Moscow Lectures

Volume 6

More information about this series at http://www.springer.com/series/15875

Serge Lvovski

Principles of Complex Analysis

NATIONAL RESEARCH
UNIVERSITY

Skolkovo Institute of Science and Technology

 Springer

Serge Lvovski
National Research University
Higher School of Economics
Moscow, Russia

Federal Science Center System
Research Institute of Russian Academy
of Sciences (FGU FNC NIISI RAN)
Moscow, Russia

Translated from the Russian by Natalia Tsilevich. Originally published as Принципы комплексного анализа by MCCME, Moscow, 2017.

ISSN 2522-0314   ISSN 2522-0322   (electronic)
Moscow Lectures
ISBN 978-3-030-59367-4   ISBN 978-3-030-59365-0   (eBook)
https://doi.org/10.1007/978-3-030-59365-0

Mathematics Subject Classification (2020): 30-01

Cover illustration: https://www.istockphoto.com/de/foto/panorama-der-stadt-moskau-gm490080014-75024685, with kind permission

This Springer imprint is published by the registered company Springer Nature Switzerland AG
The registered company address is: Gewerbestrasse 11, 6330 Cham, Switzerland

Preface to the Book Series *Moscow Lectures*

You hold a volume in a textbook series of Springer Nature dedicated to the Moscow mathematical tradition. Moscow mathematics has very strong and distinctive features. There are several reasons for this, all of which go back to good and bad aspects of Soviet organization of science. In the twentieth century, there was a veritable galaxy of great mathematicians in Russia, while it so happened that there were only few mathematical centers in which these experts clustered. A major one of these, and perhaps the most influential, was Moscow.

There are three major reasons for the spectacular success of Soviet mathematics:

1. Significant support from the government and the high prestige of science as a profession. Both factors were related to the process of rapid industrialization in the USSR.
2. Doing research in mathematics or physics was one of very few intellectual activities that had no mandatory ideological content. Many would-be computer scientists, historians, philosophers, or economists (and even artists or musicians) became mathematicians or physicists.
3. The Iron Curtain prevented international mobility.

These are specific factors that shaped the structure of Soviet science. Certainly, factors (2) and (3) are more on the negative side and cannot really be called favorable but they essentially came together in combination with the totalitarian system. Nowadays, it would be impossible to find a scientist who would want all of the three factors to be back in their totality. On the other hand, these factors left some positive and long lasting results.

An unprecedented concentration of many bright scientists in few places led eventually to the development of a unique "Soviet school". Of course, mathematical schools in a similar sense were formed in other countries too. An example is the French mathematical school, which has consistently produced first-rate results over a long period of time and where an extensive degree of collaboration takes place. On the other hand, the British mathematical community gave rise to many prominent successes but failed to form a "school" due to a lack of collaborations. Indeed, a school as such is not only a large group of closely collaborating individuals but also a group knit tightly together through student-advisor relationships. In the USA, which is currently the world leader in terms of the level and volume of mathematical research, the level of mobility is very high, and for this reason there are no US mathematical schools in the Soviet or French sense of the term. One can talk not only about the Soviet school of mathematics but also, more specifically, of the Moscow, Leningrad, Kiev, Novosibirsk, Kharkov, and other schools. In all these places, there were constellations of distinguished scientists with large numbers of students, conducting regular seminars. These distinguished scientists were often not merely advisors and leaders, but often they effectively became spiritual leaders in a very general sense.

A characteristic feature of the Moscow mathematical school is that it stresses the necessity for mathematicians to learn mathematics as broadly as they can, rather than focusing on a narrow field in order to get important results as soon as possible.

The Moscow mathematical school is particularly strong in the areas of algebra/algebraic geometry, analysis, geometry and topology, probability, mathematical physics and dynamical systems. The scenarios in which these areas were able to develop in Moscow have passed into history. However, it is possible to maintain and develop the Moscow mathematical tradition in new formats, taking into account modern realities such as globalization and mobility of science. There are three recently created centers—the Independent University of Moscow, the Faculty of Mathematics at the National Research University Higher School of Economics (HSE) and the Center for Advanced Studies at Skolkovo Institute of Science and Technology (SkolTech)—whose mission is to strengthen the Moscow mathematical tradition in new ways. HSE and SkolTech are universities offering officially licensed full-time educational programs. Mathematical curricula at these universities follow not only the Russian and Moscow tradition but also new global developments in mathematics. Mathematical programs at the HSE are influenced by those of the Independent University of Moscow (IUM). The IUM is not a formal university; it is rather a place where mathematics students of different universities can attend special topics courses as well as courses elaborating the core curriculum. The IUM was the main initiator of the HSE Faculty of Mathematics. Nowadays, there is a close collaboration between the two institutions.

While attempting to further elevate traditionally strong aspects of Moscow mathematics, we do not reproduce the former conditions. Instead of isolation and academic inbreeding, we foster global sharing of ideas and international cooperation. An important part of our mission is to make the Moscow tradition of mathematics at a university level a part of global culture and knowledge.

The "Moscow Lectures" series serves this goal. Our authors are mathematicians of different generations. All follow the Moscow mathematical tradition, and all teach or have taught university courses in Moscow. The authors may have taught mathematics at HSE, SkolTech, IUM, the Science and Education Center of the Steklov Institute, as well as traditional schools like MechMath in MGU or MIPT. Teaching and writing styles may be very different. However, all lecture notes are supposed to convey a live dialog between the instructor and the students. Not only personalities of the lecturers are imprinted in these notes, but also those of students.

We hope that expositions published within the "Moscow lectures" series will provide clear understanding of mathematical subjects, useful intuition, and a feeling of life in the Moscow mathematical school.

Moscow, Russia   Igor M. Krichever
Vladlen A. Timorin
Michael A. Tsfasman
Victor A. Vassiliev

Preface

This is a textbook aimed at students studying complex analysis from the very beginning. It reflects the author's long-term experience in teaching complex analysis at the Department of Mathematics of the Higher School of Economics (HSE) and at the *Math in Moscow* program (a joint project of HSE and the Independent University of Moscow).

The main (and essentially the only) prerequisite for reading this book is a sound knowledge of "ordinary" analysis: if the reader is comfortable with uniform convergence and such notions as openness, closedness, and compactness as applied to subsets of the plane, the rest will come. The first chapter contains, among other things, a list of facts (without proofs) from analysis that will be used in what follows. As a reference source and a textbook of analysis which can be used to fill in gaps in background knowledge, I recommend the two-volume textbook by V. A. Zorich [5, 6].

The concluding Chapter 13, devoted to Riemann surfaces, requires more background knowledge than the main part of the book. I proceed from the premise that complex analysis is usually studied along with such disciplines as multivariable real analysis, including analysis on manifolds, and the beginnings of topology (at any rate, this is the usual practice at the Department of Mathematics of HSE). The specific prerequisites for understanding this chapter are listed at its beginning.

Each chapter ends with a set of exercises, both theoretical and computational. Though they may be sufficient to help the reader master the material, they will definitely not suffice to prepare a whole lecture course with all the tests, exams, etc.: this book cannot replace a problem book in complex analysis. Such problem books are in abundance, and virtually any can be used: all of them contain sufficiently many exercises to practice computational skills specific to complex analysis.

I have sought to give a rigorous presentation of the material, yet not allowing the discussion of fundamentals to obscure the main subject. It is for the reader to judge whether or not I have succeeded in striking the right balance. Also, I have made every effort to avoid introducing abstract notions in the cases where the book's material does not allow me to demonstrate how these notions actually "work." That is why, the book contains, say, the definition of a quasiconformal map, but not that of a sheaf.

The presentation of the "homotopic" version of Cauchy's theorem is borrowed from S. Lang's textbook [2]; the derivation of the cotangent series expansion, from [1]; the proof of Weierstrass' theorem on the existence of a holomorphic function with prescribed zeros, from B. V. Shabat's textbook [4]. The presentation of Bloch's and Landau's theorems mostly follows I. I. Privalov's textbook [3]; hopefully, though, I have managed to shorten the path to Picard's great theorem as compared with Privalov.

The book owes its title to a rather curious point that a number of important theorems of complex analysis are traditionally called principles (as the reader can see by looking at the index).

I am grateful to A. V. Zabrodin and V. A. Poberezhny, with whom I have for several years run a seminar on complex analysis at the Department of Mathematics of HSE.

I am grateful to Alexei Penskoi for taking the trouble to read and comment on the manuscript; to Vladimir Medvedev for carefully reading the first draft of the text and pointing out numerous mistakes and inaccuracies; to Tatiana Korobkova for thoughtful editing; to Victor Shuvalov for formatting; and to Natalia Tsilevich for the excellent translation. All the remaining shortcomings are solely my responsibility.

Finally, my deep gratitude goes to Victoria, Jacob, and Mark by whom I have been constantly distracted, constantly helped, and without whom this book would never have been written.

S. Lvovski

Contents

Chapter 1
Preliminaries

This introductory chapter includes material needed in what follows yet not belonging to complex analysis proper. To find out how well you are acquainted with the necessary background, you can look at the exercises at the end of the chapter.

If $z = x + iy$ is a complex number ($x, y \in \mathbb{R}$, $i^2 = -1$), then x and y are called the *real* and *imaginary parts* of z, respectively. Notation: $x = \operatorname{Re} z$, $y = \operatorname{Im} z$.

A complex number z is represented by the point in the plane with coordinates $(\operatorname{Re} z, \operatorname{Im} z)$; correspondingly, we will identify the set of all complex numbers (denoted by $\mathbb{C}$) with the coordinate plane; the plane whose points are regarded as complex numbers is called the *complex plane*.

If z is a complex number, then its *absolute value* (or *modulus*) is the distance from z (more exactly, from the corresponding point in the complex plane; in what follows, we will no longer make such distinctions) to 0 (i.e., the origin): $|z| = \sqrt{x^2 + y^2}$ where $x = \operatorname{Re} z$, $y = \operatorname{Im} z$. The triangle inequality implies that $|z + w| \leq |z| + |w|$. The distance between complex numbers z and w is equal to $|z - w|$.

The Ox and Oy axes in the complex plane are called the *real* and *imaginary axes*, respectively. Given $z \in \mathbb{C}$, $z \neq 0$, the *argument* of z is the angle between the vector from 0 to z and the positive real axis (so, for instance, $\pi/4$ is the argument of $1 + i$). The argument of a complex number z is denoted by $\arg z$.

The argument of a complex number is defined up to an integer multiple of 2π: for example, the assertions $\arg(1 + i) = \pi/4$ and $\arg(1 + i) = -7\pi/4$ are equally true. If $|z| = r \neq 0$ and $\arg z = \varphi$, then

$$z = r(\cos \varphi + i \sin \varphi). \tag{1.1}$$

Expression (1.1) is called the *polar form* of the complex number z. To find the product of two complex numbers, we multiply their absolute values and add their arguments.

If $z = x + iy$ with $x, y \in \mathbb{R}$, then the number $x - iy$ is called the *complex conjugate* of z and denoted by $\bar{z}$. The point $\bar{z}$ is the reflection of z in the real axis, and $z\bar{z} = |z|^2$.

Limits of functions of a complex variable and limits of sequences of complex numbers are defined in the same way as in the real case. For example, $\lim_{z \to a} f(z) = b$

© Springer Nature Switzerland AG 2020
S. Lvovski, *Principles of Complex Analysis*, Moscow Lectures 6,
https://doi.org/10.1007/978-3-030-59365-0_1

means that for every $\varepsilon > 0$ there exists $\delta > 0$ such that $0 < |z - a| < \delta$ implies $|f(z) - b| < \varepsilon$. The theorems on the limit of a sum, difference, product, and quotient of two functions are still valid in the complex case, as well as Cauchy's convergence criterion ("a sequence converges if and only if it is Cauchy").

The notation $\lim\limits_{z \to a} f(z) = \infty$ means that $\lim\limits_{z \to a} |f(z)| = +\infty$ (and similarly for limits of sequences); $\lim\limits_{|z| \to \infty} f(z) = b$ means that "for every $\varepsilon > 0$ there exists $M > 0$ such that $|z| > M$ implies $|f(z) - b| < \varepsilon$."

Derivatives and integrals of functions of a complex variable are a delicate matter (the entire book is devoted to it), but complex-valued functions of a real variable hold no surprises. Namely, if f is a function with complex values defined on an interval of a (real) axis, and if $f(x) = u(x) + iv(x)$ (where u and v are real-valued), then the derivative of f is

$$f'(a) = \lim_{h \to 0} \frac{f(a + h) - f(a)}{h} = u'(a) + iv'(a);$$

these derivatives enjoy all the elementary properties of derivatives of real-valued functions (the derivative of a sum, product, and difference, the derivative of a composite function if the "inner" function is real-valued), with the same proofs. The integral of such a function f is defined by the formula

$$\int_a^b f(x)\,dx = \int_a^b u(x)\,dx + i \int_a^b v(x)\,dx;$$

it can also be defined in terms of Riemann sums (or, for that matter, as a Lebesgue integral). As in the case of real-valued functions, it satisfies the inequality

$$\left| \int_a^b f(x)\,dx \right| \leq \int_a^b |f(x)|\,dx; \tag{1.2}$$

to prove this, it suffices to apply the inequality "the absolute value of a sum is not greater than the sum of the absolute values" to the Riemann sums and take the limit.

1.1 Absolute and Uniform Convergence

Let X be an arbitrary set (you lose nothing by assuming that X is a subset of the complex plane) and $\{f_n\}_{n \in \mathbb{N}}$ be a countable family of *bounded* functions on X with values in $\mathbb{C}$.

Definition 1.1 We say that a series $\sum\limits_{n=1}^{\infty} f_n$ converges *absolutely and uniformly* on X if the series $\sum\limits_{n=1}^{\infty} \sup\limits_{x \in X} |f_n(x)|$ converges.

Proposition 1.2 (Weierstrass M-test) *Let* $\sum\limits_{n=1}^{\infty} f_n$ *be a series of bounded functions on* X. *If there exists a positive integer* N *such that* $\sup\limits_{x \in X} |f_n(x)| \leq a_n$ *for all* $n \geq N$, *and if the series* $\sum\limits_{n=1}^{\infty} a_n$ *converges, then the series* $\sum\limits_{n=1}^{\infty} f_n$ *converges absolutely and uniformly.*

See [6, Chap. XVI, Sec. 2].

Proposition 1.3 *If a series of bounded functions* $\sum f_n$ *converges on* X *absolutely and uniformly, then it converges on* X *uniformly. Moreover, the series obtained from* $\sum f_n$ *by any rearrangement of its terms also converges on* X *absolutely and uniformly, and to the same function.*

For the case of series with constant terms, see [5, Chap. V, Sec. 5, Proposition 4]; the modifications necessary for the case of functional series are left to the reader.

Let $\{f_{m,n}\}_{m,n \in \mathbb{N}}$ be a family of bounded functions defined on X and indexed by two positive integers. Then the formal sum $\sum\limits_{m,n \in \mathbb{N}} f_{m,n}$ is called a *double series*. We say that a double series converges absolutely and uniformly if it converges absolutely and uniformly for some (and hence, by Proposition 1.3, any) ordering of its terms.

Proposition 1.4 *Let* $\{f_{m,n}\}_{m,n \in \mathbb{N}}$ *be a family of bounded functions on* X. *Then the following two conditions are equivalent.*

(1) *The double series* $\sum\limits_{m,n} f_{m,n}$ *converges absolutely and uniformly.*

(2) *For every* $m \in \mathbb{N}$ *the series* $\sum\limits_{n=1}^{\infty} f_{m,n}$ *converges absolutely and uniformly, and, denoting the sum of this series by* f_m, *the series* $\sum\limits_{m=1}^{\infty} f_m$ *also converges absolutely and uniformly.*

Moreover, if the equivalent conditions (1) *and* (2) *are satisfied, then the sum of the series* $\sum\limits_{m,n \in \mathbb{N}} f_{m,n}$ *coincides with the sum of the series* $\sum\limits_{m=1}^{\infty} f_m$.

The reader may either prove these statements as an exercise, or find the proofs in the literature, or, finally, appropriately modify the proofs of the corresponding facts for series with constant terms (they are easier to find in textbooks).

1.2 Open, Closed, Compact, Connected Sets

In this section, we deal with subsets of the Euclidean space $\mathbb{R}^n$ for arbitrary n, but our applications mainly involve the case $n = 2$ (the complex plane $\mathbb{C}$ identified with $\mathbb{R}^2$) and $n = 1$ (the real line). If v is a point in $\mathbb{R}^n$, by $|v|$ we denote its Euclidean norm (the square root of the sum of the squares of its coordinates). The distance between points $v_1, v_2 \in \mathbb{R}^n$ is equal to $|v_1 - v_2|$. If $n = 2$ and we identify $\mathbb{R}^2$ with $\mathbb{C}$ in the usual way, then $|v|$ is the absolute value of the complex number v.

Recall that the *ε-neighborhood* of a point $a \in \mathbb{R}^n$ is the set

$$\{z \in \mathbb{R}^n : |z - a| < \varepsilon\}$$

(here ε is a positive real number).

Definition 1.5 A subset $U \subset \mathbb{R}^n$ is said to be *open* if for every point $a \in U$ there is an ε-neighborhood of a, for some ε, that is contained in U.

Definition 1.6 A subset $F \subset \mathbb{R}^n$ is said to be *closed* if its complement $\mathbb{R}^n \setminus F$ is open.

Proposition 1.7 (1) *The union of an arbitrary family of open sets is open. The intersection of an arbitrary finite family of open sets is open.*

(2) *The intersection of an arbitrary family of closed sets is closed. The union of an arbitrary finite family of closed sets is closed.*

Proposition 1.8 *A subset $F \subset \mathbb{R}^n$ is closed if and only if it satisfies the following property: if $\{a_k\}$ is a sequence of points of F and $\lim_{k \to \infty} a_k = a$, then $a \in F$.*

(The limit of a sequence in $\mathbb{R}^n$ is defined in the same way as in $\mathbb{C}$, that is, $a_k \to a$ if $\lim_{k \to \infty} |a_k - a| = 0$.)

Definition 1.9 The *closure* of a subset $X \subset \mathbb{R}^n$ is the intersection of all closed sets containing X.

Proposition 1.10 *The closure of a subset $X \subset \mathbb{R}^n$ coincides with the set of all limits of convergent sequences $\{a_k\}$ with $a_k \in X$.*

The closure of a set X is denoted by $\bar{X}$.

Definition 1.11 The *interior* of a subset $X \subset \mathbb{R}^n$ is the set of all points $a \in X$ such that some ε-neighborhood of a is contained X. The interior of a set X is denoted by $\mathrm{Int}(X)$.

The interior of a set X coincides with the union of all open sets contained in X, and also with the complement to the closure of the set $\mathbb{R}^n \setminus X$.

Definition 1.12 A subset $K \subset \mathbb{R}^n$ is said to be *compact* if it is closed and bounded.

Proposition 1.13 *The following three conditions are equivalent:*

(1) *a subset $K \subset \mathbb{R}^n$ is compact;*

(2) *for every sequence $a_m \in K$ there exists a subsequence $\{a_{m_k}\}$ such that the limit $\lim_{k \to \infty} a_{m_k}$ exists and lies in K;*

(3) *for every family of open sets $\{U_\alpha\}_{\alpha \in I}$ satisfying the property $K \subset \bigcup_\alpha U_\alpha$ there exists a finite collection $\alpha_1, \ldots, \alpha_l \in I$ such that*

$$K \subset U_{\alpha_1} \cup \ldots \cup U_{\alpha_l}.$$

A family of sets $\{U_\alpha\}$ satisfying condition (3) of the proposition is called an *open cover* of the set K, and condition (3) itself can be briefly stated as "every open cover has a finite subcover."

For all of the above, see [5, Chap. 7, Sec. 1].

Proposition 1.14 *If $K \subset \mathbb{R}^n$ is a compact set, then every continuous function $f: K \to \mathbb{R}$ attains its maximum and minimum on K.*

See [5, Chap. 7, Sec. 1].

Finally, we need the notion of connectedness. To save space, we define it only for open sets (it will not be encountered in other situations).

Definition 1.15 An open subset $U \subset \mathbb{R}^n$ is said to be *disconnected* if it can be represented as the union of two disjoint nonempty open sets.

An open subset $U \subset \mathbb{R}^n$ is said to be *connected* if it is not disconnected.

Proposition 1.16 *An open subset in $\mathbb{R}$ is connected if and only if it is an interval $(a; b)$ where a and b are real numbers, $+\infty$, or $-\infty$.*

Connected open subsets in $\mathbb{R}^n$ with $n > 1$ have no such simple characterization, but we will state and prove one important connectedness criterion.

Proposition 1.17 *An open subset $U \subset \mathbb{R}^n$ is connected if and only if for any two points $a, b \in U$ there exists a continuous map $\gamma: [0; 1] \to U$ such that*

$$\gamma(0) = a, \quad \gamma(1) = b.$$

In other words, an open set in the plane is connected if and only if any two its points can be joined by a curve.

Proof Assume that any two points of an open set $U \subset \mathbb{R}^n$ can be joined by a curve; we will prove that U is connected. Assume to the contrary that $U = U_1 \cup U_2$ where U_1 and U_2 are open, nonempty, and disjoint. Pick points $a \in U_1$, $b \in U_2$, and let $\gamma: [0; 1] \to U$ be a curve joining a and b (with $\gamma(0) = a$, $\gamma(1) = b$). Define a function $f: [0; 1] \to \mathbb{R}$ as follows: $f(t) = 1$ if $\gamma(t) \in U_1$, and $f(t) = 2$ if $\gamma(t) \in U_2$. We claim that f is continuous. Indeed, if, say, $\gamma(t) \in U_1$, then for some $\eta > 0$ there is an η-neighborhood $V \ni \gamma(t)$ lying in U_1 (see Definition 1.5); hence, since γ is continuous, there exists $\delta > 0$ such that $|t' - t| < \delta$ implies $\gamma(t') \in V \subset U_1$. Therefore, $f(t') = f(t) = 1$; in particular, for every $\varepsilon > 0$ we have

$$|t' - t| < \delta \Rightarrow \gamma(t') = \gamma(t) = 1 \Rightarrow |f(t') - f(t)| = 0 < \varepsilon,$$

which proves that f is continuous. Since a function defined on an interval that takes only two values 1 and 2 cannot be continuous, we arrive at a contradiction.

Conversely, let U be connected; we will show that any two points of U can be joined by a curve. Pick an arbitrary point $a \in U$; it suffices to show that it can be joined by a curve to every point $b \in U$. To this end, set

$$U_1 = \{z \in U : a \text{ and } z \text{ can be joined by a curve}\},$$
$$U_2 = U \setminus U_1.$$

We claim that U_1 and U_2 are open. Indeed, if $z \in U_1$, i.e., a can be joined by a curve to z, then a can be joined by a curve to every point from an ε-neighborhood of z contained in U (Fig. 1.1), so this neighborhood is contained in U_1; we have shown that U_1 is open. If, on the other hand, $z \in U_2$, i.e., z cannot be joined by a curve to a, then every point from an ε-neighborhood $V \ni z$ contained in U cannot be joined by a curve to a: otherwise, the curve joining a to a point $z' \in V$ could be extended by the line segment between z' and z. Therefore, $V \subset U_2$ and U_2 is also open. It remains to observe that $U_1 \ni a$ (the point a can be joined to itself by a "curve," namely, the constant map), so U_1 is nonempty; since U is connected, we obtain $U_2 = \varnothing$, $U = U_1$, and the point a can be joined by a curve to every point from U, as required. $\square$

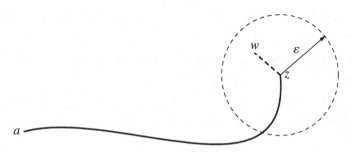

Fig. 1.1 For the proof of Proposition 1.17

Note also that in complex analysis, connected open subsets in $\mathbb{C}$ are often called *domains*.

1.3 Power Series

Consider a power series

$$c_0 + c_1(z - a) + c_2(z - a)^2 + \ldots + c_n(z - a)^n + \ldots \qquad (1.3)$$

(all coefficients c_j and the number a are complex numbers, the variable z is also assumed to be complex).

Proposition 1.18 (1) *There exists $R \in [0; +\infty]$ such that the series* (1.3) *converges absolutely for $|z - a| < R$ and diverges (its terms do not tend to zero) for $|z - a| > R$.*
    (2) *We have*

$$R = 1/\overline{\lim} \sqrt[n]{|c_n|}. \qquad (1.4)$$

In (1.4) it is meant that $1/(+\infty) = 0, 1/0 = +\infty$. This formula is called the *Cauchy–Hadamard theorem*, and the "number" R is called the *radius of convergence* of the series (1.3). (The word "number" is in quotation marks because R can be infinite).

For the Cauchy–Hadamard theorem, see [5, Chap. V, Sec. 5].

If R is the radius of convergence of a series (1.3), then the set

$$\{z \in \mathbb{C}: |z - a| < R\}$$

is called the *disk of convergence* of this series (if $R = +\infty$, then the disk of convergence coincides with the entire plane).

Proposition 1.19 *A series* (1.3) *converges absolutely and uniformly on every compact subset of its disk of convergence.*

Note that in many interesting cases, a power series does not converge uniformly on the whole disk of convergence. Actually, uniform convergence on every compact subset of a given open set U (but not necessarily on the whole set U) is a typical situation in complex analysis.

Proof Let the radius of convergence be equal to $R > 0$, and let K be a compact subset of the disk of convergence. The continuous function $z \mapsto |z - a|$ attains its maximum on K; denote this maximum by r. We have $r < R$ and

$$K \subset \bar{D}_r = \{z: |z - a| \le r\}.$$

Pick a real number r' such that $r < r' < R$ and a number z_0 such that $|z_0 - a| = r'$. By Proposition 1.18, the series (1.3) converges absolutely for $z = z_0$, i.e., the series $\sum |c_n| r'^n$ converges; in particular, all terms of this series are bounded, i.e., there is a constant $C > 0$ such that

$$|c_n| r'^n \le C \Leftrightarrow |c_n| \le \frac{C}{r'^n} \quad \text{for all } n.$$

If now $z \in \bar{D}_r$, i.e., $|z - a| \le r$, then

$$|c_n(z - a)^n| \le |c_n| r^n = |c_n| r'^n \left(\frac{r}{r'}\right)^n \le C \left(\frac{r}{r'}\right)^n.$$

Therefore, on the set $\bar{D}_r$ the terms of the series (1.3) are uniformly bounded by the terms of the convergent geometric series $C \cdot \sum (r/r')^m$, so on $\bar{D}_r$ our power series converges uniformly by the Weierstrass M-test (Proposition 1.2). $\quad\square$

Corollary 1.20 *The sum of a power series is a continuous function of z on its disk of convergence.*

Proof Indeed, in terms of the previous proof, it suffices to check that the function is continuous on every set $\bar{D}_r$ for $0 < r < R$, and this is obvious by the uniform convergence. $\quad\square$

1.4 The Exponential Function

Proposition-Definition 1.21 The series

$$1 + \frac{z}{1!} + \frac{z^2}{2!} + \ldots + \frac{z^n}{n!} + \ldots \tag{1.5}$$

converges absolutely for every $z \in \mathbb{C}$. Its sum is denoted by e^z or $\exp(z)$, and the function $z \mapsto e^z$ is called the *exponential function*, or *exponential*.

The absolute convergence of the series (1.5) for every z is well known (it follows, for example, from d'Alembert's ratio test). Since the series has infinite radius of convergence, the exponential is continuous on the entire complex plane $\mathbb{C}$. If $z \in \mathbb{R}$, then, of course, the exponential of z is equal to e^z in the usual sense.

Note that at the moment we do not define the function $z \mapsto a^z$ for any a different from e.

Setting $z = i\varphi$ with $\varphi \in \mathbb{R}$ in (1.5), we obtain the well-known Euler's formula

$$e^{i\varphi} = \cos \varphi + i \sin \varphi; \tag{1.6}$$

therefore, $e^{i\varphi}$ has absolute value 1 and argument φ, while a complex number with absolute value r and argument φ can be written as $re^{i\varphi}$.

Substituting $-\varphi$ for φ in (1.6) and then summing and subtracting the resulting equations, we arrive at the well-known formulas for sines and cosines:

$$\cos \varphi = \frac{e^{i\varphi} + e^{-i\varphi}}{2}, \quad \sin \varphi = \frac{e^{i\varphi} - e^{-i\varphi}}{2i}. \tag{1.7}$$

Here is the main property of the exponential.

Proposition 1.22 *For any $z, w \in \mathbb{C}$ we have $e^{z+w} = e^z e^w$.*

Sketch of the proof It is well known (see [5, Chap. V, Sec. 5, Proposition 5]) that if $a_0 + a_1 + \ldots + a_n + \ldots$ and $b_0 + b_1 + \ldots + b_n + \ldots$ are absolutely convergent series with sums A and B, respectively, then the series

$$a_0 b_0 + (a_0 b_1 + a_1 b_0) + \ldots + (a_0 b_n + a_1 b_{n-1} + \ldots + a_n b_0) + \ldots$$

is also absolutely convergent and its sum is equal to AB. Applying this to the series (1.5) for e^z and e^w, we see that the proposition will follow from the equation

$$1 \cdot \frac{w^n}{n!} + \frac{z}{1!} \cdot \frac{w^{n-1}}{(n-1)!} + \frac{z^2}{2!} \cdot \frac{w^{n-2}}{(n-2)!} + \ldots + \frac{z^n}{n!} \cdot 1 = \frac{(z+w)^n}{n!},$$

which is nothing else than the binomial theorem for $(z + w)^n$. □

Euler's formula implies that $e^{2\pi i} = 1$; then it follows from Proposition 1.22 that $e^{z+2\pi i} = e^z$ for all z. In other words, the exponential is a periodic function with period $2\pi i$.

1.5 Necessary Background From Multivariable Analysis

Let $U \subset \mathbb{R}^n$ be an open set. A map $F \colon U \to \mathbb{R}^m$ is said to be *differentiable*, or, for clarity, *real differentiable*, at a point $a \in U$ if there exists a linear map $L \colon \mathbb{R}^n \to \mathbb{R}^m$ such that

$$F(a+h) - F(a) = L(h) + \varphi(h) \quad \text{where} \lim_{h \to 0} \frac{|\varphi(h)|}{|h|} = 0.$$

See [5, Chap. VIII, Sec. 2].

The map L is called the *derivative* (or sometimes, for clarity, *real derivative*) of the map F at the point a (in [5, 6], the term "differential" is used).

If F is given by the formula

$$(x_1, \ldots, x_n) \longmapsto (f_1(x_1, \ldots, x_n), \ldots, f_m(x_1, \ldots, x_n))$$

and if all the partial derivatives $\frac{\partial f_j}{\partial x_i}$ exist and are continuous on the whole set U (in this case, F is said to be *of class* C^1), then F is real differentiable at every point $a \in U$ and its derivative, as a linear map from $\mathbb{R}^n$ to $\mathbb{R}^m$, is represented by the matrix $\left(\frac{\partial f_j}{\partial x_i} \right)$, called the *Jacobian matrix* of F; if $m = n$ (i.e., the Jacobian matrix is square), then its determinant is called the *Jacobian* of F (at the given point).

In most of the book, the above-mentioned results will be applied in the case where $m = n = 2$, $\mathbb{R}^m = \mathbb{R}^n = \mathbb{C}$, and the Euclidean norm is nothing else than the absolute value of a complex number.

We will also need some information on the area (or measure) of open sets in the plane. Since the reader is not assumed to be familiar with Lebesgue measure and integral, we adopt the following approach. Recall that the *support* of a function $\varphi \colon \mathbb{R}^n \to \mathbb{R}$ is the closure of the set of all points where it does not vanish. If $\varphi \colon \mathbb{R}^n \to \mathbb{R}$ is a continuous function with compact support, then by $\int_{\mathbb{R}^n} \varphi \, dx$ we mean the integral of φ over a parallelepiped containing its support (see [6, Chap. XI, Sec. 1, Definition 7]).

Definition 1.23 Let $U \subset \mathbb{R}^n$ be an open set. A *partition of unity with compact supports* on U is a countable family of continuous functions $\{\varphi_i \colon \mathbb{R}^n \to [0; +\infty)\}$ satisfying the following properties:

(1) the support of every function φ_i is compact and lies in U;

(2) every point $x \in U$ has an ε-neighborhood $V \ni x$, $V \subset U$ such that the supports of all but finitely many functions φ_i are disjoint with V;

(3) for every point $x \in U$ we have $\sum_i \varphi_i(x) = 1$ (this sum is finite by condition (2)).

Cf. [6, Chap. XV, Sec. 2, Definition 18].

Definition 1.24 Let $U \subset \mathbb{R}^n$ be an open set and $h \colon U \to [0; +\infty)$ be a continuous function. The *integral* of h *over* U is the number

$$\sum_i \int_{\mathbb{R}^n} h \cdot \varphi_i \, dx_1 \dots dx_n \qquad (1.8)$$

where $\{\varphi_i\}$ is a partition of unity with compact supports on U (if the series in the left-hand side of (1.8) diverges, then the integral is defined to be $+\infty$).

One can (easily) check that the integral does not depend on the choice of a partition of unity.

Definition 1.25 Let $U \subset \mathbb{R}^n$ be an open set. The *measure* of U is the integral $\mu(U) = \int_U 1 \, dx_1 \dots dx_n$ (a nonnegative number or $+\infty$).

If U is a bounded set, then its measure is finite.

Let U and V be open subsets in $\mathbb{R}^n$; a *diffeomorphism* between U and V is a bijective map $F: U \to V$ of class C^1 whose inverse is also of class C^1.

Proposition 1.26 *Let* $U, V \subset \mathbb{R}^n$ *be open sets and* $F: U \to V$ *be a diffeomorphism of class* C^1. *Then*

$$\mu(V) = \int_U |J(F)(x_1, \dots, x_n)| \, dx_1 \dots dx_n$$

where $|J(F)(x_1, \dots, x_n)|$ *is the determinant of the Jacobian matrix of* F *at the point* $x_1, \dots, x_n$.

This proposition can be deduced from the change of variable formula [6, Chap. XI, Sec. 5, Theorem 1].

1.6 Linear Fractional Transformations

Definition 1.27 Let a, b, c, d be complex numbers such that the matrix $\begin{pmatrix} a & b \\ c & d \end{pmatrix}$ is nondegenerate. Then the map from $\mathbb{C}$ to $\mathbb{C}$ given by the formula

$$z \mapsto \frac{az + b}{cz + d} \qquad (1.9)$$

is called a *linear fractional map*, or *linear fractional transformation*.

The nondegeneracy condition guarantees that the numerator is not proportional to the denominator, i.e., the map is not constant.

Definition 1.27 is stated with a (deliberate) carelessness: if $c \neq 0$, then for $z = -d/c$ the denominator vanishes and the map (1.9) is not defined, so, strictly speaking, it cannot be called a map from $\mathbb{C}$ to $\mathbb{C}$. To avoid repeating this caveat, it is convenient to proceed as follows. Consider the set $\overline{\mathbb{C}} = \mathbb{C} \cup \{\infty\}$ where ∞ is a symbol (called, sure enough, "infinity") to be dealt with according to the following rules.

First, we set $a \pm \infty = \infty$ and $a/\infty = 0$ for every complex number a, and also $a \cdot \infty = \infty$ and $a/0 = \infty$ for every *nonzero* complex number a (the expressions $0 \cdot \infty$, $\infty \pm \infty$, and $0/0$ are still not defined).

Second, if $f(z) = (az + b)/(cz + d)$ is a linear fractional map, we set

$$f(\infty) = \lim_{|z| \to \infty} f(z) = \frac{a}{c}.$$

With these conventions, every linear fractional transformation becomes a one-to-one map from $\overline{\mathbb{C}}$ onto itself.

The set $\overline{\mathbb{C}}$ is called the *extended complex plane*, or *Riemann sphere*.

Sets of the form $\{z \in \mathbb{C}: |z| > R\}$ will be regarded as "punctured neighborhoods of infinity" on the Riemann sphere; and the same sets with the point ∞ added, as neighborhoods of infinity (without puncture). For the reader familiar with the corresponding definitions, I should mention that this definition of neighborhoods of infinity endows $\overline{\mathbb{C}}$ with the structure of a topological space homeomorphic to the two-dimensional sphere (see Chap. 13).

The main property of linear fractional transformations is that they take lines and circles to lines and circles. More exactly, for every line $\ell \subset \mathbb{C}$ the set $\ell \cup \{\infty\} \subset \overline{\mathbb{C}}$ will be called a "line on the Riemann sphere" (it is the closure of the set $\ell \subset \mathbb{C}$ with respect to the above-mentioned topology on $\overline{\mathbb{C}}$). By a circle on the Riemann sphere we will mean a usual circle in $\mathbb{C}$ (a circle is a bounded set, it "does not go to infinity," so there is no need to add ∞). Now we introduce the following term.

Definition 1.28 A *generalized circle* is a subset in $\overline{\mathbb{C}}$ that is either a line on the Riemann sphere or a circle.

Proposition 1.29 *Every linear fractional transformation takes generalized circles to generalized circles.*

Proof Every linear fractional transformation $z \mapsto (az + b)/(cz + d)$ is easily seen to be a composition of transformations of the form $z \mapsto Az$ $(A \neq 0)$, $z \mapsto z + B$, and $z \mapsto 1/z$ (to prove this, it suffices to divide the polynomial $az + b$ by $cz + d$ with remainder). Transformations of the first two types (dilations and translations) take lines to lines and circles to circles, so it remains to consider only the case of $1/z$.

Lemma 1.30 *The equation of any generalized circle has the form*

$$pz\bar{z} + Az + \bar{A}\bar{z} + q = 0 \quad where \ A \in \mathbb{C}, \ p, q \in \mathbb{R}. \tag{1.10}$$

Proof Let $\ell \subset \mathbb{C}$ be a line given by an equation $px + qy + r = 0$ $(p, q, r \in \mathbb{R})$. Substituting $x = (z + \bar{z})/2$, $y = (z - \bar{z})/2i$, we see that the equation of ℓ has the form

$$\frac{p - iq}{2} z + \frac{p + iq}{2} \bar{z} + r = 0,$$

or, denoting $(p - iq)/2 = A$,

$$Az + \bar{A}\bar{z} + r = 0 \quad where \ A \in \mathbb{C}, r \in \mathbb{R}.$$

In a similar way, the equation of any circle in the plane has the form

$$p(x^2 + y^2) + qx + ry + s = 0 \quad \text{where } p, q, r, s \in \mathbb{R}.$$

Substituting $x = (z + \bar{z})/2$, $y = (z - \bar{z})/2i$, we obtain the equation

$$pz\bar{z} + Bz + \bar{B}\bar{z} + s \quad (B \in \mathbb{C}, s \in \mathbb{R}),$$

which is again an equation of the form (1.10).   □

Returning to the action of the transformation $z \mapsto 1/z$ on generalized circles, observe that if $w = 1/z$, then $z = 1/w$; substituting $1/w$ for z in (1.10) yields an equation for w which also has the form (1.10), and we are done.   □

Proposition 1.31 *Let $\{a_1, a_2, a_3\}$ and $\{b_1, b_2, b_3\}$ be two triples of distinct points on the Riemann sphere. Then there is a unique linear fractional transformation that sends each a_i to the corresponding b_i.*

Sketch of the proof We verify this for the case where $b_1 = 0$, $b_2 = 1$, $b_3 = \infty$, and all a_j are finite (leaving the rest as an exercise for the reader).

If a linear fractional transformation $z \mapsto (pz + q)/(rz + s)$ sends a_1 to 0 and a_3 to ∞, then $pz + q$ must vanish for $z = a_1$ and $rz + s$ must vanish for $z = a_3$. Therefore, the map can be rewritten as

$$f(z) = c\frac{z - a_1}{z - a_3}, \quad c \neq 0;$$

the condition $f(a_2) = 1$ fixes the coefficient c.   □

An important property of linear fractional maps is that they are *conformal*: linear fractional transformations preserve angles between curves. Later in this section, we will need this property for angles between (generalized) circles, which can be proved by quite elementary methods; however, we postpone the proof until Chap. 2, where this property will be deduced from a more general fact, and also the general definition of a conformal map will be given.

We also need the notion of symmetry (reflection) with respect to a circle.

Definition 1.32 Let $C \subset \bar{\mathbb{C}}$ be a generalized circle. Points $p_1, p_2 \in \bar{\mathbb{C}}$ are said to be *symmetric* with respect to C if they do not lie on C but every generalized circle that passes through p_1 and p_2 is orthogonal to C.

If p is a point on C, then it is symmetric to itself.

If C is a line, then symmetry with respect to C in the sense of Definition 1.32 is equivalent to symmetry in the ordinary sense (see Fig. 1.2).

Proposition 1.33 *For every generalized circle C and every point $p \in \bar{\mathbb{C}}$ there is a unique point symmetric to p with respect to C.*

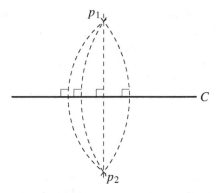

Fig. 1.2 Symmetry with respect to a line from the point of view of Definition 1.32

Proof If p and p' are symmetric points with respect to C and $A : \overline{\mathbb{C}} \to \overline{\mathbb{C}}$ is a linear fractional transformation, then (in view of Definition 1.32 and the preservation of angles) $A(p)$ and $A(p')$ are symmetric points with respect to $A(C)$, and vice versa. Applying a linear fractional transformation that takes C to a line reduces the question to the unique existence of a point symmetric to a given point with respect to a line. □

Exercises

1.1. Write the number $-1 + i\sqrt{3}$ in polar form.

1.2. Simplify the expression $(-\sqrt{3} - i)^{2017}$.

1.3. Let $\zeta_1, \ldots, \zeta_n$ be all roots of the equation $z^n = 1$ where n is a positive integer. Find the sum
$$\zeta_1^k + \zeta_2^k + \ldots + \zeta_n^k$$
for every $k \in \mathbb{Z}$.

1.4. (a) Find (draw) the images of the lines $\operatorname{Re} z = 1/2$, $\operatorname{Re} z = 1$, and $\operatorname{Re} z = 3/2$ under the map $z \mapsto z^2$; what are these curves called?
(b) The same question for the lines $\operatorname{Re} z = -1/2$, $\operatorname{Re} z = -1$, and $\operatorname{Re} z = -3/2$ (before you start calculating, consider whether this is worth doing).
(c) The same question for the lines $\operatorname{Im} z = 1/2$, $\operatorname{Im} z = 1$, and $\operatorname{Im} z = 3/2$ (and the same warning).

1.5. Give an "epsilon-delta" proof that the relation $\lim_{n \to \infty} z_n = a$ is equivalent to $\lim_{n \to \infty} \operatorname{Re} z_n = \operatorname{Re} a$ and $\lim_{n \to \infty} \operatorname{Im} z_n = \operatorname{Im} a$.

1.6. For each of the following subsets in $\mathbb{C}$, determine whether it is open, or closed, or neither.
(a) $\{z : \operatorname{Re}(z) > 2, \operatorname{Im}(z) \leq 1\}$.
(b) $\{z = x + iy : \sin x + \cos y > 2017\}$.
(c) $\{z : \operatorname{Im}(z) \geq 1, \operatorname{Re}(z) \geq -1\}$.
(d) The set of $z \in \mathbb{C}$ such that either $\operatorname{Im} z \neq 0$, or $\operatorname{Im} z = 0$ and $\operatorname{Re} z$ is rational.

1.7. Let U, V be open subsets in $\mathbb{C}$ and $f: U \to V$ be a map. Show that f is continuous in the "epsilon-delta" sense if and only if for every open subset $V_1 \subset V$ the set $f^{-1}(V_1)$ is also open.

1.8. Let U, V be open subsets in $\mathbb{C}$ and $f: U \to V$ be a continuous bijective map for which the inverse map $f^{-1}: V \to U$ is also continuous (such maps are called *homeomorphisms*). Let $b \in V$, and let φ be a function defined on a punctured neighborhood of b. Show that

$$\lim_{w \to b} \varphi(w) = \lim_{z \to f^{-1}(b)} \varphi(f(z)).$$

1.9. Show that every open set $U \subset \mathbb{C}$ can be represented as a union of pairwise disjoint connected open subsets. (*Hint.* Call two points equivalent if they can be joined by a path.)

These open subsets are called the *connected components* of U.

1.10. Let $[a; b] \subset \mathbb{R}$ be a closed interval contained in a (possibly infinite) union of open intervals $I_j \subset \mathbb{R}$. Show that there exist $a = a_0 < a_1 < a_2 < \ldots < a_n = b$ such that every closed interval $[a_k; a_{k+1}]$ is contained in at least one of I_j.

1.11. Let $U \subset \mathbb{C}$ be an open set and $K \subset U$ be a compact subset of U. Show that

$$\inf_{\substack{z \in K \\ w \in \mathbb{C} \setminus U}} |z - w| > 0.$$

1.12. Find the radius of convergence of the series $\sum\limits_{n=1}^{\infty} n! x^{n!}$.

1.13. The series

$$1 + x + \frac{x^2}{2!} + \ldots + \frac{x^n}{n!} + \ldots$$

converges for every $x \in \mathbb{R}$. Is this convergence uniform on $\mathbb{R}$?

1.14. Let $e^{z+T} = e^z$ for all $z \in \mathbb{C}$. Show that $T = 2\pi i n$ for some integer n.

1.15. Find the images of the vertical and horizontal lines under the map $z \mapsto e^z$.

1.16. Let $f: z \mapsto \frac{a_1 z + b_1}{c_1 z + d_1}$ and $g: z \mapsto \frac{a_2 z + b_2}{c_2 z + d_2}$ be linear fractional transformations. Show that their composition $g \circ f$ is a linear fractional transformation of the form $z \mapsto \frac{a_3 z + b_3}{c_3 z + d_3}$ where

$$\begin{pmatrix} a_3 & b_3 \\ c_3 & d_3 \end{pmatrix} = \begin{pmatrix} a_2 & b_2 \\ c_2 & d_2 \end{pmatrix} \begin{pmatrix} a_1 & b_1 \\ c_1 & d_1 \end{pmatrix}.$$

1.17. Fill in the gaps in the proof of Proposition 1.31.

1.18. Show that every sequence of points on the Riemann sphere contains a subsequence converging to a complex number or to ∞.

1.19. Let C be the circle of radius R centered at the origin. Show that the point symmetric to a point z with respect to C is $R^2 / \bar{z}$.

1.20. Show that the composition of two reflections with respect to generalized circles is a linear fractional transformation.

Chapter 2
Derivatives of Complex Functions

Complex analysis begins with the following definition.

Definition 2.1 Let $U \subset \mathbb{C}$ be an open subset. A function $f: U \to \mathbb{C}$ is said to be *complex differentiable* at a point $a \in U$ if the limit

$$\lim_{h \to 0} \frac{f(a+h) - f(a)}{h} \tag{2.1}$$

exists. In this case, the limit (2.1) is called the *derivative* of f at a and denoted by $f'(a)$.

Exercise 2.2 If n is a positive integer, then a calculation familiar to the reader shows that the complex derivative of the function $f(z) = z^n$ at a point a exists and is equal to na^{n-1}; as usual, a brief notation for this fact is $(z^n)' = nz^{n-1}$.

Though the definitions coincide word for word, complex differentiability has a quite different meaning than differentiability of functions of one real variable familiar to the reader from the first-year analysis course. The differences are, if you like, the subject matter of the entire book, but for the present we state several properties of complex derivatives completely analogous to the corresponding properties of derivatives of functions of a real variable. The proofs are omitted, since they coincide word for word with the proofs known to the reader from the analysis course.

Proposition 2.3 *If a function f is complex differentiable at a point $a \in \mathbb{C}$, then it is continuous at this point.*

Proposition 2.4 *If functions f and g are complex differentiable at a point $a \in \mathbb{C}$, then the same is true for the functions $f + g$, $f - g$, and fg, with*

$$(f \pm g)'(a) = f'(a) \pm g'(a), \quad (fg)'(a) = f'(a)g(a) + f(a)g'(a).$$

If, besides, $g(a) \neq 0$, then the function f/g is also complex differentiable at a, with

$$\left(\frac{f}{g}\right)'(a) = \frac{f'(a)g(a) - f(a)g'(a)}{(g(a))^2}.$$

© Springer Nature Switzerland AG 2020
S. Lvovski, *Principles of Complex Analysis*, Moscow Lectures 6,
https://doi.org/10.1007/978-3-030-59365-0_2

The proof of the next proposition (chain rule) is also essentially the same as in the real case, but I reproduce it, just to be safe.

Proposition 2.5 *Let $U \subset \mathbb{C}$ be an open set, $f: U \to \mathbb{C}$ be a function such that $f(U) \subset V$ where $V \subset \mathbb{C}$ is an open set, and $g: V \to \mathbb{C}$ be another function. If f is complex differentiable at a point $a \in U$ and g is complex differentiable at the point $f(a)$, then the composite function $g \circ f$ is complex differentiable at a, and $(g \circ f)'(a) = g'(f(a)) \cdot f'(a)$.*

Proof Set

$$\varphi(h) = f(a + h) - f(a) - f'(a) \cdot h,$$
$$\psi(k) = g(f(a) + k) - g(f(a)) - g'(f(a)) \cdot k.$$

By the definition of derivative, we have $\lim_{h \to 0} \varphi(h)/h = 0$, $\lim_{k \to 0} \psi(k)/k = 0$. Setting $\omega(k) = \sup_{0 < h \le k} |\psi(h)/h|$, we can rewrite the second relation as follows: there exists a function ω defined on an interval $[0; \varepsilon_0]$ and monotone decreasing to 0 such that $|\psi(k)| \le \omega(k)|k|$ for $|k| \le \varepsilon_0$.

Now we can write

$$g(f(a + h)) - g(f(a)) = g(f(a) + f'(a) \cdot h + \varphi(h)) - g(f(a))$$
$$= g'(f(a)) \cdot (f'(a) \cdot h + \varphi(h)) + \psi(f'(a) \cdot h + \varphi(h));$$

it follows that

$$\frac{g(f(a + h)) - g(f(a))}{h} - g'(f(a)) \cdot f'(a)$$
$$= g'(f(a)) \cdot \frac{\varphi(h)}{h} + \frac{\psi(f'(a) \cdot h + \varphi(h))}{h}. \quad (2.2)$$

To prove the proposition, we must verify that the right-hand side of (2.2) converges to zero as $h \to 0$. For the first term, this is obvious; as to the second term, observe that the relation $\lim_{h \to 0} (\varphi(h)/h) = 0$ implies that there exist constants $C > 0$, $\delta > 0$ such that $|f'(a) \cdot h + \varphi(h)| \le Ch$ for $|h| \le \delta$ (in other words, $f'(a) \cdot h + \varphi(h) = O(|h|)$). Now, for all sufficiently small h we have

$$|\psi(f'(a) \cdot h + \varphi(h))| \le \omega(|f'(a) \cdot h + \varphi(h)|)|f'(a) \cdot h + \varphi(h)| \le \omega(C|h|) \cdot C|h|;$$

dividing by $|h|$ yields

$$\left| \frac{\psi(f'(a) \cdot h + \varphi(h))}{h} \right| \le C \cdot \omega(C|h|) \to 0,$$

as required.   □

We can now introduce the fundamental definition of complex analysis.

Definition 2.6 Let $U \subset \mathbb{C}$ be an open set. A function $f : U \to \mathbb{C}$ is said to be *holomorphic on U* if it is complex differentiable at every point $a \in U$.

It follows immediately from the definitions and Proposition 2.4 that every polynomial is holomorphic on the entire complex plane $\mathbb{C}$, and every rational function (a quotient of two polynomials) is holomorphic on the entire complex plane except for the zeros of the denominator.

Here is another important example of a holomorphic function.

Proposition 2.7 *The function* $z \mapsto e^z$ *is holomorphic on the entire complex plane* $\mathbb{C}$; *its derivative at a point a is equal to* e^a. *In other words,* $(e^z)' = e^z$.

Proof By Proposition 1.22, we have

$$\lim_{h \to 0} \frac{e^{a+h} - e^a}{h} = \lim_{h \to 0} \frac{e^a \cdot e^h - e^a}{h} = e^a \cdot \lim_{h \to 0} \frac{e^h - 1}{h}.$$

Thus, to prove the proposition, it suffices to establish the relation $\lim_{h \to 0} (e^h - 1)/h = 1$. Note that for every $h \neq 0$ we have

$$\frac{e^h - 1}{h} = 1 + \frac{h}{2!} + \frac{h^2}{3!} + \ldots + \frac{h^n}{(n+1)!} + \ldots . \tag{2.3}$$

Since the series in the right-hand side of (2.3) converges for every h, its radius of convergence is infinite; hence, by Corollary 1.20, it converges to a function that is continuous on the entire complex plane $\mathbb{C}$. In particular, the limit of this function as $h \to 0$ coincides with its value at 0, that is, with 1, which completes the proof. $\square$

Now we define the sine and cosine of an arbitrary complex number.

Definition 2.8 For every $z \in \mathbb{C}$, set

$$\cos z = \frac{e^{iz} + e^{-iz}}{2}, \quad \sin z = \frac{e^{iz} - e^{-iz}}{2i}.$$

Formulas (1.7) show that if $z \in \mathbb{R}$, then $\cos z$ and $\sin z$ in the sense of Definition 2.8 coincide with the usual sine and cosine.

Proposition 2.9 *The sine and cosine functions are holomorphic on the entire complex plane, with* $(\sin z)' = \cos z$ *and* $(\cos z)' = -\sin z$.

Proof Immediately follows from Proposition 2.7 and the chain rule. $\square$

Definition 2.8 and the exponential series immediately imply that the sine and cosine functions have the following power series expansions (for all z):

$$\sin z = z - \frac{z^3}{3!} + \frac{z^5}{5!} - \ldots + (-1)^{n-1} \frac{z^{2n-1}}{(2n-1)!} + \ldots ,$$

$$\cos z = 1 - \frac{z^2}{2!} + \frac{z^4}{4!} - \ldots + (-1)^n \frac{z^{2n}}{(2n)!} + \ldots .$$

2.1 Inverse Functions, Roots, Logarithms

Now we discuss what can be said about the derivative of an inverse function.

Proposition 2.10 *Let $U, V \subset \mathbb{C}$ be open subsets, and let $f : U \to V$ be a bijective map satisfying the following properties:*
  (1) f is a holomorphic function on U;
  (2) the derivative of f does not vanish at any point of U;
  (3) the inverse map $g = f^{-1} : V \to U$ is continuous.
  Then the inverse map $g : V \to U$ is a holomorphic function on V, and for every point $b \in V$ we have $g'(b) = 1/f'(g(b))$.

Remark 2.11 Actually, if f is holomorphic and bijective, then conditions (2) and (3) follow automatically. We will establish this in Chap. 9.

Proof Using the result of Exercise 1.8, we have

$$\lim_{w \to b} \frac{g(w) - g(b)}{w - b} = \lim_{z \to f^{-1}(b)} \frac{g(f(z)) - g(b)}{f(z) - b} = \lim_{z \to g(b)} \frac{z - g(b)}{f(z) - f(g(b))}$$
$$= \frac{1}{f'(g(b))}.$$

□

Now we put the proposition to work and try to define complex logarithm, the inverse function of the exponential. Since $|e^z| = e^{\operatorname{Re} z} \neq 0$ for any z, the best we can hope for is to define logarithms only for nonzero numbers. So, let $z = re^{i\varphi}$, $r \neq 0$. It is natural to define a logarithm of z as a number $w = x + iy$ such that $z = e^w = e^x \cdot e^{iy}$. Comparing the absolute values, we see that $x = \operatorname{Re} w$ is uniquely determined: $x = \log |z|$ (the logarithm of a real number in the usual sense). However, the imaginary part of w is no longer uniquely determined: $y = \varphi$ will do, but $y = \varphi + 2\pi in$ will do just as well for every integer n. If we want to define logarithm as a function, we must somehow choose one of these values for the argument.

On the whole set $\mathbb{C} \setminus \{0\}$, however, logarithm cannot be defined even as a continuous function. Indeed, let $\log 1 = 0$ (we could start with any other value). Then, if we want the logarithm function to be continuous, for small $\varepsilon > 0$ we must set $\log e^{i\varepsilon} = i\varepsilon$. Continuing in the same fashion, we see that $\log e^{it} = it$ for $0 \leq t < 2\pi$. But then, by continuity, $\log 1 = \log e^{2\pi i} = 2\pi i \neq 0$, a contradiction.

So, we cannot define logarithm on the whole of $\mathbb{C} \setminus \{0\}$; nevertheless, we can define it on various smaller open sets. A complete description of such "good" sets will be obtained in Chap. 6, and for the present we restrict ourselves to the following. Let $\ell_\alpha = \{re^{i\alpha} : r \geq 0\}$, where $\alpha \in \mathbb{R}$, be the ray starting at the origin and making angle α with the real axis. Set $V_\alpha = \mathbb{C} \setminus \ell_\alpha$ and $U_\alpha = \{z \in \mathbb{C} : \alpha < \operatorname{Im} z < \alpha + 2\pi\}$. Then the holomorphic function $z \mapsto e^z$ defines a bijection from U_α onto V_α (Fig. 2.1), and the derivative of this function does not vanish. Instead of U_α, though, we could consider the strip $\{z \in \mathbb{C} : \alpha + 2\pi n < \operatorname{Im} z < \alpha + 2\pi(n+1)\}$ for arbitrary $n \in \mathbb{Z}$: the

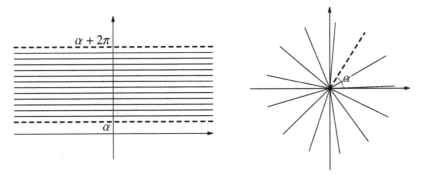

Fig. 2.1 The exponential function defines a bijection between the strip of height 2π and the complex plane cut along a ray. The horizontal lines in the left figure go to the rays in the right figure. The dotted lines do not belong to the sets.

exponential function induces a bijection of this strip onto its image in just the same way. Applying now Proposition 2.10, we obtain the following.

Proposition 2.12 *On the set* V_α, *obtained by removing the ray starting at the origin and making angle* α *with the real axis from the complex plane, for every integer n one can define a holomorphic function* log *by the formula*

$$\log(re^{it}) = \log r + it, \quad \alpha + 2\pi n < t < \alpha + 2\pi(n+1).$$

We have $e^{\log z} = z$ *and* $(\log z)' = 1/z$.

Each of the holomorphic functions from the statement of Proposition 2.12 is called a *single-valued branch of the logarithm*. Different branches of the logarithm differ by an additive constant (namely, $2\pi i k$ for some integer k); therefore, to fix a branch, it suffices to fix the value of the logarithm at one point.

The situation with the "nth root" function is very similar. First, $\sqrt[n]{\cdot}$ cannot be defined as a holomorphic (and actually even continuous) function on any open set containing the origin (for the simpler holomorphic case, see Exercise 2.9). Second, if $z = re^{i\varphi}$, then $\sqrt[n]{z} = \sqrt[n]{r} \cdot e^{i\varphi/n}$, but the choice of φ is not unique, and the value of the root depends on this choice: if w_0 is one of the nth roots of $z \neq 0$, then, adding multiples of 2π to the argument of z, we see that $w_0 e^{2\pi i/n}$, $w_0 e^{2 \cdot 2\pi i/n}, \ldots,$ $w_0 e^{2\pi i(n-1)/n}$ are also roots (then the values begin to repeat). On the whole of $\mathbb{C} \setminus \{0\}$, a single-valued "$n$th root" function still cannot be defined, but it can be defined on the complex plane cut along a ray starting at the origin.

Proposition 2.13 *Let* $n > 1$ *be a positive integer, and let* V_α, *with* $\alpha \in \mathbb{R}$, *denote the same open set as in Proposition 2.12. Then for every integer* $k \in [0; n-1]$ *one can define a holomorphic function* $\sqrt[n]{z}$ *on* V_α *by the formula*

$$\sqrt[n]{re^{i\varphi}} = \sqrt[n]{r}e^{i(\varphi+2\pi k)/n}, \quad \alpha < \varphi < \alpha + 2\pi.$$

We have $(\sqrt[n]{z})^n = z$, $(\sqrt[n]{z})' = \frac{1}{n(\sqrt[n]{z})^{n-1}}$.

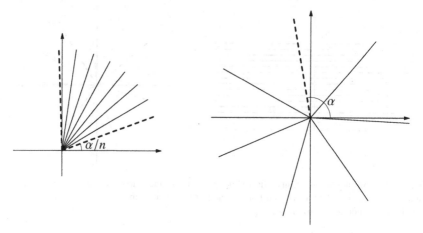

Fig. 2.2 The map $z \mapsto z^n$ (in our case, $n = 5$) takes an open sector of width $2\pi/n$ to the plane cut along a ray. Rays through the origin go to rays through the origin.

This proposition also directly follows from Proposition 2.10; for an open set U_α bijectively mapped onto V_α by the function $z \mapsto z^n$, take the sector

$$\left\{ re^{i\varphi} : \frac{\alpha}{n} + \frac{2\pi k}{n} < \varphi < \frac{\alpha}{n} + \frac{2\pi k}{n} + \frac{2\pi}{n} \right\}$$

(Fig. 2.2).

A single-valued branch of the root, as well as a single-valued branch of the logarithm, is also uniquely determined by its value at one point.

2.2 The Cauchy–Riemann Equations

Let us look at complex differentiability from another direction. If, for a moment, we turn a blind eye to the fact that complex numbers can be multiplied, then the complex plane $\mathbb{C}$ is indistinguishable from $\mathbb{R}^2$, and a function $f \colon U \to \mathbb{C}$ where U is an open set is indistinguishable from a map $f \colon U \to \mathbb{R}^2$. If we set $u = \operatorname{Re} f$, $v = \operatorname{Im} f$, then f can be written in the form

$$f \colon x + iy \mapsto u(x, y) + iv(x, y). \tag{2.4}$$

Proposition 2.14 *If a map f given by (2.4) is complex differentiable at some point, then at this point we have*

$$\begin{aligned}
\partial u/\partial x &= \partial v/\partial y, \\
\partial u/\partial y &= -\partial v/\partial x.
\end{aligned} \tag{2.5}$$

If the map f (or, equivalently, the functions u and v) is of class C^1, then f is holomorphic on U if and only if equations (2.5) hold on the whole of U.

Proof As we established in the proof of Proposition 2.5, a function f is complex differentiable at a point a with derivative equal to c if and only if

$$f(a+h) - f(a) = c \cdot h + \varphi(h) \quad \text{where } \lim_{h \to 0} |\varphi(h)|/|h| = 0.$$

Thus, the complex differentiability of f is equivalent to the fact that f is real differentiable (see Sec. 1.5) and its derivative, regarded as a linear map from $\mathbb{C}$ to $\mathbb{C}$, has the form $z \mapsto cz$ for some complex number c. Since the matrix of the linear operator "multiplication by a complex number $p + qi$" (in the basis $\{1, i\}$) has the form $\begin{pmatrix} p & -q \\ q & p \end{pmatrix}$, and the Jacobian matrix of the map (2.4) has the form

$$\begin{pmatrix} \partial u/\partial x & \partial u/\partial y \\ \partial v/\partial x & \partial v/\partial y \end{pmatrix},$$

the proposition clearly follows. $\qquad\square$

Equations (2.5) are called the *Cauchy–Riemann equations*.

The connection between complex and real derivatives described above implies an important property of holomorphic functions: they preserve angles between curves (at points where the derivative does not vanish).

To begin with, we introduce appropriate definitions. Let $[A; B] \subset \mathbb{R}$ be a closed interval.

Definition 2.15 A *smooth path* in the complex plane is a map $\gamma \colon [A; B] \to \mathbb{C}$ that is continuously differentiable on $[A; B]$ (in particular, has one-sided derivatives at the endpoints of the interval).

A smooth path $\gamma \colon [A; B] \to \mathbb{C}$ is called a *smooth curve* if γ is a one-to-one map onto its image and $\gamma'(t) \neq 0$ for all $t \in [A; B]$. The image $\gamma([A; B])$ is also sometimes called a smooth curve.

Definition 2.16 Let $\gamma \colon [A; B] \to \mathbb{C}$ be a smooth path. The *velocity vector* of γ at a point $t_0 \in [A; B]$ is the complex number

$$v = \gamma'(t_0) = \lim_{h \to 0} \frac{\gamma(t_0 + h) - \gamma(t_0)}{h}.$$

Proposition 2.17 *Let $f \colon U \to \mathbb{C}$ be a holomorphic function, where $U \subset \mathbb{C}$ is an open set. If for a point $a \in U$ we have $f'(a) \neq 0$, then the angle between any two curves passing through a is equal to the angle between their images under f.*

By the angle between two curves γ_1 and γ_2 passing through a we mean the angle between their tangent (velocity) vectors γ_1' and γ_2' at a.

Proof As noted above, the real derivative of f at a is the operator of multiplication by the nonzero complex number $f'(a)$, and the multiplication by a complex number, being a similarity transformation, preserves angles.

For the reader not yet fully comfortable with derivatives of multivariable functions, here is a more down-to-earth argument. If a curve passing through a is parametrized as $t \mapsto \gamma(t)$ where γ is a smooth map defined on an interval $(-\varepsilon; \varepsilon)$ with $\gamma(0) = a$, then the tangent vector to this curve at a is $\gamma'(0)$. The image of the curve under f is parametrized as $t \mapsto f(\gamma(t))$, so the tangent vector to the image is

$$(f(\gamma(t)))'(0) = f'(\gamma(0))\gamma'(0) = f'(a)\gamma'(0)$$

(Proposition 2.5 remains valid, with the same proof, if the "inner" function is a smooth map to $\mathbb{C}$ from a real interval). Thus, the tangent vector to the image of γ is obtained by rotating the tangent vector to γ by the angle $\arg(f'(a))$, so the angle between the tangent vectors to two curves does not change.   □

If $f'(a) = 0$, then the angle between curves passing through a is no longer preserved: for instance, the holomorphic map $z \mapsto z^2$ takes the pair consisting of the "positive" rays of the real and imaginary axes (starting at the origin) to the pair consisting of the "positive" and "negative" rays of the real axis (see also Fig. 2.2). Nevertheless, in this case the angle between curves changes in a controllable way, and later we will find out how exactly.

With regard to the angle-preserving property, it is appropriate to introduce the following definition.

Definition 2.18 Let $U, V \subset \mathbb{C}$ be open sets. A map $f \colon U \to V$ is said to be *conformal* if it is holomorphic and bijective.

As we observed above (see Remark 2.11), this condition implies that the derivative of f does not vanish and the inverse map $f^{-1} \colon V \to U$ is also holomorphic (though we cannot prove this right now).

The term "conformal" stems from a Latin word meaning "having the same form," apparently because infinitesimally[1] such a map is a similarity transformation.

There is another way to write the Cauchy–Riemann equations, different from (2.5). To describe it, first note that the derivative of every real differentiable map $f \colon U \to \mathbb{C}$ is a map $L \colon \mathbb{C} \to \mathbb{C}$ satisfying the *real linearity* condition: for any $\lambda, \mu \in \mathbb{R}$ and $z_1, z_2 \in \mathbb{C}$,

$$L(\lambda z_1 + \mu z_2) = \lambda L(z_1) + \mu L(z_2).$$

Proposition 2.19 *A map* $L \colon \mathbb{C} \to \mathbb{C}$ *is real linear if and only if there exist* $A, B \in \mathbb{C}$ *such that*

$$L(z) = Az + B\bar{z} \tag{2.6}$$

for every z.

---

[1] Saying that "a map has a property infinitesimally," one usually means that the property holds for the derivative of this map.

Proof The real linearity of the map defined by (2.6) is obvious. Conversely, if a map $L\colon \mathbb{C} \to \mathbb{C}$ is real linear, then, identifying complex numbers $z = x + iy$ with real column vectors $\binom{x}{y}$, this map takes the form

$$L\colon \binom{x}{y} \mapsto \begin{pmatrix} a & b \\ c & d \end{pmatrix}\binom{x}{y}$$

with $a, b, c, d \in \mathbb{R}$. Substituting $x = (z + \bar{z})/2$, $y = (z - \bar{z})/2i$, we obtain $L(x + iy) = L(z) = Az + B\bar{z}$, where

$$A = \frac{a + d + i(c - b)}{2}, \quad B = \frac{a - d + i(b + c)}{2}. \tag{2.7}$$

Now suppose that we are given an open set $U \subset \mathbb{C}$ and a function $f\colon U \to \mathbb{C}$ (no other open sets and functions will be encountered in the rest of this section).

Proposition-Definition 2.20 If f is real differentiable at a point $a \in U$, then its real derivative at this point has the form

$$h \mapsto \frac{\partial f}{\partial z} \cdot h + \frac{\partial f}{\partial \bar{z}} \cdot \bar{h}$$

(all partial derivatives are taken at a) where

$$\frac{\partial f}{\partial z} = \frac{1}{2}\left(\frac{\partial f}{\partial x} - i\frac{\partial f}{\partial y}\right), \quad \frac{\partial f}{\partial \bar{z}} = \frac{1}{2}\left(\frac{\partial f}{\partial x} + i\frac{\partial f}{\partial y}\right). \tag{2.8}$$

Proof Writing $f(x + iy) = u(x, y) + iv(x, y)$, the real derivative of f is the operator of multiplying the matrix

$$\begin{pmatrix} \partial u/\partial x & \partial u/\partial y \\ \partial v/\partial x & \partial v/\partial y \end{pmatrix}$$

by a column vector. Now everything follows from (2.7). □

Equations (2.8) are definitions of the operations (differential operators) $\partial/\partial z$ and $\partial/\partial \bar{z}$.

Now we can give an alternative statement of the Cauchy–Riemann equations promised above.

Proposition 2.21 *A function f is complex differentiable at a point a if and only if $(\partial f/\partial \bar{z})(a) = 0$.*

If f is of class C^1, then it is holomorphic on U if and only if $\partial f/\partial \bar{z} = 0$ on the whole of U.

Proof Indeed, a map of the form (2.6) is the multiplication by a complex number if and only if $B = 0$; in our case, $B = \partial f/\partial \bar{z}$. □

The equation $\partial f/\partial \bar{z} = 0$ is also called the Cauchy–Riemann equation. It is completely equivalent to the system of equations (2.5), but not useless: many important formulas look much simpler and much more natural in terms of $\partial f/\partial z$ and $\partial f/\partial \bar{z}$. Some examples can be found in the exercises at the end of this chapter, as well as in Chap. 10.

Exercises

2.1. Draw the images of the rays starting at the origin under the map $z \mapsto e^z$.
2.2. Show that the function $f(z) = \bar{z}$ is not complex differentiable at any point.
2.3. At what points is the function $f(z) = |z|^2$ complex differentiable?
2.4. Show that the function

$$f: re^{i\varphi} \mapsto re^{2i\varphi}$$

is not holomorphic on $\mathbb{C} \setminus \{0\}$.
2.5. Show that the sine and cosine are periodic functions with period 2π (i.e., that $\sin(z + 2\pi) = \sin z$ and $\cos(z + 2\pi) = \cos z$ for every $z \in \mathbb{C}$).
2.6. Is it true that $|\sin z| \leq 1$ for every $z \in \mathbb{C}$?
2.7. Prove the following identities (for arbitrary complex z and w):

$$\sin^2 z + \cos^2 z = 1;$$
$$\sin(z + w) = \sin z \cos w + \cos z \sin w;$$
$$\cos(z + w) = \cos z \cos w - \sin z \sin w.$$

You are supposed to solve this exercise using Definition 2.8. In Chap. 5 we will see how these identities can be proved in one line without any calculations.
2.8. Solve in $\mathbb{C}$ the simplest trigonometric equations: $\sin z = 0$, $\cos z = -1$, etc.
2.9. Let $n > 1$ be a positive integer and U be an open set containing 0. Show that on U one cannot define a holomorphic "nth root" function, i.e., a holomorphic function f satisfying the identity $(f(z))^n = z$.
2.10. Find all real a and b for which the function

$$f(x + iy) = ax^2 + by^2 + ixy$$

is holomorphic on the entire complex plane $\mathbb{C}$.
2.11. Assume that a function $f: \mathbb{C} \to \mathbb{C}$ is holomorphic and satisfies the relation

$$\operatorname{Im} f(z) = -\operatorname{Re} f(z)$$

for all z. Show that f is constant.
2.12. Let $U \subset \mathbb{C}$ be an open set and $f: U \to \mathbb{C}$ be a function of class C^1.
(a) Show that if f is holomorphic, then its Jacobian (the determinant of the Jacobian matrix) at a point $a \in U$ is equal to $|f'(a)|^2$.
(b) Show that in the general case the Jacobian of f at a is equal to

$$\left|\frac{\partial f}{\partial z}(a)\right|^2 - \left|\frac{\partial f}{\partial \bar{z}}(a)\right|^2.$$

(*Hint.* See formulas (2.7).)

Chapter 3
A Tutorial on Conformal Maps

This chapter contains neither complicated definitions, nor deep theorems: its main purpose is to give examples of how one can explicitly (i.e., in terms of elementary functions) construct a conformal map between given two subsets of the complex plane, to help the reader better grasp the geometric meaning of elementary functions of a complex variable.

Definition 3.1 If there is a conformal map between open sets $U, V \subset \mathbb{C}$, then they are said to be *conformally isomorphic* (or simply isomorphic).

A conformal map from an open set $U \subset \mathbb{C}$ onto itself is called a *conformal automorphism* of U.

A conformal map between given two open sets exists quite often, but not always; and if it does exist, it is by no means always expressible in terms of elementary functions. Let us learn to do what little we can.

3.1 Linear Fractional Transformations

Example 3.2 Let us find the image of the unit circle (the circle of radius 1 centered at the origin) under the linear fractional transformation $z \mapsto (z-1)/(z+1)$.

Let z be a point of the unit circle. By a well-known theorem of plane geometry ("an inscribed angle subtended by a diameter is a right angle"), the triangle formed by the points 1, z, and -1 is right-angled (Fig. 3.1); thus, if z lies on the unit circle and in the upper half-plane, then the argument of $(z-1)/(z+1)$ is equal to $\pi/2$ (and if z lies in the lower half-plane, then the argument of the same number is equal to $-\pi/2$); in any case, $(z-1)/(z+1)$ lies on the imaginary axis. Hence, the transformation $z \mapsto (z-1)/(z+1)$ takes the unit circle to the imaginary axis.

Since linear fractional transformations are continuous one-to-one maps from the Riemann sphere to itself, it follows that the unit disk, i.e., the set $\{z : |z| < 1\}$, gets mapped onto one of the two half-planes into which the imaginary axis divides the complex plane. To find out which of the two, it suffices to look at the image of

© Springer Nature Switzerland AG 2020
S. Lvovski, *Principles of Complex Analysis*, Moscow Lectures 6,
https://doi.org/10.1007/978-3-030-59365-0_3

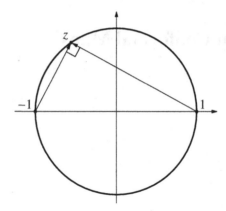

Fig. 3.1 The map $z \mapsto (z-1)/(z+1)$ takes the unit circle to the imaginary axis

one point. The point 0 is mapped to the point $(0-1)/(0+1) = -1$, which lies in the left half-plane $\{z: \operatorname{Re} z < 0\}$; if we then multiply by $-i$ (i.e., rotate by 90° clockwise), then the left half-plane will be mapped onto the upper one. So, we obtain the following result.

The function $z \mapsto i(1-z)/(1+z)$ is a conformal map from the unit disk $\{z: |z| < 1\}$ onto the upper half-plane $\{z: \operatorname{Im} z > 0\}$.

The constructed linear fractional transformation from the unit disk onto the upper half-plane is by no means unique (soon we will find out the degree of nonuniqueness).

Example 3.3 Let us, on the contrary, construct a conformal map from the upper half-plane onto the unit disk. Of course, one of the corresponding linear fractional transformations can be constructed by inverting the map from Example 3.2: write the equation $z = i(1-w)/(1+w)$ and solve it for w. However, it is more fun to construct this map from geometric considerations too.

Namely, observe that the real axis consists of the points equidistant from i and $-i$, i.e., if $\operatorname{Im} z = 0$, then $|z-i| = |z+i|$, or

$$\left|\frac{z-i}{z+i}\right| = 1.$$

This means that the map $z \mapsto (z-i)/(z+i)$ takes the real axis to the unit circle (the point $\infty \in \overline{\mathbb{C}}$ is mapped to 1). Thus, it takes the upper half-plane either to the interior of the circle (the disk $\{z: |z| < 1\}$), or to its exterior $\{z: |z| > 1\}$. One easily sees that it is the first of these possibilities that occurs: for example, nothing from the upper half-plane is mapped to infinity, while in the lower half-plane the point $-i$ is mapped to infinity. Another way is to observe that the point i from the upper half-plane is mapped to the point 0 from the interior of the unit circle. Let us summarize.

The function $z \mapsto (z-i)/(z+i)$ is a conformal map from the upper half-plane $H = \{z: \operatorname{Im} z > 0\}$ onto the unit disk $U = \{z: |z| < 1\}$.

We emphasize that, again, this linear fractional transformation from the upper half-plane onto the unit disk is not unique.

Well, let us look into this nonuniqueness. If $f_1, f_2 \colon H \to U$ are two isomorphisms, then $f_1 = f_2 \circ g$ where $g = f_2^{-1} \circ f_1$ is a linear fractional automorphism of the upper half-plane H, and also $f_2 = h \circ f_1$ where $h = f_1 \circ f_2^{-1}$ is a linear fractional automorphism of the unit disk U.

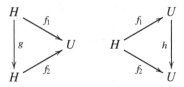

Thus, to determine the degree of nonuniqueness of linear fractional isomorphisms between the unit disk and the upper half-plane, we must describe the linear fractional automorphisms of these sets.

Proposition 3.4 *The linear fractional automorphisms of the upper half-plane* $H = \{z \colon \operatorname{Im} z > 0\}$ *are the maps of the form*

$$z \mapsto \frac{az+b}{cz+d}, \quad a, b, c, d \in \mathbb{R}, \quad \begin{vmatrix} a & b \\ c & d \end{vmatrix} = 1.$$

The group of linear fractional automorphisms of the upper half-plane is isomorphic to $\mathrm{SL}_2(\mathbb{R})/\{\pm I\}$, *where* $\mathrm{SL}_2(\mathbb{R})$ *is the group of real matrices with determinant 1 and I is the identity matrix.*

Proof A linear fractional transformation is an automorphism of the upper half-plane H if it takes its boundary (i.e., the real axis) to itself and sends at least one point of H to a point of H. It is clear from the proof of Proposition 1.31 that a linear fractional transformation sends real numbers to real numbers if and only if it can be written in the form $z \mapsto (az+b)/(cz+d)$ with real a, b, c, d; every such linear fractional transformation takes H either to itself or to the lower half-plane. To distinguish between these two cases, look at the imaginary part of the image of the point $i \in H$:

$$\operatorname{Im} \frac{ai+b}{ci+d} = \operatorname{Im} \frac{(ai+b)(-ci+d)}{c^2+d^2} = \frac{ad-bc}{c^2+d^2}.$$

The right-hand side is positive if and only if the determinant of the matrix $\begin{pmatrix} a & b \\ c & d \end{pmatrix}$ is positive. If we multiply all elements of a matrix by a real number $\lambda \neq 0$, then the linear fractional transformation corresponding to this matrix remains unchanged, while the determinant gets multiplied by λ^2, so by such a multiplication every matrix with positive determinant can be transformed into a matrix with determinant 1; this proves the first assertion.

To prove the second one, recall that composing linear fractional transformations corresponds to multiplying the corresponding matrices (see Exercise 1.16); it is easy to see that a matrix corresponds to the identity linear fractional transformation if and only if it is a multiple of the identity matrix, and that in $SL_2(\mathbb{R})$ only I and $-I$ are multiples of the identity matrix; this proves the second assertion.   □

In principle, a description of the linear fractional automorphisms of the unit disk could be obtained from Proposition 3.4: if $\varphi\colon H \to U$ is a linear fractional isomorphism, then every automorphism of U has the form $\varphi \circ f \circ \varphi^{-1}$ where f is an automorphism of the upper half-plane (see the diagram (3.1)).

$$
\begin{array}{ccc}
H & \xrightarrow{\;\varphi\;} & U \\
\downarrow{\scriptstyle f} & & \downarrow{\scriptstyle \varphi\circ f\circ\varphi^{-1}} \\
H & \xrightarrow{\;\varphi\;} & U
\end{array}
\qquad (3.1)
$$

However, it takes some doing to obtain a clear-cut description along these lines; it is more convenient to describe the automorphisms of the unit disk independently.

Proposition 3.5 *The linear fractional automorphisms of the unit disk $U = \{z\colon |z| < 1\}$ are the maps of the form*

$$
z \mapsto e^{i\theta}\frac{z - a}{1 - \bar{a}z}, \qquad \theta \in \mathbb{R},\ |a| < 1. \qquad (3.2)
$$

Proof First, we show that every map of the form (3.2) is an automorphism of the disk. Since the multiplication by $e^{i\theta}$ with $\theta \in \mathbb{R}$ is a rotation around 0, it suffices to check that the map $z \mapsto (z-a)/(1-\bar{a}z)$ takes the unit circle to itself and sends some point of the unit disk to a point of the unit disk. The second assertion is obvious, since the point a of the unit disk is mapped to 0. As to the first one, let z be a point of the unit circle, i.e., $|z| = 1$; then $z^{-1} = \bar{z}$. Now we have

$$
\left|\frac{z-a}{1-\bar{a}z}\right| = |z|\cdot\left|\frac{1-az^{-1}}{1-\bar{a}z}\right| = \left|\frac{1-a\bar{z}}{1-\bar{a}z}\right| = \frac{|1-a\bar{z}|}{|1-\bar{a}z|} = 1,
$$

since the absolute values of conjugate numbers coincide.

Conversely, let $f\colon U \to U$ be a linear fractional automorphism, and let $a \in U$ be the point mapped to 0. Then the point $1/\bar{a}$, which is symmetric to a with respect to the unit circle (see Exercise 1.19), is mapped to the point symmetric to 0 with respect to the unit circle, i.e., to ∞ (now we regard f as a map from $\overline{\mathbb{C}}$ to itself). Thus, the map has the form

$$
f(z) = c\cdot\frac{z-a}{z-1/\bar{a}} = c_1\cdot\frac{z-a}{1-\bar{a}z}, \qquad c_1 = -c\bar{a}.
$$

By the calculations above,

$$
|z| = 1 \Rightarrow \left|c_1\cdot\frac{z-a}{1-\bar{a}z}\right| = |c_1|,
$$

and since f is an automorphism of the unit circle, we have $|f(z)| = 1$ for $|z| = 1$, whence $|c_1| = 1$, so $c_1 = e^{i\theta}$, $\theta \in \mathbb{R}$. Thus, f has the form (3.1), and the proof is completed.   □

Later we will show that all conformal automorphisms of the upper half-plane and the unit disk are necessarily linear fractional and thus have the form described in Propositions 3.4 and 3.5.

3.2 More Complicated Maps

Bringing in maps other than linear fractional transformations, we can increase the stock of sets that can be conformally mapped onto the upper half-plane by elementary functions.

We are already acquainted with some conformal maps which are not linear fractional. For example, in Fig. 2.1 we have seen that the function $z \mapsto e^z$ establishes a conformal bijection between a horizontal strip of height 2π and the complex plane cut along a ray; if $\varphi < 2\pi$, then the same function is a conformal map between the strip $U_1 = \{z: \alpha < \operatorname{Im} z < \alpha + \varphi\}$ of height φ and the sector $U_2 = \{z: \alpha < \arg z < \alpha + \varphi\}$ (indeed, if $z = x + iy$ where x is arbitrary and $\alpha < y < \alpha + \varphi$, then $e^z = e^x e^{iy}$ where $|z| = e^x$ is an arbitrary positive number and $\arg e^z = y$ lies between α and $\alpha + \varphi$). In particular, if $\varphi = \pi$, then the function $z \mapsto e^z$ is a conformal map from the horizontal strip $\{z: 0 < \operatorname{Im} z < \pi\}$ onto the upper half-plane ("a sector of width π") $\{z: \operatorname{Im} z > 0\}$. To obtain conformal maps of other strips onto the upper half-plane, one should first apply a translation (add a constant), rotation, and dilation centered at the origin (multiply by a constant).

Further, the interior of every sector can be mapped onto the upper half-plane: if, for instance, the sector is of width π/a and its vertex is at the origin, then the map $z \mapsto z^a = e^{a \log z}$ takes this sector to some half-plane (and if we want it to be the upper half-plane, we will possibly need to rotate it, i.e., multiply by a complex number with an appropriate argument).

Here is a somewhat less trivial example.

Example 3.6 Let us map the half-disk $U = \{z: |z| < 1,\ \operatorname{Im} z > 0\}$ (Fig. 3.2a) onto the upper half-plane. The boundary of U consists of a straight line and a semicircle. The idea is to first "straighten" the curvilinear part of the boundary. For this, we can use a linear fractional transformation that sends some point of the circle to infinity. In order for the rectilinear part of the boundary to remain rectilinear rather than become a circle, we must choose this point so as to lie on the rectilinear part of the boundary too. There are two such points: 1 and -1; for definiteness, take the point 1 and consider the linear fractional transformation

$$z \mapsto z_1 = \frac{1}{z - 1}$$

(any other linear fractional transformation whose denominator vanishes at 1 would also do). Under this transformation, the segment $[-1; 1]$ of the real axis is mapped to the ray $(-\infty; -1/2]$. The semicircle $\{e^{it} : 0 \leq t \leq \pi\}$ is mapped, obviously, to a subset of a line (since 1 goes to ∞); clearly, it is a ray starting at $-1/2$; since linear fractional transformations preserve angles, this ray is perpendicular to the image of the segment $[-1; 1]$, i.e., vertical. To see whether it is directed upwards or downwards, we can look at the orientation of the angle: as viewed from the vertex -1, the direction of the shortest rotation from the rectilinear part of the boundary to the curvilinear one is counterclockwise, and this direction must be preserved under a conformal map. However, to avoid confusion, it is easier to find the image of another point of the arc. Clearly, the point i goes to the point $(-1 - i)/2$ in the lower half-plane. Therefore, the image of the semicircle is the downward ray starting at $-1/2$, and the whole half-disk is mapped onto the interior of the right angle with vertex at $-1/2$ (see Fig. 3.2b).

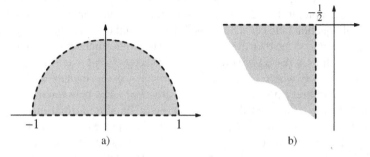

a)   b)

Fig. 3.2 The map $z \mapsto z_1 = \frac{1}{z-1}$ takes a half-disk (a) to the interior of a right angle (b)

The rest is easy: apply the shift $z_1 \mapsto z_1 + \frac{1}{2}$ to move the vertex of the right angle to the origin, then squaring will map it onto the upper half-plane (see Fig. 3.3). Collecting everything together, we see that a conformal map from the half-disk onto the half-plane is given by the formula

$$z \mapsto \left(\frac{1}{z-1} + \frac{1}{2}\right)^2 = \frac{1}{4}\left(\frac{z+1}{z-1}\right)^2.$$

Of course, the coefficient $\frac{1}{4}$ may be omitted.

Example 3.7 Let us look at how the map $f : z \mapsto z + \frac{1}{z}$ acts on the set $U = \{z : |z| > 1\}$ (the exterior of the unit disk).

The equality $a = f(z)$ means that z is a root of the quadratic equation

$$z^2 - az + 1 = 0. \tag{3.3}$$

By Vieta's formulas, the product of the two roots of this equation is equal to 1, so if one of them is less than 1 in absolute value, then the other one is greater than 1

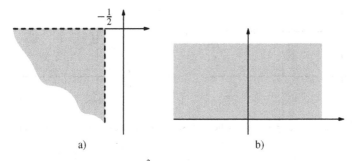

Fig. 3.3 The map $z_1 \mapsto z_2 = \left(z_1 + \frac{1}{2}\right)^2$ takes the interior of a right angle (a) onto the upper half-plane (b)

in absolute value. Thus, the only case where a point $a \in \mathbb{C}$ does not lie in the image of the exterior of the unit disk is when both roots of equation (3.3) have absolute value 1. Since the product of these roots is equal to 1, they have the form $e^{\pm i\theta}$ with $\theta \in \mathbb{R}$, so, again by Vieta's formulas,

$$a = e^{i\theta} + e^{-i\theta} = 2\cos\theta, \quad \theta \in \mathbb{R}.$$

Now, $a = 2\cos\theta$, $\theta \in \mathbb{R}$, if and only if a is real and $-2 \le a \le 2$, hence the image of the exterior of the unit disk under the map f is $\mathbb{C} \setminus [-2; 2]$, i.e., the complex plane with the segment $[-2; 2]$ of the real axis removed. Since for $a \notin [-2; 2]$ exactly one root of equation (3.3) lies in the exterior of the unit disk, we arrive at the following conclusion: *the function $f(z) = z + \frac{1}{z}$ is a conformal map between the exterior of the unit disk and the complex plane with the segment $[-2; 2]$ of the real axis removed.*

In applied mathematics, the function f from this example is called the *Joukowsky transform.*

Example 3.8 Let $U = \{z \colon \operatorname{Re} z > 0, 0 < \operatorname{Im} z < 1\}$ (semi-strip, Fig. 3.4a), and let us map U onto the upper half-plane. Applying the exponential function as in the case of an entire strip, we see that the function $z \mapsto z_1 = e^{\pi z}$ maps U onto a part of the half-plane, namely, the part consisting of the numbers greater than 1 in absolute value (since $\operatorname{Re}(\pi z) > 0$, we have $|e^{\pi z}| > 1$, see Fig. 3.4). Now, the resulting half-plane with a semicircular cavity can be mapped onto the upper half-plane by straightening the circular part of the boundary as in Example 3.6. However, we can also use a ready-made formula: it is easy to see that the Joukowsky transform $z_1 \mapsto z_2 = z_1 + (1/z_1)$ sends points lying in the upper half-plane outside the unit disk to the upper half-plane, so it is a conformal map from the half-plane with a cavity onto the entire half-plane. Taking the composition of the two maps, we see that the function

$$z \mapsto e^{\pi z} + e^{-\pi z} = 2\cos\pi i z$$

maps the semi-strip onto the upper half-plane. Of course, the coefficient 2 may be omitted.

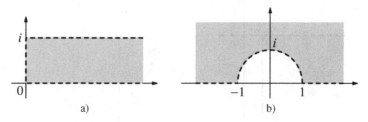

Fig. 3.4 The map $z \mapsto z_1 = e^{\pi z}$ takes the semi-strip (a) onto the upper half-plane with a cavity (b)

Finally, we give an example of a conformal map from a "cut domain," i.e., an open set with some line segments or circular arcs removed.

Example 3.9 Let $U = \{z \colon \mathrm{Im}\, z > 0\} \setminus \{it \colon 0 < t \le 1\}$, that is, U is the upper half-plane with the line segment between the points 0 and i removed (Fig. 3.5a).

Fig. 3.5 The map $z \mapsto z_1 = z^2$ takes the half-plane cut along the segment $[0; i]$ (a) onto the plane cut along the ray $[-1; +\infty)$ (b)

To construct a conformal map from U onto the upper half-plane, we (partially) straighten its boundary by applying the map $z \mapsto z_1 = z^2$. Then the segment between 0 and i goes to the segment $[-1; 0]$ of the real axis, the rays $(-\infty; 0]$ and $[0; +\infty)$ «merge» into the ray $[0; +\infty)$ (these segment and ray do not belong to the image of U), and the resulting domain is the plane with the ray $[-1; +\infty)$ removed (in the classical terminology, the plane cut along a ray, Fig. 3.5b). The rest is easy. If we apply the map $z_1 \mapsto z_2 = z_1 + 1$, then the plane cut along the ray $[-1; +\infty)$ goes to the plane cut along the ray $[0; +\infty)$, and the latter domain can be mapped onto the upper half-plane by the function $z_2 \mapsto z_3 = \sqrt{z_2}$. Collecting everything together, we see that the open set U is conformally mapped onto the upper half-plane by the function $z \mapsto \sqrt{z^2 + 1}$.

It only remains to specify which "branch" of the square root we use. Since the square root was applied to map $\mathbb{C} \setminus [0; +\infty)$ onto the upper half-plane and not the lower one, this is, obviously, the branch for which $\sqrt{-1} = i$ (and not $-i$).

Exercises

3.1. Find and draw the image of the line $\{z\colon \operatorname{Re} z + \operatorname{Im} z = 1\}$ under the map $z \mapsto 1/z$.

3.2. The same for the line $\{z\colon \operatorname{Re}(z) = -1\}$ and the map $z \mapsto z/(z-2)$.

3.3. Find and draw the image of the set $\{z\colon \operatorname{Im}(z) < 0, |z| < 2\}$ under the map $z \mapsto z^2$.

3.4. Construct a conformal automorphism of the unit disk $U = \{z\colon |z| < 1\}$ that sends $1/2$ to $2/3$.

3.5. Construct a conformal automorphism of the set $U = \{z\colon \operatorname{Re} z > 0, \operatorname{Im} z > 0\}$ that sends $1 + i$ to $1 + i\sqrt{3}$.

For each of the open sets U listed below, construct a conformal map from U onto the upper half-plane.

3.6. $U = \{z\colon 1 < \operatorname{Re}(z) + \operatorname{Im}(z) < 2\}$.

3.7. $U = \{z\colon (\operatorname{Re} z)^2 \leq (\operatorname{Im} z/2)^2 - 1\}$ (*hint*: see Exercise 1.4).

3.8. $U = \left\{z\colon |z - \sqrt{3}| > 2, \left|z - \frac{\sqrt{3}}{3}\right| < 1\right\}$.

3.9. $U = D \setminus [0; 1]$ (here $D = \{z\colon |z| < 1\}$ is the unit disk, $[0; 1]$ is the segment of the real line between the points 0 and 1).

3.10. $U = D \setminus [1/2; 1]$ (here $D = \{z\colon |z| < 1\}$ is the unit disk, $[1/2; 1]$ is the segment of the real line between the points $1/2$ and 1).

3.11. $U = \mathbb{C} \setminus \{it\colon t \in \mathbb{R}, t \geq 1\}$.

3.12. $U = \mathbb{C} \setminus ((-\infty; -1] \cup [1; +\infty))$ (here $(-\infty; -1]$ and $[1; +\infty)$ are rays of the real line).

3.13. $U = \{z\colon \operatorname{Re} z > 0, 0 < \operatorname{Im} z < \pi\} \setminus \{t + \pi i/2 : 0 \leq t \leq 1\}$ (semi-strip with a cut).

3.14. $U = H \cup D$ where $H = \{z\colon \operatorname{Im} z > 0\}$ is the upper half-plane, $D = \{z\colon |z| < 1\}$ is the unit disk.

3.15. $U = \{z\colon \operatorname{Re} z < (\operatorname{Im} z)^2, \operatorname{Im} z > 0\}$.

3.16. Show that the map $z \mapsto z + (1/z)$ takes circles centered at the origin (with radius different from 1) to ellipses with foci at the points 2 and -2.

3.17. Find the images of the rays starting at the origin under the map $z \mapsto z + (1/z)$.

3.18. Construct a conformal map from the set

$$V = \left\{z\colon \frac{(\operatorname{Re} z)^2}{a^2} + \frac{(\operatorname{Im} z)^2}{b^2} > 1\right\}, \quad a, b > 0$$

(the exterior of an ellipse) onto the set $W = \{z\colon |z| > 1\}$ (the exterior of the unit disk).

Chapter 4
Complex Integrals

4.1 Basic Definitions

In complex analysis, functions are usually integrated over paths. In this course, I will not try to define integrals in greatest generality: integrals over piecewise smooth paths will quite suffice for our modest purposes. We slightly generalize Definition 2.15.

Definition 4.1 A continuous map $\gamma\colon [A; B] \to \mathbb{C}$, where $[A; B] \subset \mathbb{R}$ is a closed interval, is called a *piecewise smooth path* in $\mathbb{C}$ if there exists a finite partition $A = A_0 < A_1 < \ldots < A_n = B$ such that the restriction of γ to each segment $[A_j; A_{j+1}]$ is a smooth path.

In particular, a piecewise smooth path necessarily has one-sided derivatives at every point. In what follows, the word "path" always means "piecewise smooth path" unless otherwise stated.

Definition 4.2 Let $\gamma\colon [A; B] \to \mathbb{C}$ be a piecewise smooth path and f be a continuous complex-valued function defined on the set $\gamma([A; B]) \subset \mathbb{C}$ (or on an open set containing $\gamma([A; B])$). Then the *integral* of f over γ is the number

$$\int_{\gamma} f(z)\, dz := \int_{A}^{B} f(\gamma(t))\gamma'(t)\, dt.$$

(Motivation: if $z = \gamma(t)$, then $dz = \gamma'(t)\, dt$, as in the change of variable formula.)

Definition 4.2 involves a parametrization of the path, but, as the following proposition shows, the dependence on the choice of such a parametrization is not so strong.

Proposition 4.3 *Assume that the conditions of Definition 4.2 are satisfied and $\varphi\colon [A_1; B_1] \to [A; B]$ is a bijective differentiable function whose derivative is everywhere positive. Then*

© Springer Nature Switzerland AG 2020
S. Lvovski, *Principles of Complex Analysis*, Moscow Lectures 6,
https://doi.org/10.1007/978-3-030-59365-0_4

$$\int_{\gamma \circ \varphi} f(z)\, dz = \int_{\gamma} f(z)\, dz.$$

If the derivative of φ is everywhere negative, then

$$\int_{\gamma \circ \varphi} f(z)\, dz = -\int_{\gamma} f(z)\, dz.$$

Proof This follows immediately from the change of variable formula for definite integrals. For example, if φ is increasing, i.e., $\varphi(A_1) = A$ and $\varphi(B_1) = B$, we have

$$\int_A^B f(\gamma(t))\gamma'(t)\, dt = \int_{A_1}^{B_1} f(\gamma(\varphi(u)))\gamma'(\varphi(u))\varphi'(u)\, dt$$

$$= \int_{A_1}^{B_1} f(\gamma \circ \varphi(u))(\gamma \circ \varphi)'(u)\, du$$

(by the chain rule formula for $\gamma \circ \varphi$). $\square$

Bearing in mind what we have just proved, we introduce another definition.

Definition 4.4 A *piecewise smooth curve* is a subset in $\mathbb{C}$ of the form $\gamma([A;B])$ where $\gamma \colon [A;B] \to \mathbb{C}$ is a piecewise smooth path, γ is one-to-one onto its image, and $\gamma'(t) \neq 0$ for all $t \in [A;B]$ except, possibly, finitely many points.

The following convention is common in complex analysis.

Convention 4.5 If $\gamma \subset \mathbb{C}$ is a piecewise smooth curve, then $\int_{\gamma} f(z)\, dz$ stands for the integral over a piecewise smooth path parameterizing γ, for an arbitrary *bijective* piecewise smooth parametrization. In view of Proposition 4.3, the value of such an integral is uniquely determined up to sign. If the curve is endowed with an orientation, then the integral is uniquely determined.

If the orientation of the curve is not clear from the context, it will be indicated explicitly.

The next two remarks concern integrals over closed curves. First, if $\gamma \subset \mathbb{C}$ is a closed piecewise smooth curve (or $\gamma \colon [A;B] \to \mathbb{C}$ is a piecewise smooth path for which $\gamma(A) = \gamma(B)$), then the integral over γ (with the chosen orientation) does not depend not only on the parametrization, but also on the choice of the initial (= final) point on the curve, i.e., the point $\gamma(A) = \gamma(B)$. For instance, in Fig. 4.1 if we choose p as the initial point and the "counterclockwise" direction, then the integral over γ is equal to the sum "the integral from p to q plus the integral from q to p"; and if we choose q as the initial point (and the same direction), then the integral over γ is equal to the sum "the integral from q to p plus the integral from p to q": order does not matter for addition!

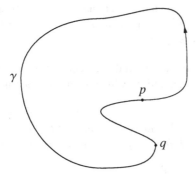

Fig. 4.1 The integral over a closed path does not depend on the choice of the initial point

Second, for closed non-self-crossing curves in the plane, the "counterclockwise" direction will be called *positive*, and the opposite one, *negative*. If the direction is not explicitly indicated, we assume it to be positive.

Remark 4.6 The pedantic reader may notice that the mathematical notions of "clockwise" and "counterclockwise" do not exist, so the "positive direction of traversing a curve" remains undefined. In justification I may say that rigorously defining an orientation of a closed curve and, most importantly, using this definition in proofs require a deep dive into topology, inappropriate for an introductory course.

However, I still mention that a closed non-self-crossing curve γ is positively oriented if its index around every point of $\mathbb{C}$ that does not lie on γ is equal to 0 or 1 (see Sec. 4.2).

Now we give an example of evaluating an integral which plays a very important role in what follows.

Example 4.7 If γ is a positively (i.e., counterclockwise) oriented circle of radius $r > 0$ centered at a point $a \in \mathbb{C}$, then

$$\int_\gamma \frac{dz}{z-a} = 2\pi i.$$

Indeed, parameterizing the circle as $\gamma(t) = a + re^{it}$, $t \in [0; 2\pi]$, we have

$$z = \gamma(t) = a + re^{it}; \qquad dz = \gamma'(t)\,dt = ire^{it}\,dt;$$

$$\frac{dz}{z-a} = \frac{ire^{it}\,dt}{re^{it}} = i\,dt; \qquad \int_\gamma \frac{dz}{z-a} = \int_0^{2\pi} i\,dt = 2\pi i.$$

For integrals over curves, an analog of the fundamental theorem of calculus holds.

Definition 4.8 Let $U \subset \mathbb{C}$ be an open set and $f \colon U \to \mathbb{C}$ be a continuous function. We say that a function $F \colon U \to \mathbb{C}$ is an *antiderivative* of f if F is holomorphic on U and $F'(z) = f(z)$ for every $z \in U$.

Remark 4.9 The continuity of f is a redundant condition: in Sec. 5.3 we will see that if a complex function has an antiderivative, then it is not only continuous, but even holomorphic!

Proposition 4.10 *Let $U \subset \mathbb{C}$ be an open set and $f \colon U \to \mathbb{C}$ be a continuous function that has an antiderivative F in U. If γ is a path in U joining points $p \in U$ and $q \in U$, then*

$$\int_\gamma f(z)\, dz = F(q) - F(p).$$

Proof Let $\gamma \colon [A; B] \to U$ be a path such that $\gamma(A) = p$, $\gamma(B) = q$. Then we have

$$\int_\gamma f(z)\, dz = \int_\gamma F'(z)\, dz = \int_A^B F'(\gamma(t))\gamma'(t)\, dt$$

$$= \int_A^B \frac{d}{dt} F(\gamma(t))\, dt = F(\gamma(t))\big|_A^B = F(\gamma(B)) - F(\gamma(A)) = F(q) - F(p).$$

Corollary 4.11 *If a function f defined on an open set $U \subset \mathbb{C}$ has an antiderivative on this set, then the integral of F over every closed path in U vanishes.*

In particular, since, obviously, every polynomial has an antiderivative (which is also a polynomial), the integral of every polynomial over every closed contour vanishes. Also, the integral of $(z - a)^n$ over a closed contour[1] vanishes if n is an integer different from -1. For $n = -1$, this is no longer the case, see Example 4.7.

Another corollary of Proposition 4.10 is a generalization of a well-known result from elementary analysis.

Corollary 4.12 *If f is a holomorphic function on a connected open set $U \subset \mathbb{C}$ and $f'(z) = 0$ for all $z \in U$, then f is constant.*

Proof Any two points $a, b \in U$ can be joined by a continuous path γ; moreover, it is easy to see that this path can be made piecewise smooth (see Exercise 4.9 below). Then

$$f(b) - f(a) = \int_\gamma f'(z)\, dz = \int_\gamma 0 \cdot dz = 0,$$

and we are done.   □

---

[1] For $n < 0$, a closed contour that does not pass through a.

Now we state and prove several simple, purely technical properties of complex integrals, to avoid being distracted by them in more serious situations.

Proposition 4.13 *Let* $\gamma\colon [A; B] \to \mathbb{C}$ *be a piecewise smooth path and* $\{f_n\}$ *be a sequence of functions defined and continuous on* $\gamma([A; B])$ *that converges uniformly on* $\gamma([A; B])$ *to a function* f*. Then*

$$\lim_{n\to\infty} \int_\gamma f_n(z)\,dz = \int_\gamma f(z)\,dz.$$

Proof If the sequence $\{f_n\}$ converges uniformly to f on $\gamma([A; B])$, then the sequence $\{f_n(\gamma(t))\}$ converges uniformly to $f(\gamma(t))$ on $[A; B]$. Since γ is a piecewise smooth map, the function $\gamma'(t)$ is bounded, hence the sequence $\{f_n(\gamma(t))\gamma'(t)\}$ converges uniformly to $f(\gamma(t))\gamma'(t)$. Therefore, this sequence can be integrated term by term over $[A; B]$, as required. □

To state the next property, we need the notion of the length of a curve (more exactly, of a path). Since we do not strive for the greatest possible generality, the most primitive definition will do: length is the integral of velocity with respect to time. More rigorously, this idea may be expressed as follows.

Definition 4.14 Let $\gamma\colon [A; B] \to \mathbb{C}$ be a piecewise smooth path. Then its *length* is the number

$$\text{length}(\gamma) = \int_A^B |\gamma'(t)|\,dt.$$

It is clear (see Proposition 4.3) that the length of a path is invariant under a monotone change of parametrization; this time, due to the absolute value sign in the integrand, a decreasing change of parametrization will also do. This property will not be stated as a numbered proposition.

We will also speak of the lengths of curves, meaning that the curves are parameterized according to Convention 4.5 (again, here we need not be concerned about the orientation of the curve).

Proposition 4.15 *Let* $\gamma\colon [A; B] \to \mathbb{C}$ *be a piecewise smooth path and* f *be a function defined and continuous on* $\gamma([A; B])$*. Then*

$$\left| \int_\gamma f(z)\,dz \right| \leq \sup_{z\in\gamma([A;B])} |f(z)| \cdot \text{length}(\gamma). \tag{4.1}$$

Proof Denoting by M the supremum of the values of f on $\gamma([A; B])$, we have

$$\left| \int_\gamma f(z)\,dz \right| = \left| \int_A^B f(\gamma(t))\gamma'(t)\,dt \right|$$

$$\leq \int_A^B |f(\gamma(t))\gamma'(t)|\,dt \leq \int_A^B M|\gamma'(t)|\,dt = M \cdot \text{length}(\gamma),$$

as required (we have used inequality (1.2)). $\square$

In some cases, the rough estimate (4.1) is not sufficient. A sharper bound uses another notion of line integral.

Definition 4.16 Let $\gamma\colon [A; B] \to \mathbb{C}$ be a piecewise smooth path and f be a continuous complex-valued function defined on the set $\gamma([A; B]) \subset \mathbb{C}$. Then the *arc length integral of f over γ* is the number

$$\int_\gamma f(z)\,|dz| = \int_A^B f(\gamma(t))|\gamma'(t)|\,dt.$$

For instance, $\int_\gamma |dz|$ is nothing else than the length of γ.

Proposition 4.17 *Let $\gamma\colon [A; B] \to \mathbb{C}$ be a piecewise smooth path and f be a function defined and continuous on $\gamma([A; B])$. Then*

$$\left| \int_\gamma f(z)\,dz \right| \leq \int_\gamma |f(z)|\,|dz|.$$

This is obvious from Definition 4.16 and inequality (1.2).

4.2 The Index of a Curve Around a Point

In this section, we completely resolve the question concerning the integral of $dz/(z-a)$ over closed paths. As we will see, this integral has a clear geometric interpretation.

Let $\gamma\colon [A; B] \to \mathbb{C}$ be a closed (which means that $\gamma(A) = \gamma(B)$) piecewise smooth path in the complex plane, and let $a \in \mathbb{C}$ be a point that does not lie on γ. Informally, the index of γ around a is the number of turns that γ makes around a.

Here is a more accurate, but still informal, definition of index.

Let γ be a closed path in the complex plane that does not pass through a point $a \in \mathbb{C}$. Consider the radius vector from a to the points of γ; as a traverses the path, we keep track of the number of complete turns this radius vector makes. The *index of the path γ around the point a* is the total number of turns of the radius vector counted with

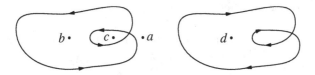

Fig. 4.2 The left curve has index 0 around the point a, index 1 around b, and index 2 around c; the right curve has index -1 around d

signs: each complete turn in the positive direction (counterclockwise) contributes $+1$, and each complete turn in the negative direction (clockwise) contributes -1. The index of γ around a is denoted by $\operatorname{Ind}_a \gamma$.

We will also speak of the *index of a closed curve*, meaning that the curve is endowed with an orientation and keeping in mind Convention 4.5.

Figure 4.2 shows several examples.

Now we introduce a rigorous definition of index. We will need the following lemma.

Lemma 4.18 *Let $\gamma: [A; B] \to \mathbb{C}$ be a piecewise smooth path and a be a point in $\mathbb{C}$ that does not lie on γ. Then there are piecewise smooth functions $r, \varphi: [A; B] \to \mathbb{R}$ such that $\gamma(t) = a + r(t)e^{i\varphi(t)}$.*

Proof There is no question about r: if such r and φ do exist, then necessarily $r(t) = |\gamma(t) - a|$, and that is how we define $r(t)$. Let us deal with φ. As in the previous chapter, denote by $\ell_\alpha = \{re^{i\alpha} : r \geq 0\}$ the ray starting at 0 that makes angle α with the positive real axis, and set $V_\alpha = \mathbb{C} \setminus \ell_\alpha$ (the plane cut along the ray ℓ_α). For every integer k there is a well-defined smooth function $\arg: V_\alpha \to \mathbb{R}$ on V_α that sends a complex number $z \in V_\alpha$ to its argument $\arg z$ satisfying the inequalities

$$\alpha + 2\pi k < \arg z < \alpha + 2\pi(k + 1).$$

Since the map $\gamma: [A; B] \to \mathbb{C}$ is continuous, every point $t \in [A; B]$ is contained in an open interval I such that $\gamma(I \cap [A; B]) \subset V_\alpha$ for some α. Then, the closed interval $[A; B]$ being compact, it follows that there is a partition

$$A = A_0 < A_1 < \ldots < A_n = B$$

such that for every k, $0 \leq k \leq n - 1$, the interval $[A_k; A_{k+1}]$ is contained in at least one such I (see Exercise 1.10). Hence, for every k there is α_k such that $\gamma(t) - a \in V_{\alpha_k}$ for every $t \in [A_k; A_{k+1}]$.

Now we construct φ as follows. First, pick one of the functions $\arg$ on V_{α_0} (for definiteness, denote it by $\arg_0$) and set

$$\varphi(t) = \arg_0(\gamma(t) - a)$$

for $t \in [A_0; A_1]$. Since $\varphi(A_1)$ is an argument of the number $\gamma(A_1) - a$, we can pick one of the functions $\arg$ on V_{α_1} (for definiteness, denote it by $\arg_1$) such that

$\arg_1(\gamma(A_1) - a) = \varphi(A_1)$. Set

$$\varphi(t) = \begin{cases} \arg_0(\gamma(t) - a), & t \in [A_0; A_1], \\ \arg_1(\gamma(t) - a), & t \in [A_1; A_2]. \end{cases}$$

Now, when φ has been defined on $[A_0; A_2]$, pick a function $\arg_2$ on V_{α_2} for which $\arg_2(\gamma(A_2) - a) = \varphi(a_2)$, use it to extend φ to the interval $[A_2; A_3]$, etc. Continuing in this vein, after finitely many steps φ will be defined on the whole interval $[A; B]$.□

Note that φ is not uniquely defined: we can certainly replace $\varphi(t)$ with $\varphi(t) + 2\pi n$ for every $n \in \mathbb{Z}$. It is intuitively clear that this is the only ambiguity that can arise, but we have not yet proved this (see Exercise 4.13).

Proposition-Definition 4.19 Let $\gamma: [A; B] \to \mathbb{C}$ be a closed path that does not pass through a point $a \in \mathbb{C}$. Writing it in the form $\gamma(t) = a + r(t)e^{i\varphi(t)}$ where $r, \varphi: [A; B] \to \mathbb{R}$ are piecewise smooth functions (such a representation does exist by Lemma 4.18), the quotient

$$\mathrm{Ind}_a \gamma = \frac{\varphi(B) - \varphi(A)}{2\pi}$$

is an integer independent of the choice of φ. This integer is called the *index of the path γ around the point a*. Moreover, we have

$$\mathrm{Ind}_a \gamma = \frac{1}{2\pi i} \int_\gamma \frac{dz}{z - a}. \tag{4.2}$$

Proof Since both $\varphi(B)$ and $\varphi(A)$ are arguments of the same complex number $\gamma(B) - a = \gamma(A) - a$, the difference $\varphi(B) - \varphi(A)$ is an integer multiple of 2π. Thus, the index is an integer. To prove that it is independent of the choice of φ, it suffices to verify (4.2). Since $z = \gamma(t) = a + r(t)e^{i\varphi(t)}$, we have

$$dz = (r(t)e^{i\varphi(t)})'\, dt = r'(t)e^{i\varphi(t)}\, dt + ir(t)e^{i\varphi(t)}\varphi'(t)\, dt,$$

whence

$$\frac{dz}{z - a} = \frac{r'(t)}{r(t)}\, dt + i\varphi'(t)\, dt$$

and

$$\int_\gamma \frac{dz}{z - a} = \int_A^B \frac{r'(t)}{r(t)}\, dt + i \int_A^B \varphi'(t)\, dt = \log r(t)\Big|_A^B + i\varphi(t)\Big|_A^B$$

$$= \log r(B) - \log r(A) + i(\varphi(B) - \varphi(A)) = 2\pi i \cdot \frac{\varphi(B) - \varphi(A)}{2\pi}$$

(here $r(B) = r(A) = |\gamma(A) - a|$ and log is the ordinary logarithm of a positive number), so we are done.   □

It is intuitively clear that if we fix a path γ and vary a point a continuously so as not to cross γ, then the index of γ around a must remain unchanged. Now we prove this rigorously.

Proposition 4.20 *If $\gamma: [A; B] \to \mathbb{C}$ is a closed path, then the index $\mathrm{Ind}_a \gamma$ is the same for all points a lying in the same connected open set $U \subset \mathbb{C} \setminus \gamma([A; B])$.*

Proof We begin with a lemma.

Lemma 4.21 *If $U \subset \mathbb{C}$ is a connected open set and $f: U \to \mathbb{R}$ is a continuous function taking only integer values, then f is constant.*

Proof Let $a, b \in U$. Since U is connected, there is a continuous path $\gamma: [A; B] \to U$ for which $\gamma(A) = a$, $\gamma(B) = b$. The function f is continuous, hence the composition $f \circ \gamma: [A; B] \to \mathbb{R}$ is continuous too, and since this continuous function takes only integer values, it is constant (e.g., by the intermediate value theorem). Therefore, $f(a) = f(\gamma(A)) = f(\gamma(B)) = f(b)$, and the lemma is proved.   □

In view of Lemma 4.21, it suffices to verify that the function $a \mapsto \mathrm{Ind}_a \gamma$ is continuous on $\mathbb{C} \setminus \gamma([A; B])$. To see this, assume that $a_n \to a \in \mathbb{C} \setminus \gamma([A; B])$. We will show that

$$\lim_{n \to \infty} \frac{1}{z - a_n} = \frac{1}{z - a} \quad \text{uniformly on } \gamma([A; B]).$$

To this end, set $c = \inf_{z \in \gamma([A;B])} |z - a|$; since a does not lie on the curve $\gamma([A; B])$, we have $c > 0$ (cf. Exercise 1.11). For all sufficiently large n, we have $|a_n - a| < c/2$, hence $|z - a_n| \geq c/2$ for all $z \in \gamma([A; B])$ and all sufficiently large n. Then

$$\left| \frac{1}{z - a_n} - \frac{1}{z - a} \right| = \frac{|a_n - a|}{|z - a| \cdot |z - a_n|} \leq \left(\frac{2}{c} \right)^2 |a_n - a|,$$

and the required uniform convergence is now obvious. Thus, by Proposition 4.13, we have

$$\frac{1}{2\pi i} \int \frac{dz}{z - a_n} \to \frac{1}{2\pi i} \int \frac{dz}{z - a};$$

in view of (4.2), this means exactly that $\mathrm{Ind}_a \gamma$ depends continuously on a.   □

There is another intuitively clear statement about index: the index of a closed curve around a point a remains unchanged if we deform the curve continuously so that it does not touch a. A rigorous proof of this statement will be given once we have a rigorous definition of a continuous deformation of a curve.

Exercises

Find the integrals in Exercises 4.1–4.7.

4.1. $\int_\gamma \bar{z}\,dz$ where γ is the line segment between the points 0 and $1+i$ (directed from 0 to $1+i$).

4.2. $\int_\gamma \bar{z}\,dz$ where γ is the counterclockwise oriented circle of radius 2 centered at 1. (If you are familiar with Stokes' theorem, you can do this problem in your head; if not, then you will have to do some calculations.)

4.3. $\int_\gamma \sin z\,dz$ where γ is the union of the line segment joining the point 1 to the point $3+i\sqrt{7}$ and the line segment joining the point $3+i\sqrt{7}$ to the point $1+i$ (with this orientation).

4.4. $\int_\gamma \frac{dz}{(z+1)^2}$ where γ is the triangle with vertices $1-i$, $-i$, and -2 (traversed counterclockwise).

4.5. $\int_\gamma \frac{z^2-3z+4}{z(z-2)^2}\,dz$ where γ is the circle defined by the equation $|z|=3$ (traversed counterclockwise).

4.6. $\int_{[1;2i]} \frac{dz}{z}$ where $[1;2i]$ is the line segment between the points 1 and $2i$.

4.7. $\int_\gamma \frac{dz}{z}$ where γ is the path shown in Fig. 4.3. Is there enough data to find the answer?

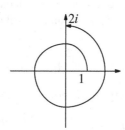

Fig. 4.3 For Exercise 4.7

4.8. Show that the integral of a continuous function f over a piecewise smooth curve γ can be defined in the spirit of "Riemann integral." Namely, if γ is a curve joining points a and b, then by a partition of this curve we mean a sequence of points $z_0 = a, z_1, \ldots, z_n = b$ lying on γ (in this order, in the direction determined by the orientation of γ). The Riemann sum corresponding to this partition is $\sum_{i=0}^{n-1} f(z_i)(z_{i+1}-z_i)$. Show that the integral $\int_\gamma f(z)\,dz$ is equal to the limit of Riemann sums over the set of all partitions as $\max_i |z_{i+1}-z_i| \to 0$.

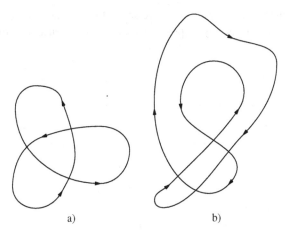

a)   b)

Fig. 4.4

4.9. Show that if $U \subset \mathbb{C}$ is a connected open set, then any two points of U can be joined by a piecewise smooth path lying in U. (*Hint*. Proceed in the same way as in the proof of the "only if" part of Proposition 1.17.)

4.10. Prove rigorously that there is no logarithm function defined on the whole of $\mathbb{C} \setminus \{0\}$. In more detail: there is no holomorphic function $f : \mathbb{C} \setminus \{0\} \to \mathbb{C}$ satisfying the identity $e^{f(z)} = z$.

4.11. The curve γ in Fig. 4.4a divides the plane into connected open sets. For each of them, find $\mathrm{Ind}_a \gamma$ where a is a point of this set.

4.12. The same question for the curve in Fig. 4.4b.

4.13. Show that the function φ whose existence is stated in Lemma 4.18 is defined up to an additive constant of the form $2\pi n$ with $n \in \mathbb{Z}$.

4.14. Let $\gamma \subset \mathbb{C}$ be a piecewise smooth closed curve endowed with an orientation (see Convention 4.5), and let $a \in \mathbb{C} \setminus \gamma$. Assume that there exists a ray $\ell \subset \mathbb{C}$ starting at a with the following properties:
(1) the intersection $\ell \cap \gamma$ is finite;
(2) γ is smooth at any intersection point of ℓ and γ;
(3) at every intersection point of ℓ and γ, the tangent vector to γ is not collinear to ℓ (in this case, ℓ and γ are said to "intersect transversally").
To each intersection point $p \in \ell \cap \gamma$ we assign multiplicity $+1$ if the quotient of complex numbers
$$\frac{\text{tangent vector to } \gamma \text{ at } p}{\text{direction vector of } \ell}$$
lies in the upper half-plane, and -1 if this quotient lies in the lower half-plane (it cannot lie on the real axis by transversality).
Show that the index $\mathrm{Ind}_a \gamma$ is equal to the sum of the multiplicities of all points from $\ell \cap \gamma$.

4.15* Try to generalize the result of Exercise 4.14 to the case where the curve γ is allowed to touch the ray ℓ (possibly, under some restrictions on the type of tangency).

Chapter 5
Cauchy's Theorem and Its Corollaries

5.1 Cauchy's Theorem

In the previous chapter, we have seen that the integral of a holomorphic function over a closed contour frequently turns out to be zero. Now we are going to investigate this phenomenon more or less systematically, and this investigation will take us quite far.

Before setting off, let us agree that by a triangle in the plane we mean a closed subset of $\mathbb{C}$ consisting of its interior, sides, and vertices. If $\Delta \subset \mathbb{C}$ is a triangle, then its *boundary* is the closed polygonal curve composed of the three sides traversed in the positive direction (i.e., counterclockwise). The boundary of Δ will be denoted by $\partial\Delta$.

Theorem 5.1 (Cauchy's theorem, version 1) *Let $U \subset \mathbb{C}$ be an open subset, $f: U \to \mathbb{C}$ be a holomorphic function, and $\Delta \subset U$ be a triangle. Then $\int_{\partial\Delta} f(z)\, dz = 0$.*

Proof First assume that $\Delta \subset \mathbb{C}$ is an arbitrary triangle and $f: \Delta \to \mathbb{C}$ is an arbitrary continuous function. The midsegments divide Δ into four congruent triangles Δ_a, Δ_b, Δ_c, and Δ_d (Fig. 5.1). Then

$$\int_{\partial\Delta} f(z)\, dz = \int_{\partial\Delta_a} f(z)\, dz + \int_{\partial\Delta_b} f(z)\, dz + \int_{\partial\Delta_c} f(z)\, dz + \int_{\partial\Delta_d} f(z)\, dz. \qquad (5.1)$$

To prove this, it suffices to look at Fig. 5.1: the integrals over the sides of small triangles lying inside Δ cancel, and the integrals over the remaining sides sum to the integral over the boundary of Δ.

Now let f and Δ denote the same objects as in the statement of the theorem. We argue by contradiction. Assume that the integral of f over the boundary of Δ does not vanish; set

$$\left| \int_{\partial\Delta} f(z)\, dz \right| = C > 0.$$

© Springer Nature Switzerland AG 2020
S. Lvovski, *Principles of Complex Analysis*, Moscow Lectures 6,
https://doi.org/10.1007/978-3-030-59365-0_5

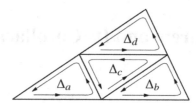

Fig. 5.1 If the boundaries of all triangles have the same orientation, then the integral over the big triangle is equal to the sum of the integrals over the four small triangles

The midsegments divide Δ into four triangles; it follows from (5.1) that for at least one of them (denote it by Δ_1)

$$\left| \int_{\partial \Delta_1} f(z)\,dz \right| \geq \frac{C}{4}. \tag{5.2}$$

Indeed, otherwise the absolute values of all terms in the right-hand side of (5.1) would be strictly less than $C/4$, and hence the absolute value of their sum would be strictly less than C, contradicting our definition of this number.

Further, the midsegments divide the triangle Δ_1 into four triangles; inequality (5.2) and the same argument as above show that the absolute value of the integral of f over the boundary of at least one of them is not less than $(C/4)/4 = C/4^2$; pick one of such triangles and denote it by Δ_2. Again, the midsegments divide Δ_2 into four triangles, and so on. We obtain a sequence of nested triangles

$$\Delta = \Delta_0 \supset \Delta_1 \supset \ldots \supset \Delta_n \supset \ldots$$

in which Δ_n is similar to Δ with ratio $1/2^n$ and

$$\left| \int_{\partial \Delta_n} f(z)\,dz \right| \geq \frac{C}{4^n}. \tag{5.3}$$

It is easy to see that there exists a (unique, but we do not need this) point $a \in U$ lying in all Δ_n: for example, observe that the sequence of vertices of all these triangles is Cauchy and take a to be its limit.

Denote the perimeter of the original triangle Δ by p; then the perimeter of Δ_n is equal to $p/2^n$. Now recall that the function f is holomorphic on U and, in particular, has a derivative at a. This idea can be expressed as follows. Define a function φ by the relation

$$f(z) = f(a) + f'(a)(z - a) + \varphi(z).$$

Then there exist a number $\varepsilon > 0$ and an increasing function $\omega \colon (0; \varepsilon] \to \mathbb{R}$ such that $|\varphi(z)| \leq \omega(|z - a|)|z - a|$ for $0 < |z - a| < \varepsilon$, see the proof of Proposition 2.5; it suffices to set

$$\omega(t) = \sup_{0<|z-a|\le t} \frac{|\varphi(z)|}{|z-a|}.$$

Note that the function $z \mapsto f(a) + f'(a)(z-a)$, being linear, has an antiderivative in U, hence its integral over the boundary of every triangle vanishes. Thus, $\int_{\partial\Delta_n} f(z)\,dz = \int_{\partial\Delta_n} \varphi(z)\,dz$; as to the integral in the right-hand side, we estimate it as follows. If n is so large that Δ_n is contained in the ε-neighborhood of a, then $|z-a| \le p/2^n$ for every $z \in \Delta_n$ (the distance between two points in a triangle does not exceed its greatest side and, a fortiori, the perimeter). Therefore,

$$|\varphi(z)| \le \omega(|z-a|)|z-a| \le \omega\left(\frac{p}{2^n}\right)\cdot\frac{p}{2^n}$$

for every $z \in \Delta_n$. Since the perimeter of Δ_n is equal to $p/2^n$, Proposition 4.15 implies that

$$\left|\int_{\partial\Delta_n} f(z)\,dz\right| = \left|\int_{\partial\Delta_n} \varphi(z)\,dz\right| \le \omega\left(\frac{p}{2^n}\right)\cdot\frac{p}{2^n}\cdot\frac{p}{2^n} = \omega\left(\frac{p}{2^n}\right)\cdot\frac{p^2}{4^n};$$

comparing this with (5.3), we obtain

$$\frac{C}{4^n} \le \omega\left(\frac{p}{2^n}\right)\cdot\frac{p^2}{4^n} \Rightarrow \omega\left(\frac{p}{2^n}\right) \ge \frac{C}{p^2}$$

for all sufficiently large n. But $\omega(t) \to 0$ as $t \to 0$, and the resulting contradiction proves the theorem. $\qquad\square$

Having proved the theorem, we now try to generalize it.

Recall that a subset of the plane is said to be *convex* if whenever it contains two points, it also contains the segment between them. To generalize Cauchy's theorem, we need the following result.

Proposition 5.2 *Let $U \subset \mathbb{C}$ be a convex open set and $f : U \to \mathbb{C}$ be a holomorphic function. Then f has an antiderivative, i.e., there is a function $F : U \to \mathbb{C}$ such that $F'(a) = f(a)$ for all $a \in \mathbb{C}$.*

Proof Let us agree on the following notation: if $a, b \in U$, then by $[a; b]$ we denote the line segment between a and b directed from a to b.

Choose, once and for all, a point $z_0 \in U$, and define a function $F : U \to \mathbb{C}$ by the formula $F(z) = \int_{[z_0;z]} f(t)\,dt$ (the right-hand side is well defined, since the segment $[z_0; z]$ is contained in U). We will show that F is an antiderivative of f.

Let $a \in U$, $a + h \in U$. Denote by Δ the triangle with vertices z_0, a, and $a + h$; by convexity, it is contained in U. Then, by Cauchy's theorem,

$$\int_{\partial\Delta} f(t)\,dt = \int_{[z_0;a]} f(t)\,dt + \int_{[a;a+h]} f(t)\,dt + \int_{[a+h;z_0]} f(t)\,dt,$$

whence

$$F(a+h) - F(a) = - \int\limits_{[a+h;z_0]} f(t)\, dt - \int\limits_{[z_0;a]} f(t)\, dt = \int\limits_{[a;a+h]} f(t)\, dt.$$

Note that

$$\int\limits_{[a;a+h]} f(t)\, dt = \int\limits_{[a;a+h]} f(a)\, dt + \int\limits_{[a;a+h]} (f(t) - f(a))\, dt$$

$$= f(a) \cdot h + \int\limits_{[a;a+h]} (f(t) - f(a))\, dt.$$

Now, using the estimate from Proposition 4.15, we obtain

$$\left| \frac{F(a+h) - F(a)}{h} - f(a) \right| = \frac{\left| \int\limits_{[a;a+h]} (f(t) - f(a))\, dt \right|}{|h|}$$

$$\leq \frac{|h| \cdot \sup\limits_{t \in [a;a+h]} |f(t) - f(a)|}{|h|} = \sup\limits_{t \in [a;a+h]} |f(t) - f(a)|.$$

Since the function f is holomorphic and hence continuous, the right-hand side tends to zero as $h \to 0$, so the derivative of F at the point a exists and is equal to $f(a)$, as required. $\qquad\square$

Theorem 5.3 (Cauchy's theorem, version 2) *Let $U \subset \mathbb{C}$ be a convex open set and $f \colon U \to \mathbb{C}$ be a holomorphic function. Then:*

(a) the integral of f over every closed path in U vanishes;

(b) if $p, q \in U$ and γ_1, γ_2 are two paths in U joining p and q, then the integrals of f over γ_1 and γ_2 coincide.

Proof By Proposition 5.2, the function f has an antiderivative in U; now everything follows from Corollary 4.11 and Proposition 4.10. $\qquad\square$

Now we turn from convex open sets to a wider class.

"Theorem" 5.4 (Cauchy's theorem, version 3) *Let $f \colon U \to \mathbb{C}$ be a holomorphic function on an open set $U \subset \mathbb{C}$.*

(a) Let $\gamma_1, \gamma_2 \subset U$ be closed curves that do not cross themselves and each other and are positively (counterclockwise) oriented. If the part of the plane between γ_1 and γ_2 lies entirely inside U, then $\int_{\gamma_1} f\, dz = \int_{\gamma_2} f\, dz$.

(b) Let γ_1 and γ_2 be two curves in U joining points $p \in U$ and $q \in U$. If the part of the plane between γ_1 and γ_2 lies entirely inside U, then $\int_{\gamma_1} f\, dz = \int_{\gamma_2} f\, dz$.

«Proof» In each of the two cases, divide the part of the plane between γ_1 and γ_2 into finitely many parts such that each of them is contained in an open disk which,

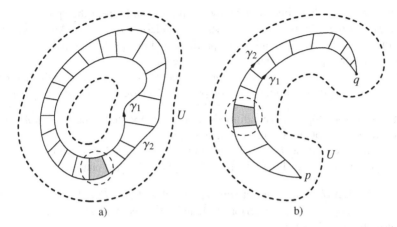

Fig. 5.2 (a) The closed curves γ_1 and γ_2 are positively oriented; (b) the curves γ_1 and γ_2 join the points p and q. The bold dotted line represents the boundary of the domain U. The part of the plane between γ_1 and γ_2 is divided into "small" parts; one of these parts is cross-hatched, and the dotted line represents a disk that contains this part and, in turn, is contained in U.

in turn, is contained in U (Fig. 5.2), and endow the boundary of each small part with the positive orientation. Denoting these boundaries by $\mu_1, \ldots, \mu_n$, by Theorem 5.3 we have

$$\int_{\mu_1} f \, dz = \ldots = \int_{\mu_n} f \, dz = 0 \Rightarrow \int_{\mu_1} f \, dz + \ldots + \int_{\mu_n} f \, dz = 0. \qquad (5.4)$$

In the right-hand side, the integrals over the parts of boundaries lying inside the domain between γ_1 and γ_2 cancel (cf. the proof of Cauchy's theorem for triangles), while the integrals over the parts of boundaries lying on γ_1 and γ_2 sum to the integral over γ_1 plus the integral over γ_2, but with signs: in case (a), the outer curve is oriented counterclockwise, as in the statement of the theorem, while the inner one is oriented clockwise, so the sum of the integrals in (5.4), which is equal to zero, is also equal to the difference of the integral of $f \, dz$ over γ_1 and the integral of $f \, dz$ over γ_2; in case (b), one of the curves has the original orientation from p to q, while the other one has the opposite orientation from q to p. Hence, the right-hand side in (5.4) is also equal to the difference of the integrals over γ_1 and γ_2. In both cases, we see that the difference of the integrals over γ_1 and γ_2 vanishes, as required.   □

We put quotes around the words "theorem" and "proof" for an obvious reason: we have not defined the notion of "the part of the plane between given curves," and it is not so easy to give it a rigorous mathematical meaning. The most general correct statement would apparently be "γ_1 and γ_2 are homologous as 1-chains in U," but the definitions, lemmas, etc. required by this approach would take too much space.

That is why, strictly speaking, Theorem 5.4 should be regarded as a recipe for constructing correct proofs. Namely, whenever we wish to prove that the integrals of a given holomorphic function over two curves coincide, we must divide the part of the plane between them into sufficiently small parts and repeat the "proof" for the boundaries of these parts instead of abstract boundaries $\mu_1, \ldots, \mu_n$. Moreover, it is a common practice not to mention this partition and this proof explicitly, but to run through the corresponding argument in one's head.

In Chap. 6 we will prove (without quotes) another version of Cauchy's theorem. Namely, we will define homotopies of paths and prove that $\int_{\gamma_1} f \, dz = \int_{\gamma_2} f \, dz$ if γ_1 and γ_2 are homotopic (this condition is stronger than γ_1 and γ_2 being homologous, so we still will not obtain the most general result).

Here are several typical examples of applying "Theorem" 5.4. All of them will be used repeatedly in this book; the first one, right in the next chapter.

Example 5.5 Let U be an open set containing a closed disk $\bar{D} = \{z : |z - z_0| \leq r\}$, $D = \{z : |z - z_0| < r\}$ be the interior of $\bar{D}$, and a be a point in D. If a function φ is holomorphic on $U \setminus \{a\}$, then

$$\int_{\partial D} \varphi(z) \, dz = \int_{|z-a|=\varepsilon} \varphi(z) \, dz,$$

where ∂D (the boundary of D) is the circle $\{z : |z - z_0| = r\}$, both circles are positively oriented, and the number ε is so small that the circle $\{z : |z - a| = \varepsilon\}$ is contained in the interior of $\bar{D}$.

Indeed, this follows from Theorem 5.4: the part of the plane between the larger and smaller circles is entirely contained in U.

Example 5.5 can be varied in different ways. For instance, it is clear that the disk $\bar{D}$ can be replaced by a half-disk, rectangle, or anything else of this kind. Generalizing it in another direction, note that there is no need to mention the set U at all, instead we may require φ to be continuous on $\bar{D} \setminus \{a\}$ and holomorphic on $D \setminus \{a\}$. Indeed, it is clear from the above that

$$\int_{|z-z_0|=r-\delta} \varphi(z) \, dz = \int_{|z-a|=\varepsilon} \varphi(z) \, dz \quad \text{for all sufficiently small } \delta > 0,$$

and it remains to observe that

$$\lim_{\delta \to 0} \int_{|z-z_0|=r-\delta} \varphi(z) \, dz = \int_{|z-z_0|=r} \varphi(z) \, dz$$

by the continuity of φ on the boundary of the disk.

Example 5.6 Let $\bar{U} \subset \mathbb{C}$ be the part of the complex plane bounded by a closed non-self-crossing piecewise smooth curve γ (the curve γ is contained in $\bar{U}$); set $\text{Int}(\bar{U}) = U$. (If you are tired of inexact statements, think that $\bar{U}$ is a closed disk, γ is the circle bounding this disk, and U is the interior of $\bar{U}$.) Let $a_1, \ldots, a_n \in U$, and let f be a function continuous on the set $\bar{U} \setminus \{a_1, \ldots, a_n\}$ and holomorphic on the

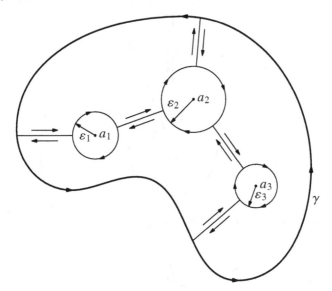

Fig. 5.3

set $U \setminus \{a_1, \ldots, a_n\}$. Take a circle of radius $\varepsilon_j > 0$ around each point a_j so that all these circles lie in U and are pairwise disjoint. Then

$$\int_\gamma f(z)\,dz = \int_{|z-a_1|=\varepsilon_1} f(z)\,dz + \ldots + \int_{|z-a_n|=\varepsilon_n} f(z)\,dz, \qquad (5.5)$$

where all curves are assumed to be positively oriented.

Formally, equality (5.5) does not follow from Theorem 5.4, but it can also be established by following the recipe from the "proof" of this theorem. Namely, if we cut the part of the plane between the curve γ and the circles of radii $\varepsilon_1, \ldots, \varepsilon_n$ into sufficiently small parts (so small that the integral of $f(z)\,dz$ over the boundary of each of them vanishes), then, as before, these integrals sum to the integral over γ traversed in the positive direction plus the sum of the integrals over the n circles traversed in the negative direction. See Fig. 5.3.

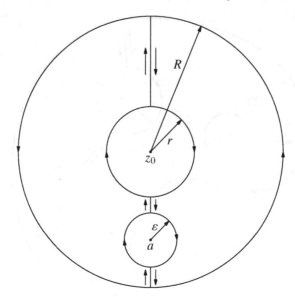

Fig. 5.4

Example 5.7 Set

$$\bar{U} = \{z \colon r \le |z - z_0| \le R\},$$

$$U = \{z \colon r < |z - z_0| < R\},$$

where $0 < r < R$. Let $a \in U$, and let $\varphi \colon \bar{U} \setminus \{a\} \to \mathbb{C}$ be a function continuous on $\bar{U} \setminus \{a\}$ and holomorphic on $U \setminus \{a\}$. If $\varepsilon > 0$ is so small that the closed disk of radius ε centered at a lies in U, then

$$\int_{|z-z_0|=R} \varphi(z)\, dz - \int_{|z-z_0|=r} \varphi(z)\, dz = \int_{|z-a|=\varepsilon} \varphi(z)\, dz,$$

where all three circles are positively oriented.

This formula can be proved in the same way as formula (5.5) from Example 5.6 if we denote by γ the circle of radius R centered at z_0 and set $a_1 = a$, $a_2 = z_0$, $\varepsilon_1 = \varepsilon$, $\varepsilon_2 = r$. See Fig. 5.4.

5.2 Cauchy's Formula and Analyticity of Holomorphic Functions

The theorem we are going to prove underlies a very large part of complex analysis.

Theorem 5.8 (Cauchy's formula) *Let $\bar{U} \subset \mathbb{C}$ be the part of the complex plane bounded by a closed non-self-crossing curve γ (the curve γ is contained in $\bar{U}$); set*

$\text{Int}(\bar{U}) = U$. If $f \colon \bar{U} \to \mathbb{C}$ is a function continuous on $\bar{U}$ and holomorphic on U, then for every $a \in U$,

$$f(a) = \frac{1}{2\pi i} \int_\gamma \frac{f(z)\, dz}{z - a}, \tag{5.6}$$

where the curve γ is oriented in the positive direction.

If you are appalled by the notion of "the part of the plane bounded by a curve" in the statement of the theorem, then this statement can also be regarded as a recipe for obtaining rigorous statements for each particular curve γ we will need (semicircle, rectangle, etc.).

Before proving Cauchy's formula, let us comprehend its meaning: it implies that all values of a holomorphic function inside a closed contour are completely determined once its values on the contour are known. Needless to say, for smooth (and even infinitely smooth) functions, that is very much not the case: a smooth function can always be "perturbed" on an arbitrarily small subset of its domain of definition so as to remain smooth.

This is one of the manifestations of a phenomenon we will encounter repeatedly: a holomorphic function is an extremely rigid object.

Proof Let $\varepsilon > 0$ be so small that the closed disk $\{z \colon |z - a| \le \varepsilon\}$ is contained in U; denote its boundary by $\gamma_\varepsilon = \{z \colon |z - a| = \varepsilon\}$. Cauchy's theorem applied to the function $z \mapsto f(z)/(z - a)$, which is holomorphic on $U \setminus \{a\}$, implies that

$$\int_\gamma \frac{f(z)\, dz}{z - a} = \int_{\gamma_\varepsilon} \frac{f(z)\, dz}{z - a} \tag{5.7}$$

(see Example 5.5 and the discussion following it). Hence, it suffices to prove that

$$\int_{\gamma_\varepsilon} \frac{f(z)\, dz}{z - a} = 2\pi i f(a); \tag{5.8}$$

we note, for future reference, that it follows from formula (5.7) (or directly from Cauchy's theorem in the form 5.4) that the integral in the left-hand side of (5.8) does not depend on ε.

Since the function f has a complex derivative at the point a, we can write

$$f(z) = f(a) + f'(a)(z - a) + \varphi(z),$$
$$\left| \frac{\varphi(z)}{z - a} \right| \le \omega(|z - a|), \quad \lim_{t \to 0} \omega(t) = 0 \tag{5.9}$$

(see the proofs of Theorem 5.1 and Proposition 2.5). Therefore,

$$\int_{\gamma_\varepsilon} \frac{f(z)\, dz}{z - a} = \int_{\gamma_\varepsilon} \frac{f(a)\, dz}{z - a} + \int_{\gamma_\varepsilon} f'(a)\, dz + \int_{\gamma_\varepsilon} \frac{\varphi(z)\, dz}{z - a}. \tag{5.10}$$

The first term in the right-hand side is equal to $2\pi i f(a)$ (see Example 4.7), and
the second one vanishes, because this is the integral of a constant over a closed
contour. Thus, to prove the theorem, we must show that the third term in the right-
hand side of (5.10) also vanishes; for this, we estimate it using Proposition 4.15 and
inequality (5.9):

$$\left| \int_{\gamma_\varepsilon} \frac{\varphi(z)\,dz}{z-a} \right| \leq 2\pi\varepsilon \cdot \sup_{|z-a|=\varepsilon} \left| \frac{\varphi(z)}{z-a} \right| \leq 2\pi\varepsilon \cdot \omega(\varepsilon);$$

as we have already observed, the left-hand side of this inequality does not depend
on ε, while the right-hand side tends to zero as $\varepsilon \to 0$, hence the left-hand side is
necessarily zero, completing the proof. $\square$

Now we are going to derive the first of the important corollaries of Cauchy's
formula. First, a definition is in order.

Definition 5.9 A function $f \colon U \to \mathbb{C}$, where $U \subset \mathbb{C}$ is an open set, is said to be
analytic if for every point $a \in U$ there exists an open disk $D = \{z \colon |z-a| < r\} \subset U$
centered at a such that for all $z \in D$ the function f can be represented as a convergent
power series

$$f(z) = c_0 + c_1(z-a) + \ldots + c_n(z-a)^n + \ldots. \tag{5.11}$$

Since, by Proposition 1.19, the series (5.11) converges uniformly on every compact subset in D, and its partial sums are continuous, it follows that every analytic
function is continuous.

Proposition 5.10 *Every holomorphic function is analytic.*

Proof Let f be a holomorphic function on $U \subset \mathbb{C}$, and let $a \in U$. Then for some
$r > 0$ the closed disk $\bar{D} = \{z \colon |z-a| \leq r\}$ is contained in U. Let $z \in \mathrm{Int}(\bar{D})$. By
Cauchy's formula applied to the disk $\bar{D}$ and the point $z \in \mathrm{Int}(\bar{D})$, we have

$$f(z) = \frac{1}{2\pi i} \int_{|z-a|=r} \frac{f(\zeta)\,d\zeta}{\zeta - z}. \tag{5.12}$$

For every ζ lying on the boundary of $\bar{D}$ (denote it by ∂D), the fraction $1/(\zeta - z)$ can
be expanded in a geometric series:

$$\frac{1}{\zeta - z} = \frac{1}{\zeta - a} \cdot \frac{1}{1 - \frac{z-a}{\zeta-a}} = \sum_{n=0}^{\infty} \frac{(z-a)^n}{(\zeta-a)^{n+1}}. \tag{5.13}$$

For $\zeta \in \partial D$ we have $|\zeta - a| = r$, hence the nth term of the series (5.13) can be
estimated as

$$\left| \frac{(z-a)^n}{(\zeta-a)^{n+1}} \right| = \frac{1}{|\zeta-a|} \left| \frac{z-a}{\zeta-a} \right|^n \leq \frac{1}{r} \cdot \left| \frac{z-a}{r} \right|^n.$$

Since $|(z-a)/r| < 1$, the right-hand side is a convergent geometric series whose
terms do not depend on ζ; thus, the series (5.13) (for a fixed z) converges on the

circle $\{\zeta : |\zeta - a| = r\}$ uniformly with respect to ζ. Multiplying both sides of (5.13) by $f(\zeta)$, we obtain the expansion

$$\frac{f(\zeta)}{\zeta - z} = \sum_{n=0}^{\infty} \frac{f(\zeta)(z - a)^n}{(\zeta - a)^{n+1}}. \tag{5.14}$$

The function f is bounded on this circle, and the series in the right-hand side of (5.13) converges uniformly with respect to $\zeta \in \partial D$, hence the series in the right-hand side of (5.14) also converges uniformly, and it can be integrated term by term (with respect to ζ) over ∂D. Multiplying by $1/2\pi i$ yields

$$f(z) = \frac{1}{2\pi i} \int_{\partial D} \frac{f(\zeta)\, d\zeta}{\zeta - z} = \frac{1}{2\pi i} \sum_{n=0}^{\infty} \left((z - a)^n \cdot \int_{\partial D} \frac{f(\zeta)\, d\zeta}{(\zeta - a)^{n+1}} \right)$$

$$= \sum_{n=0}^{\infty} c_n (z - a)^n, \quad (5.15)$$

where

$$c_n = \frac{1}{2\pi i} \int_{\partial D} \frac{f(\zeta)\, d\zeta}{(\zeta - a)^{n+1}}. \tag{5.16}$$

The proof is completed. □

Corollary 5.11 (of the proof) *Every function holomorphic on an open disk $D \subset \mathbb{C}$ centered at a point a can be represented in D by a power series $\sum_{j=0}^{\infty} c_j (z - a)^j$ which converges absolutely and uniformly on every compact subset $K \subset \mathbb{C}$.*

So, we have established that every holomorphic function is analytic. The converse is also true: every analytic function is holomorphic. This is a much simpler fact than Proposition 5.10, it can be proved by quite elementary arguments using properties of double series (Proposition 1.4; see also [5, Chap. V, Sec. 5]). We will prove it in the next section applying the same "line integration method" that we used to prove Proposition 5.10.

5.3 Infinite Differentiability. Term-By-Term Differentiability

We continue deriving important corollaries of Cauchy's formula.

Theorem 5.12 *Let $\bar{U} \subset \mathbb{C}$ be the part of the complex plane bounded by a closed non-self-crossing curve γ (the curve γ is contained in $\bar{U}$); set $\mathrm{Int}(\bar{U}) = U$. If a function $f : \bar{U} \to \mathbb{C}$ is continuous on $\bar{U}$ and holomorphic on U, then in the interior of U it has derivatives of all orders; these derivatives are also holomorphic, and they are given by the formula*

$$f^{(n)}(z) = \frac{n!}{2\pi i} \int_\gamma \frac{f(\zeta)\, d\zeta}{(\zeta - z)^{n+1}}. \qquad (5.17)$$

Proof For $n = 0$, formula (5.17) is nothing else than the ordinary Cauchy's formula (5.6). Further, it is easy to see that formula (5.17) for $n + 1$ is obtained from formula (5.17) for n by differentiating the integrand with respect to z. Thus, to prove the theorem, it suffices to justify this differentiation.

Fixing an integer $n \geq 0$ and a point $z \in U$, set $\Phi_n(\zeta, z) = f(\zeta)/(\zeta - z)^{n+1}$. If we verify that the "difference quotient" $(\Phi_n(\zeta, z+h) - \Phi_n(\zeta, z))/h$ converges uniformly with respect to $\zeta \in \gamma$ as $h \to 0$, then, integrating (with respect to ζ) the relation

$$\lim_{h \to 0} \frac{\Phi_n(\zeta, z+h) - \Phi_n(\zeta, z)}{h} = (n+1)\Phi_{n+1}(\zeta, z) \qquad (5.18)$$

over the curve γ (which is justified by the uniform convergence), we will obtain

$$\frac{d}{dz}\left(\int_\gamma \frac{f(\zeta)\, d\zeta}{(\zeta - z)^{n+1}} \right) = (n+1) \int_\gamma \frac{f(\zeta)\, d\zeta}{(\zeta - z)^{n+2}},$$

and the theorem will be proved.

It remains to verify that the convergence in (5.18) is uniform with respect to $\zeta \in \gamma$. This can be proved by boring, but straightforward calculations. Namely, expanding $(\zeta - z - h)^{n+1} = ((\zeta - z) - h)^{n+1}$ according to the binomial theorem and reducing the terms to a common denominator, we obtain

$$\frac{\Phi_n(\zeta, z+h) - \Phi_n(\zeta, z)}{h} = f(\zeta) \cdot G(\zeta, z, h),$$

where

$$G(\zeta, z, h) = \frac{n+1}{(\zeta - z - h)^{n+1}(\zeta - z)}$$
$$- \frac{\binom{n+1}{2} h(\zeta - z)^{n-1} - \binom{n+1}{3} h^2(\zeta - z)^{n-2} + \ldots \pm h^n}{(\zeta - z - h)^{n+1}(\zeta - z)^{n+1}}. \qquad (5.19)$$

Since $f(\zeta)$ is bounded on γ by the continuity of f, it suffices to show that

$$\lim_{h \to 0} G(\zeta, z, h) = \frac{n+1}{(\zeta - z)^{n+2}} \quad \text{uniformly with respect to } \zeta \in \gamma.$$

Note that, since z does not lie on γ and $\gamma \subset \mathbb{C}$ is a compact subset, there exists a constant $c > 0$ such that $|\zeta - z| > c$ for all $\zeta \in \gamma$; hence, $|\zeta - z - h| > c/2$ for all $\zeta \in \gamma$ and all sufficiently small h. Now the inequality

$$\left| \frac{1}{\zeta - z - h} - \frac{1}{\zeta - z} \right| = \left| \frac{h}{(\zeta - z)(\zeta - z - h)} \right| \leq \frac{|h|}{c \cdot c/2},$$

which holds for all sufficiently small h, implies that the fraction $1/(\zeta-z-h)$ converges uniformly to $1/(\zeta - z)$. Thus, it follows from the boundedness of $(n + 1)/(\zeta - z)$ and the rule "the limit of a product is equal to the product of the limits" that the first term in the right-hand side of (5.19) converges uniformly to the desired expression $(n + 1)/(\zeta - z)^{n+2}$. It remains to verify that the second term in the right-hand side of (5.19) converges uniformly to zero. This follows from the fact that the absolute value of its denominator is bounded from below by the strictly positive (for all sufficiently small h) constant $c^{2n+2}/2^{n+1}$ and the absolute value of each term in the numerator is bounded from above by const $\cdot |h|^k$ with $k \geq 1$. □

Corollary 5.13 (of the theorem) *Every holomorphic function is "infinitely complex differentiable": if f is holomorphic on an open set U, then the function $z \mapsto f'(z)$ is holomorphic on the same set.*

Corollary 5.14 (of the proof) *Let $\gamma \subset \mathbb{C}$ be a piecewise smooth curve and $f : \gamma \to \mathbb{C}$ be a continuous function. Then the function*

$$F : z \mapsto \frac{1}{2\pi i} \int_\gamma \frac{f(\zeta)\, d\zeta}{\zeta - z} \tag{5.20}$$

is holomorphic on $\mathbb{C} \setminus \gamma$; its nth derivative is equal to

$$F^{(n)}(z) = \frac{n!}{2\pi i} \int_\gamma \frac{f(\zeta)\, d\zeta}{(\zeta - z)^{n+1}}.$$

Proof The argument in the proof of Theorem 5.12 involved only the values of f on γ and its continuity on γ, so it works without change. □

Warning 5.15 Assume that γ in the statement of Corollary 5.14 is just the circle $\{z \colon |z| = 1\}$. Then the function defined by (5.20) is holomorphic on the disk $\{z \colon |z| < 1\}$. But it by no means follows, and is generally false, that the value of $F(z)$ approaches $u(z)$ as z approaches the boundary of the disk. See Exercise 5.17.

Another corollary of Theorem 5.12 can be regarded as an inversion of Cauchy's theorem for triangles.

Proposition 5.16 (Morera's theorem) *Let $U \subset \mathbb{C}$ be an open set. If a continuous function $f : U \to \mathbb{C}$ has the property that $\int_{\partial\Delta} f(z)\, dz = 0$ for every triangle $\Delta \subset U$, then f is holomorphic on U.*

Proof Since whether or not a function is complex differentiable at a given point $a \in U$ depends only on its behavior in an arbitrarily small neighborhood of a, it suffices to show that the restriction of f to an arbitrary open disk contained in U is holomorphic. Thus, without loss of generality we may assume that U is an open disk. Now we proceed as in the proof of Proposition 5.2. Namely, choose, once and for all, a point $z_0 \in U$ and set $F(z) = \int_{[z_0;z]} f(t)\, dt$ for every $z \in U$ (by $[z_0; z]$ we denote a line segment). The same argument as in the proof of Proposition 5.2 establishes that $F'(z) = f(z)$ for all z (this argument uses only the continuity of f and the fact that the integral of f over the boundary of every triangle vanishes). Thus, f is the derivative of the holomorphic function F, and we have just established that the derivative of a holomorphic function is holomorphic too.   □

It is known that the term-by-term differentiation of uniformly convergent series (or, equivalently, sequences) of smooth (and even infinitely differentiable) functions of a real variable is not always possible, and the sum of a uniformly convergent series of infinitely differentiable functions on an interval can be an arbitrary continuous function, even nowhere differentiable. Now we will see that for holomorphic functions, no such disgraceful things may occur.

Proposition 5.17 *Let $U \subset \mathbb{C}$ be an open set and $\sum\limits_{n=1}^{\infty} f_n$ be a series of holomorphic functions on U that converges uniformly on every compact subset $K \subset U$. Then the sum of this series (denote it by $f(z)$) is a holomorphic function on U, we have $f'(z) = \sum\limits_{n=1}^{\infty} f_n'(z)$, and the series in the right-hand side converges uniformly on every compact subset $K \subset U$, too.*

Proof Since the original series converges uniformly on every compact subset in U, its sum f is at least continuous. If now $\Delta \subset U$ is an arbitrary triangle, then the series $\sum f_n$ converges uniformly on its boundary $\partial\Delta$, the latter being compact. Thus, it follows from the first version of Cauchy's theorem that

$$\int\limits_{\partial\Delta} f(z)\, dz = \sum\limits_{n=1}^{\infty} \int\limits_{\partial\Delta} f(z)\, dz = 0,$$

so the function f satisfies the assumptions of Morera's theorem and hence is holomorphic.

To prove that the series of derivatives converges uniformly on every compact subset $K \subset U$, we proceed as follows. Let $K \subset U$ be a compact subset. Every point $a \in K$ is contained in some closed disk $\bar{D}_a \subset U$. The interiors of these disks cover K, and this cover has a finite subcover; moreover, K is contained in the union of finitely many closed disks contained in U. The uniform convergence on each of these disks will imply (since the cover is finite) the uniform convergence on the whole set K.

Now let $\bar{D} = \{z \colon |z - a| \le r\}$ be a closed disk of radius r contained in U. Then for some $r_1 > r$ the closed disk $\bar{D}_1 = \{z \colon |z - a| \le r_1\}$ is also contained in U. Denote by $S_n = \sum\limits_{j=1}^{n} f_n$ the partial sum of the series $\sum f_n$, and by γ, the boundary of the larger disk $\bar{D}_1$; then for every $z \in \bar{D}$, by (5.17),

$$f'(z) - S_n'(z) = \frac{1}{2\pi i} \int\limits_{\gamma} \frac{(f(\zeta) - S_n(\zeta))\, d\zeta}{(\zeta - z)^2}.$$

Since for $z \in \bar{D}$ and $\zeta \in \gamma$ we have $|\zeta - z| \geq r_1 - r$, estimating the integral by Proposition 4.15, we see that for every $z \in \bar{D}$,

$$|f'(z) - S_n'(z)| \leq \frac{1}{2\pi} \cdot \frac{\sup\limits_{\zeta \in \gamma} |f(\zeta) - S_n(\zeta)|}{(r_1 - r)^2} \cdot \text{length}(\gamma)$$

$$= \text{const} \cdot \sup_{\zeta \in \gamma} |f(\zeta) - S_n(\zeta)|. \quad (5.21)$$

Since $\sum f_n$ converges to f uniformly on the compact set γ, the right-hand side of (5.21) tends to zero as $n \to \infty$. Therefore, $\sum f_n'$ converges to f' uniformly on $\bar{D}$, as required. $\qquad \square$

Now we can systematize and complement what we know about the interrelations between the notions of "holomorphic function" and "analytic function."

Proposition 5.18 *A function of a complex variable is analytic if and only if it is holomorphic. If f is a holomorphic function on an open disk $\{z : |z - a| < r\}$, then in this disk it can be represented by the following power series:*

$$f(z) = f(a) + f'(a)(z - a) + \frac{f''(a)}{2!}(z - a)^2 + \ldots + \frac{f^{(n)}(a)}{n!}(z - a)^n + \ldots . \quad (5.22)$$

As in the real case, the expansion (5.22) is called the Taylor series of f at a.

Proof Every holomorphic function is analytic by Proposition 5.10. To prove the converse, recall that an analytic function can be represented by a convergent power series in a neighborhood of every point. Since a power series converges uniformly on every compact subset of its disk of convergence (Proposition 1.19), the fact that an analytic function is holomorphic now follows from Proposition 5.17, because polynomials are holomorphic.

The fact that a holomorphic function on a disk centered at a can be represented in this disk by a series in powers of $z - a$ is established in the proof of Proposition 5.10. Finally, formulas for the coefficients of the series (5.22) can be obtained in at least two ways. First, we may observe that the coefficient of $(z - a)^n$ in (5.15) is exactly $n!$ times smaller than the expression for $f^{(n)}(a)$ in (5.17). Second, we may observe that, by Proposition 5.17 and the uniform convergence of power series on compact subsets, the expansion

$$f(z) = c_0 + c_1(z - a) + c_2(z - a)^2 + \ldots + c_n(z - a)^n + \ldots ,$$

whose existence is stated in Proposition 5.10, can be differentiated any number of times; hence, we can apply the same argument as one (informally) uses in calculus courses to guess the coefficients of Taylor series: substitute $z = a$ to obtain that $c_0 = f(a)$, then differentiate term by term and substitute $z = a$ (to obtain that $c_1 = f'(a)$), etc. $\qquad \square$

Our considerations imply useful upper bounds on the coefficients of the Taylor series of a holomorphic function, which are called Cauchy's estimates.

Proposition 5.19 (Cauchy's estimates) *Let f be a function holomorphic on a disk* $D = \{z \colon |z - a| < R\}$ *and*

$$f(z) = c_0 + c_1(z - a) + \ldots + c_n(z - a)^n + \ldots$$

be its power series expansion in this disk. Then for every n we have

$$|c_n| \leq \frac{\displaystyle\sup_{z \in D} |f(z)|}{R^n}. \tag{5.23}$$

(Of course, these inequalities are meaningful only if f is bounded in D.)

Proof Set $\sup_{z \in D} |f(z)| = M$. If $M = +\infty$, then there is nothing to prove, so we may assume that M is finite. Choose a number $\varepsilon > 0$. By (5.16), we have

$$c_n = \frac{1}{2\pi i} \int\limits_{|z-a|=R-\varepsilon} \frac{f(\zeta)\, d\zeta}{(\zeta - a)^{n+1}},$$

whence, by (4.1),

$$|c_n| \leq \frac{1}{2\pi} \frac{M}{(R - \varepsilon)^{n+1}} \cdot 2\pi(R - \varepsilon) = \frac{M}{(R - \varepsilon)^n}.$$

Letting $\varepsilon \to 0$, we obtain the desired inequalities. □

Proposition 5.18 has also an important "qualitative" corollary.

Corollary 5.20 *If f is a holomorphic function in a neighborhood of a point a and $f(a) = 0$, then either f is identically zero in a neighborhood of a, or f does not vanish in some punctured neighborhood of a.*

Proof If all coefficients of the Taylor series of f at the point a vanish, then f is identically zero in a neighborhood of a. Otherwise, the Taylor series expansion of f has the form

$$f(z) = c_k(z - a)^k + c_{k+1}(z - a)^{k+1} + \ldots + c_n(z - a)^n + \ldots$$

where $k > 0$, $c_k \neq 0$, and the series converges in a disk D centered at a. Set

$$g(z) = c_k + c_{k+1}(z - a) + \ldots + c_{k+n}(z - a)^n + \ldots;$$

the series in the right-hand side also converges in D, so the function g is well defined, holomorphic, and hence continuous on D. Since $g(a) = c_k \neq 0$, by continuity there exists a neighborhood $V \ni a$ on which g does not vanish. Therefore, the function $f(z) = (z - a)^k g(z)$ does not vanish on $V \setminus \{a\}$, as required. □

Corollary 5.20 has an important global counterpart called the uniqueness theorem or the principle of analytic continuation.

Proposition 5.21 *Let $U \subset \mathbb{C}$ be a connected open set and $S \subset U$ be a subset that has an accumulation point in U. If $f, g \colon U \to \mathbb{C}$ are holomorphic functions that coincide on the subset $S \subset U$, then $f(z) = g(z)$ for all $z \in U$.*

Proof Set $f(z) - g(z) = \varphi(z)$. We know that $\varphi(z) = 0$ for all $z \in S$ and must prove that $\varphi(z) = 0$ for all $z \in U$.

Let $a \in U$ be an accumulation point of the set S. By continuity, we have $\varphi(a) = 0$, and Corollary 5.20 shows that φ is identically zero in a neighborhood of a. Now denote by Z the set of all points of U at which the function φ and its derivatives of all orders vanish. It follows from the above that $Z \ni a$, so Z is not empty. Further, Z is open: if $b \in Z$, then all coefficients of the Taylor series of φ at the point b are zero, so the function φ is identically zero in some neighborhood of b, and this neighborhood, obviously, lies in Z. Finally, $U \setminus Z$ is also open. Indeed, if $b \in U \setminus Z$, then $\varphi^{(k)}(b) \neq 0$ for some k; since the kth derivative of a holomorphic function is also holomorphic and, in particular, continuous, $\varphi^{(k)}$ does not vanish on some neighborhood $V \ni b$, whence $V \subset U \setminus Z$. So, the set U is the union of two disjoint open subsets Z and $U \setminus Z$. Since U is connected, one of these sets is empty and the other one coincides with U. Since $Z \neq \varnothing$, we have $Z = U$, i.e., the function φ is identically zero. $\qquad\square$

Of course, Proposition 5.21 fails for a disconnected set U. If, for example, $U = U_0 \cup U_1$ where U_0 and U_1 are disjoint open sets, then it suffices to consider the function $f \colon U \to \mathbb{C}$ that is identically equal to 0 and the function g that is equal to 0 on U_0 and to 1 on U_1. The condition that S has an accumulation point that lies in U cannot be removed as well (see the exercises at the end of the chapter).

As the first application of Proposition 5.21, we explain how it can be used to prove, without any calculations, the standard trigonometric identities for functions of a complex variable (cf. Exercise 2.7). Assume, for instance, that we must prove the identity $\sin^2 z + \cos^2 z = 1$. Set $f(z) = \sin^2 z + \cos^2 z$ and $g(z) = 1$; both functions are holomorphic on the entire complex plane $\mathbb{C}$, and we know from calculus that they coincide on the subset $\mathbb{R} \subset \mathbb{C}$. Hence, f and g satisfy the conditions of Proposition 5.21, so $f(z) = g(z)$ for all $z \in \mathbb{C}$. To prove identities involving more than one variable, a few more words are needed. For instance, to verify without calculations that

$$\sin(z + w) = \sin z \cos w + \cos z \sin w, \qquad (5.24)$$

we can first fix an arbitrary $z \in \mathbb{R}$ and regard both sides of (5.24) as holomorphic functions of w; since they coincide for all $w \in \mathbb{R}$, the principle of analytic continuation shows that they coincide also for all $w \in \mathbb{C}$. Thus, identity (5.24) is verified for all $z \in \mathbb{R}$, $w \in \mathbb{C}$. Now for every $w \in \mathbb{C}$ we regard both sides of (5.24) as holomorphic functions of z; we already know that they coincide for all $z \in \mathbb{R}$, hence they coincide also for all $z \in \mathbb{C}$.

Exercises

5.1. If f and g are continuous functions on an open subset $U \subset \mathbb{R}^2$ and $\gamma: [p;q] \to U$, $\gamma(t) = (x(t), y(t))$, is a piecewise smooth path, then one defines

$$\int_\gamma (f \, dx + g \, dy) = \int_p^q (f(\gamma(t))x'(t) + g(\gamma(t))y'(t)) \, dt.$$

Now let f and g be differentiable at every point of U in the sense of Sec. 1.5 with $\partial f/\partial y = \partial g/\partial x$ identically on U. Imitating the proof of Theorem 5.1, show that if $\Delta \subset U$ is a triangle, then $\int_{\partial \Delta} (f \, dx + g \, dy) = 0$.

(We do not assume that the partial derivatives are continuous, so Green's theorem does not work!)

5.2. Let U, f, and g be the same as in the previous exercise, but assume that U is convex. Show that there exists a function $F: U \to \mathbb{R}$ such that $\partial F/\partial x = f$, $\partial F/\partial y = g$.

5.3. Let f be a function holomorphic on an open set containing a closed disk $\{z: |z - a| \le r\}$. Show that the value of f at the center of the disk is equal to the average of its values on the boundary:

$$f(a) = \frac{1}{2\pi} \int_0^{2\pi} f(a + re^{i\varphi}) \, d\varphi.$$

(*Hint*: Cauchy's formula.)

5.4. Let f be a function holomorphic on an open set containing a closed disk $\{z: |z - a| \le r\}$. Show that the value of f at the center of the disk is equal to the average of its values on the disk:

$$f(a) = \frac{1}{\pi r^2} \int_{\{z: \, |z-a| \le r\}} f(x + iy) \, dx \, dy.$$

5.5. Set

$$f(z) = \begin{cases} \frac{\sin z}{z}, & z \ne 0, \\ 1, & z = 0. \end{cases}$$

Show that the function f is holomorphic on the entire complex plane $\mathbb{C}$.

For each of the following functions, find its power series representation.

5.6. $f(z) = \cos^2 z$ in a neighborhood of 0.

5.7. $f(z) = z/(z - 1)^2$ in a neighborhood of 0.

5.8. $f(z) = \log \frac{1+z}{1-z}$ in a neighborhood of 0 (we use the branch of the logarithm for which $\log 1 = 0$).

5.9. $f(z) = z/(z^2 + 2z + 5)$ in a neighborhood of the point $z = -1$.

5.10. Find the power series expansion of the function $f(z) = \sqrt{\cos z}$ in a neighborhood of 0 up to z^4. We use the branch of the square root for which $\sqrt{1} = 1$.

5.11. Find $f'''(0)$ if $f(z) = e^{\sin z}$ (try to find a more sensible way to do this than using the chain rule).

5.12. Let f be a function holomorphic in a neighborhood of a point a. Show that it satisfies "Taylor's formula with Peano's remainder": for every positive integer n,

$$f(z) = f(a) + f'(a)(z-a) + \frac{f''(a)}{2!}(z-a)^2 + \ldots + \frac{f^{(n)}(a)}{n!}(z-a)^n + o((z-a)^n).$$

5.13. Does there exist a function f that is holomorphic in the unit disk, is not identically zero, and satisfies the condition $f(1 - 1/n) = 0$ for all integers $n \geq 2$?

5.14. Does there exist a function f that is holomorphic in the unit disk and satisfies the condition $f(1/n) = (-1)^n/n^2$ for all integers $n \geq 2$?

5.15. A function f is holomorphic in the unit disk $\{z: |z| < 1\}$; it is known that $f(1/n) = (2n + 1)/(3n + 1)$ for every integer $n \geq 2$. What are the possible values of $f(3/4)$?

5.16. Show that there is no holomorphic function f in the unit disk $\{z: |z| < 1\}$ such that $f(1/n) = (2 + n)/(3 + n)$ for every integer $n \geq 2$.

5.17. Let $\gamma = \{z \in \mathbb{C}: |z| = 1\}$ be the unit circle and $f: \gamma \to C$ be the function given by $f(z) = z + \bar{z}$. For every $z \notin \gamma$, set

$$F(z) = \frac{1}{2\pi i} \int_\gamma \frac{f(\zeta)\, d\zeta}{\zeta - z}.$$

(a) Find an explicit formula for the function $z \mapsto F(z)$.

(b) Verify that the restriction of F to the open unit disk $\{z: |z| < 1\}$ extends to a continuous function on its closure, and that the restriction of F to the exterior of the unit disk $\{z: |z| > 1\}$ also extends to a continuous function on its closure. Compare the function f on the circle with the limits of $F(z)$ as z approaches the circle from the interior and from the exterior.

5.18. Let $U \subset \mathbb{C}$ be an open connected set and $f, g: U \to \mathbb{C}$ be holomorphic functions that are not identically zero. Show that their product fg is not identically zero.

5.19. Again, let $\bar{D} = \{z: |z| \leq 1\}$ be the closed unit disk, $\partial D = \{z: |z| = 1\}$ be the unit circle, and f be a function continuous on $\bar{D}$ and holomorphic on the interior of $\bar{D}$. Show that if f is identically zero on some arc in ∂D, then it is identically zero everywhere. (*Hint.* The entire circle can be covered by finitely many translations of the arc.)

Chapter 6
Homotopy and Analytic Continuation

6.1 Homotopy of Paths

We have already mentioned that the integral of a holomorphic function over a path connecting given two points is unchanged under deformations of the path as long as it remains within the domain of definition of the function. In this chapter, among other things, we will rigorously prove this statement. To begin with, we give a precise definition of a deformation of a path.

So far, we have dealt mainly with piecewise smooth maps, but in this chapter we need arbitrary continuous paths (i.e., continuous maps from intervals to the complex plane). Whenever a path must be piecewise smooth, this will be stated explicitly.

Let $U \subset \mathbb{C}$ be an open set and $\gamma_0, \gamma_1 \colon [A; B] \to U$ be two continuous paths connecting points $p \in U$ and $q \in U$ (in other words, $\gamma_0(A) = \gamma_1(A) = p$ and $\gamma_0(B) = \gamma_1(B) = q$). Saying that the paths γ_0 and γ_1 can be deformed into each other, we mean that γ_0 and γ_1 can be included in a family $\{\gamma_s\}$ with $s \in [0; 1]$. We must only explain somehow what it means for such a deformation to be continuous. To this end, set $F(t, s) = \gamma_s(t)$; then F is a map from the rectangle $[A; B] \times [0; 1] \subset \mathbb{R}^2$ to U, where

$$[A; B] \times [0; 1] = \{(t, s) \colon A \le t \le B, \, 0 \le s \le 1\}.$$

As a definition of the continuity of our deformation, we require F to be continuous. We have arrived at the following definition.

Definition 6.1 Let $\gamma_0, \gamma_1 \colon [A; B] \to U$ be continuous paths in an open set $U \subset \mathbb{C}$ connecting points p and q. They are said to be *homotopic* if there exists a continuous map $F \colon [A; B] \times [0; 1] \to U$ such that $F(t, 0) = \gamma_0(t)$ and $F(t, 1) = \gamma_1(t)$ for all $t \in [A; B]$, and also $F(A, s) = p$ and $F(B, s) = q$ for all $s \in [0; 1]$.

The map F is called a *homotopy* between the paths γ_0 and γ_1.

The condition "$F(A, s) = p$ and $F(B, s) = q$ for all s" means that each of the "intermediate" paths $\gamma_s \colon t \mapsto F(t, s)$ also joins p to q. This is all illustrated in Fig. 6.1.

S. Lvovski, *Principles of Complex Analysis*, Moscow Lectures 6,
https://doi.org/10.1007/978-3-030-59365-0_6

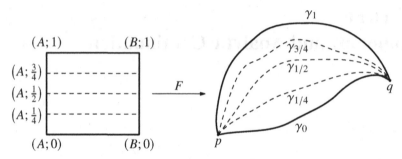

Fig. 6.1 A homotopy between paths γ_0 and γ_1 connecting points p and q

A homotopy from Definition 6.1 will sometimes be called a *homotopy with fixed endpoints*.

We can also define a homotopy of closed paths.

Definition 6.2 Let $\gamma_0, \gamma_1 \colon [A;B] \to U$ be continuous closed paths in an open set $U \subset \mathbb{C}$. They are said to be *homotopic* if there exists a continuous map $F \colon [A;B] \times [0;1] \to U$ such that $F(t,0) = \gamma_0(t)$ and $F(t,1) = \gamma_1(t)$ for all $t \in [A;B]$, and also $F(A,s) = F(B,s)$ for all $s \in [0;1]$.

The map F is called a *homotopy* (more exactly, a homotopy of closed paths) between γ_0 and γ_1.

The condition "$F(A,s) = F(B,s)$ for all s" means, of course, that all paths γ_s are closed.

Examples of non-homotopic paths will be given a little later: for instance, it turns out that two closed piecewise smooth paths in $\mathbb{C} \setminus \{0\}$ are homotopic if and only if their indices around the origin coincide. Now we consider a simple, but important example of a situation in which we do have a homotopy.

Proposition 6.3 *Let $U \subset \mathbb{C}$ be a convex open set and $\gamma_0, \gamma_1 \colon [A;B] \to U$ be either two paths in U connecting the same pair of points p and q, or two closed paths. Then γ_0 and γ_1 are homotopic: in the first case, as paths with fixed endpoints, in the second case, as closed paths.*

Proof In both cases, a homotopy

$$F \colon [A;B] \times [0;1] \to U$$

we are seeking is given by the formula $F(t,s) = (1-s)\gamma_0(t) + s\gamma_1(t)$. Since the set of points of the form $(1-s)\gamma_0(t) + s\gamma_1(t)$, $s \in [0;1]$, is nothing else than the line segment between $\gamma_0(t)$ and $\gamma_1(t)$, and the set U is convex by assumption, we have $F([A;B] \times [0;1]) \subset U$, so F is indeed a map to U. If γ_0 and γ_1 are closed paths, then $\gamma_0(A) = \gamma_0(B)$ and $\gamma_1(A) = \gamma_1(B)$, whence $F(A,s) = F(B,s)$ for all s. If γ_0 and γ_1 connect points p and q, then $F(A,s) = (1-s)p + sp = p$ for every s, and similarly for $F(B,s)$. Thus, in both cases F is a homotopy between γ_0 and γ_1. $\square$

The homotopy constructed in this proof is called a *linear homotopy*.

If there is a homotopy $F: [A; B] \times [0; 1] \to U$ between two piecewise smooth paths $\gamma_0, \gamma_1: [A; B] \to U$, then the "intermediate" paths $\gamma_s: t \mapsto F(t, s)$, though continuous, can be arbitrarily non-smooth, to the point of being Peano curves! However, it turns out that if there is at least some homotopy between piecewise smooth paths in an open set, then one can construct a homotopy between them whose all intermediate curves are also piecewise smooth. Now we prove this assertion (actually, even a stronger one). It will be needed in Sec. 6.3 below.

Lemma 6.4 (homotopy enhancement) *Let $U \subset \mathbb{C}$ be an open subset, and let $\gamma_0, \gamma_1: [A; B] \to U$ be two piecewise smooth paths that either connect the same pair of points p and q, or both are closed. If γ_0 and γ_1 are homotopic (as paths with fixed endpoints in the first case, and as closed paths in the second case), then there exists a homotopy $G: [A; B] \times [0; 1] \to U$ between them (which is also a homotopy of paths with fixed endpoints in the first case and a homotopy of closed paths in the second case) such that the path $\gamma_s: [A; B] \to U$, $\gamma_s(t) = G(t, s)$, is piecewise smooth for every $s \in [0; 1]$, and also the path $\mu_t: [0; 1] \to U$, $\mu_t(s) = G(t, s)$, is piecewise smooth for every $t \in [A; B]$.*

In the proof of this lemma (and not just in it), we need a general analytic fact; to begin with, we state and prove it.

Proposition 6.5 (Lebesgue number lemma) *Let $K \subset \mathbb{C} = \mathbb{R}^2$ be a compact set and $\{V_\alpha\}$ be a family of open subsets in $\mathbb{C}$ such that $K \subset \bigcup_\alpha V_\alpha$. Then there exists $\varepsilon > 0$ such that the ε-neighborhood of every point $z \in K$ is entirely contained in one of the sets V_α.*

In this lemma, a compact subset $K \subset \mathbb{C}$ can be replaced with a compact subset $K \subset \mathbb{R}^n$ for arbitrary n, or even with an arbitrary compact metric space (and an open cover of this space). The proof in these more general cases is word for word the same as that given below. By a *Lebesgue number* of an open cover $\{V_\alpha\}$ we will mean any $\varepsilon > 0$ satisfying the assumptions of the proposition.

Proof We argue by contradiction. If there is no Lebesgue number for $\{V_\alpha\}$, then, in particular, no number of the form $1/n$ where n is a positive integer does qualify. Thus, for every positive integer n there exists $z_n \in K$ such that the $(1/n)$-neighborhood of z_n is not contained in any V_α. Since K is compact, the sequence $\{z_n\}$ has a subsequence $\{z_{n_k}\}$ converging to a point $z \in K$. The sets V_α cover the whole of K, hence z is contained in some set V_β; since V_β is open, there exists $\delta > 0$ such that the δ-neighborhood of z is contained in V_β. Now let k be so large that $1/n_k < \delta/2$ and $|z_{n_k} - z| < \delta/2$. It is clear from the triangle inequality and the inequality $|z_{n_k} - z| < \delta/2$ that the $(\delta/2)$-neighborhood of z_{n_k} is contained in the δ-neighborhood of z, and hence in the set V_β; since $1/n_k < \delta/2$, the $(1/n_k)$-neighborhood of z_{n_k} is certainly contained in V_β, a contradiction. $\qquad\square$

Proof of Lemma 6.4 Let $F: [A; B] \times [0; 1]$ be a homotopy between the paths γ_0 and γ_1. Since the set U is open and the map F is continuous, for every point

$p \in [A; B] \times [0; 1]$ there is an open set $V_p \subset \mathbb{R}^2$ containing p such that $F(V_p)$ is contained in some open disk $U_p \subset U$ containing $F(p)$. Apply Lebesgue's number lemma (Proposition 6.5) to the compact set $[A; B] \times [0; 1] \subset \bigcup V_p$, and let ε be a Lebesgue number of the cover $\{V_p\}$. Now divide the rectangle $[A; B] \times [0; 1]$ into rectangles of diameter less than ε by drawing lines parallel to the coordinate axes. In more detail, choose numbers $A = t_0 < t_1 < \ldots < t_m = B$ and $0 = s_0 < s_1 < \ldots < s_n = 1$ in such a way that the diagonal of each of the rectangles

$$\Pi_{jk} := [t_j, t_{j+1}] \times [s_k; s_{s+1}] = \{(t, s) : t_j \leq t \leq t_{j+1}, s_k \leq s \leq s_{k+1}\}$$

is less than ε. By construction, for any j and k the set $F(\Pi_{jk})$ is contained in an open disk which, in turn, is contained in U. Therefore, for any points $p, q \in \Pi_{jk}$ the line segment between $F(p)$ and $F(q)$ is also contained in U; we will repeatedly use this observation.

Now we construct a homotopy G. First, we must set $G(t, 0) = \gamma_0(t) = F(t, 0)$ and $G(t, 1) = \gamma_1(t) = F(t, 1)$, here we have no choice. At the second step, we define $G(t, s_k)$ for all $t \in [A; B]$ and $0 < k < n$ (the inequalities are strict, since $G(t, s_0) = G(t, 0)$ and $G(t, s_n) = G(t, 1)$ are already defined and not to be tampered with). Effectively, this means that each of the paths

$$\gamma_{s_k} : t \mapsto F(t, s_k),$$

which is only guaranteed to be continuous, will be replaced by a polygonal curve. Specifically, this can be done as follows: first, we set $G(t_j, s_k) = F(t_j, s_k)$ (in other words, the "internal" vertices of the rectangles into which the large rectangle $[A; B] \times [0; 1]$ is divided are mapped to the same points as before); and if $t_j \leq t \leq t_{j+1}$, then

$$G(t, s_k) = F(t_j, s_k) + \frac{t - t_j}{t_{j+1} - t_j}(F(t_{j+1}, s_k) - F(t_j, s_k)); \qquad (6.1)$$

in other words, as t grows from t_j to t_{j+1}, the point $F(t, s_j)$ moves uniformly along the straight line from $F(t_j, s_k)$ to $F(t_{j+1}, s_k)$. Since, as we have observed earlier, the line segment between the images of any two points of the rectangle Π_{jk} lies in U, all points $G(t, s_k)$ do indeed lie in U.

It remains to define $G(t, s)$ for s other than the division points $s_0, \ldots, s_n$. For this, we use the construction of a linear homotopy. Namely, if $s_k \leq s \leq s_{k+1}$ and $t \in [A; B]$ is arbitrary, set

$$G(t, s) = G(t, s_k) + \frac{s - s_k}{s_{k+1} - s_k}(G(t, s_{k+1}) - G(t, s_k)). \qquad (6.2)$$

Again, as s grows from s_k to s_{k+1}, the point $G(t, s)$ moves uniformly along the straight line from $G(t, s_k)$ to $G(t, s_{k+1})$. The points $G(t, s)$ thus defined do indeed lie in U for the same reason as above.

Obviously, the map G defined in this way is continuous. Now we verify that the maps $t \mapsto G(t, s)$ are piecewise smooth for every fixed $s \in [0; 1]$. For $s = s_0 = 0$

or $s = s_n = n$, they coincide with γ_0 or γ_1 and hence are piecewise smooth by assumption. For $s = s_k$, $0 < k < m$, these maps are given by (6.1) and hence are piecewise smooth (and even piecewise linear). Finally, if s does not coincide with any s_j, then these maps, as follows from (6.2), are linear combinations of two piecewise smooth maps and hence are also piecewise smooth.

The piecewise smoothness of the maps $s \mapsto G(t, s)$ for every fixed $t \in [A; B]$ can be verified in the same way and even more easily: it follows from (6.2) that all these maps are even piecewise linear. Finally, it is clear from construction that the map G is a homotopy with fixed endpoints (respectively, a homotopy of closed paths) if F is. The lemma is proved.   □

6.2 Analytic Continuation

In Chap. 2 we discovered that if there were a continuous function log such that $\log 1 = 0$, then, considering the values $\log(e^{it})$ for $t \in [0; 2\pi]$, we would obtain, by continuity, that $\log 1$ must be equal not to 0, but to $2\pi i$. Now we are going to examine this kind of phenomena more systematically. We begin with an important abstract definition.

Given a point $p \in \mathbb{C}$, consider the set of all pairs (U, f) where $U \ni p$ is a neighborhood and $f: U \to \mathbb{C}$ is a holomorphic function. Two pairs (U_1, f_1) and (U_2, f_2) are considered equivalent if the functions f_1 and f_2 coincide on some neighborhood of p.

Definition 6.6 A *germ* of a holomorphic function at a point p is an equivalence class of pairs (U, f) with respect to this equivalence relation; the equivalence class of a pair (U, f) is called the *germ of the function f at the point p*.

The set of germs of holomorphic functions at a point p is denoted by O_p.

Informally, a germ at a point p is a function defined in a neighborhood of p for which we care about nothing but the behavior in arbitrarily small neighborhoods of p.

A definition of germs completely analogous to Definition 6.6 can be given also for other classes of functions: smooth, continuous, and even quite arbitrary (in some cases, this may be useful too). However, specific properties of holomorphic functions ensure that their germs have a simple explicit description.

Proposition 6.7 *There exists a natural one-to-one correspondence between the germs of holomorphic functions at a point $p \in \mathbb{C}$ and the power series of the form $\sum c_k(z - p)^k$ with positive radius of convergence.*

Proof This is just the assertion that a function is holomorphic if and only if it is analytic: if two holomorphic functions coincide in a neighborhood of p, then they have the same power series at p, because the coefficients of this series for a function f are expressed in terms of the derivatives of f at p; in this way, with every germ we associate a convergent power series. Conversely, every power series with positive radius of convergence is a holomorphic function in its disk of convergence, and now we can consider its germ.   □

Definition 6.8 The set of all germs of holomorphic functions at a point $p \in \mathbb{C}$ is called the *local ring* of holomorphic functions and denoted by O_p.

The union of O_p over all $p \in \mathbb{C}$ is denoted by O.

The set O_p is called a ring, because it can be easily equipped with an addition and multiplication satisfying all the ring axioms (but we will usually regard various O_p merely as sets).

In the definition of O, it is understood that the sets $O_p \subset O$ for different p are disjoint. More precisely,

$$O = \{(p, f) : p \in \mathbb{C}, \, f \in O_p\}.$$

Definition 6.9 Let $U \subset \mathbb{C}$ be an open set and $f : U \to \mathbb{C}$ be a holomorphic function. The subset $\mathcal{U} \subset O$ consisting of the germs of f at all points $p \in U$ is called a *neighborhood* in O. If $f \in O$ and $\mathcal{U} \subset O$ is a neighborhood containing f, then $\mathcal{U}$ is called a *neighborhood of the germ f*.

Remark 6.10 For the readers familiar with the corresponding definitions, I mention that the set of neighborhoods in O defined above is a base of a topology on O (moreover, this topology is Hausdorff, which follows from the principle of analytic continuation). The set O equipped with this topology is called the "sheaf of holomorphic functions on $\mathbb{C}$." Sheaf techniques are not used in this book, so I will not further elaborate on this point.

The map $O \to \mathbb{C}$ that sends a germ $f \in O_p$ to the point $p \in \mathbb{C}$ has properties resembling those of a covering map, and the reader familiar with coverings will see that the statements and proofs of Propositions 6.14 and 6.15 have much in common with the well-known "path lifting theorem" and "homotopy lifting theorem." Still, this map is not a covering map (it is only a local homeomorphism), so we cannot deduce these results just by referring to the corresponding facts about coverings.

Now we are finally in a position to give a definition of analytic continuation.

Definition 6.11 Let $\gamma : [A; B] \to \mathbb{C}$ be a continuous path in the complex plane (the case $\gamma(A) = \gamma(B)$ is not excluded). Set $\gamma(A) = p$, $\gamma(B) = q$, and let $f \in O_p$. An *analytic continuation* of the germ f along the path γ is a map $\Gamma : [A; B] \to O$ with the following properties:

(1) $\Gamma(t) \in O_{\gamma(t)}$ for every $t \in [A; B]$;

(2) the map Γ is "continuous" in the following sense: if $t \in [A; B]$ and $\mathcal{U} \ni \Gamma(t)$ is a neighborhood of the germ $\Gamma(t) \in O_{\gamma(t)}$ in the sense of Definition 6.9, then there exists $\delta > 0$ such that $\Gamma(t') \in \mathcal{U}$ whenever $|t' - t| < \delta$.

If Γ is an analytic continuation of a germ $f \in O_p$, then the germ $\Gamma(B) \in O_q$ is called the *analytic continuation of f along γ*.

Of course, it is not true that every germ can be analytically continued along every path (see Exercise 6.1).

The above definition of analytic continuation is so abstract that it looks impracticable. However, it has an elementary and "down-to-earth" reformulation.

Proposition 6.12 *Let* $\gamma\colon [A; B] \to \mathbb{C}$ *be a continuous path in the complex plane. Set* $\gamma(A) = p$, $\gamma(B) = q$, *and let* f *be a germ of a holomorphic function at the point* p *and* g *be a germ of a holomorphic function at the point* q. *Then the following two conditions are equivalent.*

(1) *The germ* g *is obtained by analytic continuation of* f *along* γ.

(2) *There exist a partition* $A = t_1 < t_2 < \ldots < t_n < t_{n+1} = B$ *of the interval* $[A; B]$, *open subsets* $U_1, \ldots, U_n \subset \mathbb{C}$, *and holomorphic functions* $f_j\colon U_j \to \mathbb{C}$ *with the following properties*:

(a) $\gamma([t_j; t_{j+1}]) \subset U_j$ *for every* j;
(b) *on every intersection* $U_j \cap U_{j+1}$, *the functions* f_j *and* f_{j+1} *coincide*;
(c) *the germ of* f_1 *at* p *is* f, *the germ of* f_n *at* q *is* g.

In somewhat less precise terms, Proposition 6.12 says the following: the analytic continuation of a germ f to a germ g along a curve γ can always be constructed by covering γ with a chain of open sets with holomorphic functions defined on them that agree on the intersections.

Proof First, we prove the implication (1) $\Rightarrow$ (2). Let $\Gamma\colon [A; B] \to O$ be the map that defines the analytic continuation of f to g along γ. For every $t \in [A; B]$, consider the germ $\Gamma(t) \in O_{\gamma(t)}$. It is given by a holomorphic function f_t on an open set $U_t \ni \gamma(t)$; replacing U_t, if necessary, by a smaller set, we may assume that it is an open disk centered at $\gamma(t)$: this does not affect the germ. Now let $\mathcal{U}_t \subset O$ be the set of germs of f_t at all points of U_t. The set $\mathcal{U}_t$ is a neighborhood of the germ $\Gamma(t)$ in the sense of Definition 6.9; hence, by the "continuity" of Γ, there exists a neighborhood $V_t \ni t$, $V_t \subset [A; B]$, such that $\Gamma(V_t) \subset \mathcal{U}_t$.

Choosing a finite subcover of the cover of $[A; B]$ by these neighborhoods V_t, we obtain a partition $A = t_1 < t_2 < \ldots < t_n < t_{n+1} = B$ and a collection of neighborhoods $\mathcal{U}_1, \ldots, \mathcal{U}_n \subset O$ for which $\Gamma([t_j, t_{j+1}]) \subset \mathcal{U}_j$ and $\mathcal{U}_j$ itself is the set of germs of a holomorphic function $f_j\colon U_j \to \mathbb{C}$ at all points of an open disk $U_j \supset [t_j, t_{j+1}]$. The latter observation implies, in particular, that $f = \Gamma(t_1) = \Gamma(A)$ is the germ of f_1 at the point $p = \gamma(A)$ and, analogously, $g = \Gamma(B)$ is the germ of f_n at the point $q = \Gamma(B)$. Thus, we have proved properties (a) and (c); to prove property (b), observe that at the points $\gamma(t_{j+1})$ the germs of the functions f_j and f_{j+1} coincide (with the germ $\Gamma(t_{j+1}) \in O_{\gamma(t_{j+1})}$), so the functions $f_j, f_{j+1}\colon U_j \cap U_{j+1} \to \mathbb{C}$ coincide on some neighborhood of the point $\gamma(t_{j+1}) \in U_j \cap U_{j+1}$. Since U_j and U_{j+1} are disks, the intersection $U_j \cap U_{j+1}$ is convex and, therefore, connected (any two points can be joined by a line segment). Thus, by the principle of analytic continuation, f_j coincides with f_{j+1} on $U_j \cap U_{j+1}$, and condition (b) is also satisfied. We have established that (1) $\Rightarrow$ (2).

The implication (2) $\Rightarrow$ (1) is easier to prove. Namely, assume that we are given t_j, U_j, and f_j satisfying conditions (a), (b), and (c). We construct a map $\Gamma\colon [A; B] \to O$ by setting

$$\Gamma(t) = (\text{the germ of } f_j \text{ at } \gamma(t)) \quad \text{if } t \in [t_j; t_{j+1}].$$

Note that Γ is well defined: if t belongs to two neighboring segments of the partition, i.e., $t = t_j$ for some j, then, since the functions f_{j-1} and f_j coincide on $U_{j-1} \cap U_j \ni \gamma(t_j)$, they have the same germ at $\gamma(t_j)$. It remains to check that Γ is "continuous" in the sense of Definition 6.11. To this end, assume that $t \in [t_j; t_{j+1}]$, and let $\mathcal{U} \ni \Gamma(t)$ be a neighborhood in O. If $\mathcal{U}$ is the set of germs of a function $f \colon U \to \mathbb{C}$ at all points of an open set $U \subset \mathbb{C}$, then, since $\mathcal{U} \ni \Gamma(t)$, the germ of f at the point $\Gamma(t)$ coincides with the germ of f_j at the same point; thus, there exists an open disk $D \ni \Gamma(t)$, $D \subset U \cap U_j$, on which the functions f and f_j coincide. Now if $\delta > 0$ is so small that $|t' - t| < \delta$ implies $\gamma(t') \in U$, then for all t' in the δ-neighborhood of t, the germ $\Gamma(t') \in O_{\gamma(t')}$ is the germ of the function f_j, which is the same as f, so $\Gamma(t') \in \mathcal{U}$. The proposition is proved.   □

Proposition 6.12 shows that we have already encountered various examples of analytic continuation along a path. For instance, if we take the branch of the logarithm defined in a neighborhood of 1 and determined by the condition $\log(1) = 0$ and consider its analytic continuation along the circle $\{z \colon |z| = 1\}$ oriented counterclockwise, then we obtain the branch of the logarithm defined in a neighborhood of 1 for which $\log(1) = 2\pi i$. Here, sets U_i and functions f_i can be chosen, for example, as follows:

$$U_1 = \mathbb{C} \setminus (-\infty; 0], \qquad f_1(z) = \log z, \qquad \operatorname{Im} f(z) \in (-\pi; \pi);$$
$$U_2 = \mathbb{C} \setminus [0; +\infty), \qquad f_2(z) = \log z, \qquad \operatorname{Im} f(z) \in (0; 2\pi);$$
$$U_3 = U_1, \qquad\qquad\quad f_3(z) = \log z, \qquad \operatorname{Im} f(z) \in (\pi; 3\pi).$$

In a similar way we can show that every germ of the functions $\log z$ and $\sqrt[n]{z}$ with n a positive integer (of course, at a point other than 0) admits an analytic continuation along every path that does not pass through the origin. Moreover, if f is a holomorphic function on an open set U, then every germ of $\log f$ or $\sqrt[n]{f}$ admits an analytic continuation along every path that does not pass through the points where f vanishes.

Here is another corollary of Proposition 6.12, almost obvious but nevertheless useful.

Corollary 6.13 *Let $\gamma \colon [A; B] \to \mathbb{C}$ be a continuous path connecting points p and q. Assume that a germ $f \in O_p$ can be analytically continued along γ to a germ $g \in O_q$. If V is an open set containing $\gamma([A; B])$ and $\varphi \colon V \to \mathbb{C}$ is a holomorphic function, then the germ $\varphi \circ f$ can be continued along γ to the germ $\varphi \circ g$.*

(We leave the reader to work out how to compose a function with a germ.)

Proof Indeed, if the analytic continuation of f to g involves open sets U_j and functions f_j, then the analytic continuation of $\varphi \circ f$ to $\varphi \circ g$ can be obtained by using the open sets $U_j \cap V$ and the functions $\varphi \circ f_j$.   □

Now we prove two general facts about analytic continuation.

Proposition 6.14 *Let* $\gamma\colon [A; B] \to \mathbb{C}$ *be a continuous path in the complex plane connecting points* $p = \gamma(A)$ *and* $q = \gamma(B)$. *If there exists an analytic continuation of a germ* $f \in O_p$ *along* γ, *then it is unique.*

Proof Let $\Gamma, \Gamma_1\colon [A; B] \to O$ be two analytic continuations of f along γ; we will show that $\Gamma(t) = \Gamma_1(t)$ for all $t \in [A; B]$, which, obviously, proves the theorem.

To this end, denote by $E \subset [A; B]$ the set of values C from the interval $A < C \leq B$ such that $\Gamma(t) = \Gamma_1(t)$ for all $t \in [A; C]$; by assumption, $E \ni A$, and we must prove that $E = [A; B]$.

First, we show that if $t_0 \in E$, $t_0 < C$, then there exists $\varepsilon > 0$ such that $t_0 + \varepsilon \in E$. Indeed, let $\Gamma(t_0) = \Gamma_1(t_0)$ be the germ of a function $f\colon U \to \mathbb{C}$ at the point $\gamma(t_0) \in U$. Denote by $\mathcal{U}$ the set of germs of f at all points of the set U; by the "continuity" of the maps Γ and Γ_1, there exists $\delta > 0$ such that $|t - t_0| < \delta$ implies $\Gamma(t) \in \mathcal{U}, \Gamma_1(t) \in \mathcal{U}$. Since, by the definition of $\mathcal{U}$ and of analytic continuation, both $\Gamma(t)$ and $\Gamma_1(t)$ are germs of f at the point $\gamma(t)$, we obtain $[t_0; t_0 + \delta) \subset E$, whence, say, $t_0 + \delta/2 \in E$.

Now we show that $\sup E \in E$. Indeed, by the definition of E, we have $t \in E \Rightarrow [A; t] \subset E$.

Denote $\sup E = t_0$; if $t_0 \notin E$, then it follows from the above that $[A; t_0) \subset E$. Let $\Gamma(t_0)$ and $\Gamma_1(t_0)$ be the germs at $\gamma(t_0)$ of the functions f and f_1, respectively; we may assume without loss of generality that both f and f_1 are defined on the same open disk $U \ni \gamma(t_0)$ (instead of a disk, any connected open set would do). Denote by $\mathcal{U}$ the set of germs of f at all points of U, and by $\mathcal{U}_1$ the set of germs of f_1 at all points of U. By the "continuity" of the maps Γ and Γ_1, there exists $\delta > 0$ such that $|t - t_0| < \delta$ implies $\Gamma(t) \in \mathcal{U}, \Gamma_1(t) \in \mathcal{U}_1$. Since $t_0 = \sup E$, there exists $t_1 \in E$ such that $t_1 < t$ and $|t_1 - t| < \delta$. Then

$$\Gamma(t) = \Gamma_1(t) \in \mathcal{U} \cap \mathcal{U}_1.$$

Therefore, the germs of f and f_1 at the point $\gamma(t_1) \in U$ coincide: both are equal to $\Gamma(t) = \Gamma_1(t)$. In other words, $f = f_1$ on some neighborhood of $\gamma(t)$ contained in U. Since U is connected, the principle of analytic continuation implies that $f = f_1$ on U; in particular, the germs of the functions f and f_1 at the point $\gamma(t_0)$ coincide. Therefore, $\Gamma(t_0) = \Gamma_1(t_0)$; since $[A; t_0] \subset E$, we have $t_0 \in E$.

Thus, the set E has the following properties: if $t \in E$, then $[A; t] \subset E$; if $t < B$ lies in E, then some neighborhood of t lies in E; $\sup E \in E$. It is clear from these three properties that $E = [A; B]$, and we are done. $\qquad\square$

The second property of analytic continuation concerns continuations of the same germ along homotopic paths.

Proposition 6.15 *Let* $F\colon [A; B] \times [0; 1] \to \mathbb{C}$ *be a homotopy with fixed endpoints of paths connecting points* p *and* q, *so* $F(A, s) = p$ *for all* $s \in [0; 1]$ *and* $F(B, s) = q$ *for all* $s \in [0; 1]$. *Let* $f \in O_p$ *be a germ of a holomorphic function at the point* p, *and assume that* f *admits an analytic continuation along every intermediate path* $\gamma_s\colon t \mapsto F(t, s)$.

Then the analytic continuations of f *along all paths* γ_s *(in particular, along* γ_0 *and* γ_1*) coincide.*

Proposition 6.15 is called the monodromy theorem. One usually speaks about monodromy in a situation where "translations" of something along all homotopic paths (in our case, this "something" is a germ) yield the same result.

Proof Assume that the analytic continuation of f along a path γ_s, $s \in [0; 1]$, yields a germ $g \in O_q$. We will show that there exists $\varepsilon_s > 0$ such that for every s_1 with $|s_1 - s| < \varepsilon_s$, the analytic continuation of f along γ_{s_1} also yields g. Since, by the compactness of closed intervals (known as the "Heine–Borel theorem"), the interval $[0; 1]$ can be covered by finitely many ε_s-neighborhoods of points $s \in [0; 1]$, i.e., by finitely many intervals such that within each of them the analytic continuation of f yields the same result, it follows that this result is the same for all s.

Now we prove the assertion about γ_s stated above. By Proposition 6.12, there exist a partition $A = t_1 < t_2 < \ldots < t_{n+1} = B$, open sets $U_1, \ldots, U_n$ with $U_j \supset \gamma_s([t_j; t_{j+1}])$, and holomorphic functions $f_j : U_j \to \mathbb{C}$ such that f_j and f_{j+1} coincide on $U_j \cap U_{j+1}$, f is the germ of f_1 at the point $p = F(0, s)$, and g is the germ of g_n at the point $q = F(1, s)$. We will show that for every j, $1 \le j \le n$, there is $\delta_j > 0$ such that $\gamma_{s'}([t_j; t_{j+1}]) \subset U_j$ whenever $|s' - s| < \delta_j$. This can be proved by a standard compactness argument. Namely, denote the rectangle $[A; B] \times [0; 1]$ by Π. Since the map $F : \Pi \to \mathbb{C}$ is continuous and the image of the line segment between (t_j, s) and (t_{j+1}, s) is contained in the open set U_j, for every point $(t, s) \in \Pi \subset \mathbb{R}^2$, $t_j \le t \le t_{j+1}$, there exists an open disk $D_t \subset \mathbb{R}^2$ centered at (t, s) such that $F(D_t) \subset U_j$. By compactness, the line segment between (t_j, s) and (t_{j+1}, s) is contained in a finite union $D_1 \cup \ldots \cup D_m$ of such disks; therefore, there exists $\delta_j > 0$ such that the rectangle $[A; B] \times [s - \delta_j; s + \delta_j]$ is contained in the union of D_j, whence $F([A; B] \times [s - \delta_j; s + \delta_j]) \subset U_j$. Repeat this construction for each j, $1 \le j \le n$, and set $\delta = \min(\delta_1, \ldots, \delta_n)$. Now if $|s' - s| < \delta$, then for $t \in [t_k, t_{k+1}]$, $1 \le k \le n$, we have $\gamma_{s'}(t) = F(t, s) \in U_k$, so the collection of open sets $U_1, \ldots, U_n$ and functions $f_1, \ldots, f_n$ defines an analytic continuation of f (the germ of f_1 at p) to g (the germ of g_1 at q) along the curve $\gamma_{s'}$. In other words, the analytic continuation of f along $\gamma_{s'}$ is the same for all s' lying in the δ-neighborhood of s, and this is precisely what we lacked to complete the proof of the proposition. □

A useful application of the monodromy theorem is a result about constructing holomorphic functions on simply connected sets. First, we introduce a definition.

Definition 6.16 An open set $U \subset \mathbb{C}$ is said to be *simply connected* if the following two conditions are satisfied:

(1) U is connected;

(2) if $p, q \in U$ and $\gamma_0, \gamma_1 : [A; B] \to U$ are two paths connecting p and q, then γ_0 and γ_1 are homotopic as paths with fixed endpoints.

The construction of a linear homotopy shows that every convex open set is simply connected. Further, if there is a «homeomorphism» (that is, a continuous bijective map whose inverse is also continuous) $\varphi : U_1 \to U_2$ between open sets U_1 and U_2 and one of these sets is simply connected, then the other one is simply connected too: if $F : [A; B] \times [0; 1] \to U_1$ is a homotopy between paths γ_0 and γ_1 in U_1, then $\varphi \circ F$ is a homotopy between the paths $\varphi \circ \gamma_1$ and $\varphi \circ \gamma_2$ in U_2.

Proposition 6.17 *Let $U \subset \mathbb{C}$ be a simply connected open set, $p \in U$, and $f \in O_p$ be a germ of a holomorphic function at the point p. If f admits an analytic continuation along every path in U, then there exists a unique holomorphic function $f \colon U \to \mathbb{C}$ whose germ at p coincides with f.*

Proof For each point $q \in U$, join p to q by a path γ and consider the analytic continuation of f along γ. Since U is simply connected, the monodromy theorem says that the result does not depend on γ; set $f(q)$ to be the value of the resulting germ at the point q. It is clear by construction that f is holomorphic, and the uniqueness follows from the connectedness of U and the principle of analytic continuation. $\quad\square$

6.3 Cauchy's Theorem Revisited

We return one last time to Cauchy's theorem. Now we are at last ready to state precisely what it means that "the integral of a holomorphic function is unchanged under deformations of the path," and to prove this statement.

Theorem 6.18 (Cauchy's theorem, version 4) *Let $f \colon U \to \mathbb{C}$ be a holomorphic function, where $U \subset \mathbb{C}$ is an open set. Assume that $\gamma_0, \gamma_1 \colon [A; B] \to U$ are piecewise smooth paths and either γ_0 and γ_1 connect the same pair of points $p, q \in U$, or both are closed. If γ_0 and γ_1 are homotopic (in the first case, as paths with fixed endpoints; in the second case, as closed paths), then $\int_{\gamma_0} f(z)\,dz = \int_{\gamma_1} f(z)\,dz$.*

Proof By the homotopy enhancement lemma (Lemma 6.4), we may assume that the homotopy $F \colon [A; B] \times [0; 1] \to U$ between γ_0 and γ_1 has the following properties: every path $\gamma_s \colon t \mapsto F(t, s)$, $s \in [0; 1]$, is piecewise smooth, and every path $s \mapsto F(t, s)$, $t \in [A; B]$, is also piecewise smooth. Denote the rectangle $[A; B] \times [0; 1]$ by $\Pi \subset \mathbb{R}^2$, and let $\mu \colon [P; Q] \to \partial \Pi$, where $[P; Q] \subset \mathbb{R}$ is an interval, be a bijective piecewise linear parametrization of the boundary of Π that traverses it counterclockwise. By the properties of F, the composition $F \circ \mu \colon [P; Q] \to U$ is a piecewise smooth path in U. I claim that

$$\int_{F \circ \mu} f(z)\,dz = \int_{\gamma_0} f(z)\,dz - \int_{\gamma_1} f(z)\,dz. \tag{6.3}$$

Indeed, the restriction of $F \circ \mu$ to the lower base of the rectangle Π (i.e., to $[A; B] \times \{0\}$) is nothing else than the path γ_0 (with an increasing linear change of parametrization), so the integral of $f(z)\,dz$ over this part of the path $F \circ \mu$ is equal to $\int_{\gamma_0} f(z)\,dz$. In a similar way, the integral of $f(z)\,dz$ over the restriction to the upper base of Π is equal to $-\int_{\gamma_1} f(z)\,dz$, because this part of $F \circ \mu$ is the path γ_1 traversed in the opposite direction (formally, it is the path γ_1 with a *decreasing* linear change of parametrization). To establish (6.3), it suffices to verify that the sum of the integrals over the parts of $F \circ \mu$ corresponding to the lateral sides of the rectangle vanishes. In the case where F is a homotopy with fixed endpoints, this is true even for each term separately: F maps each of the lateral sides to a single point, so the restrictions of $F \circ \mu$ to the lateral sides are "constant" paths, and integrals over such paths are always zero. If F is a homotopy of closed paths, then $F(0, s) = F(1, s)$ for every $s \in [0; 1]$, so the two parts of $F \circ \mu$ corresponding to the lateral sides are the same path traversed in opposite directions, and the integrals of the same function over a path and the same path traversed in the opposite direction sum to zero.

Thus, we have established (6.3); now, to prove the theorem, it suffices to verify that the left-hand side of (6.3) is zero. To this end, observe that, by the continuity of F, for every point $x \in \Pi$ there exists a neighborhood $V_x \ni x$, $V_x \subset \mathbb{R}^2$, such that $F(V_x \cap \Pi)$ is contained in an open disk $D_x \subset U$. By Proposition 6.5, the cover $\Pi \subset \bigcup_x V_x$ has a "Lebesgue number" $\varepsilon > 0$ such that the ε-neighborhood of every point $y \in \Pi$ is contained in some V_x. Now divide the rectangle Π into finitely many rectangles of diameter less than ε by drawing lines parallel to its sides. If Π' is any of these small rectangles, then Π' is contained in the ε-neighborhood of some[1] its vertex, and thus in some of the neighborhoods V_x; therefore, $F(\Pi')$ is contained in some open disk D_x which, in turn, is contained in U. Let $\mu' \colon [P'; Q'] \to \partial\Pi'$ be a bijective piecewise linear parametrization of the boundary of Π' that traverses it in the positive direction. Since the homotopy F is "enhanced," the composition $F \circ \mu'$ is a closed piecewise linear path, so the integral $\int_{F \circ \mu'} f(z) \, dz$ is well defined. Since the image of this path lies in the open disk $D_x \subset U$, we have $\int_{F \circ \mu'} f(z) \, dz = 0$ by Theorem 5.3 (on a convex set, the integral of a holomorphic function over a closed path vanishes). Take the sum of the integrals $\int_{F \circ \mu'} f(z) \, dz$ over all small rectangles Π'. We have already proved that each of the terms is equal to zero; on the other hand, these integrals sum to $\int_{F \circ \mu} f(z) \, dz$: the integral over each side of a small rectangle lying strictly in the interior of Π appears twice with opposite signs, and the integrals over the parts corresponding to the other sides sum to the integral over $F \circ \mu$. Thus, the left-hand side of (6.3) is zero, and the theorem is proved. $\square$

6.4 Indices of Curves Revisited

As an application of the above results, we will obtain a homotopic classification of closed paths (in the simplest nontrivial case).

Definition 6.19 Let $a \in \mathbb{C}$ and $0 \le r < R \le +\infty$. The set $\{z \colon r < |z - a| < R\}$ is called an *annulus* centered at a.

In particular, special cases of an annulus are the plane with a point removed, a punctured disk, and the complement to a closed disk.

---

[1] In fact, any, but it does not matter.

Proposition 6.20 *Let $U \subset \mathbb{C}$ be an annulus centered at a, and let γ_0 and γ_1 be piecewise smooth closed paths in U. Then the following two conditions are equivalent:*

(1) *γ_0 and γ_1 are homotopic as closed paths;*

(2) *$\mathrm{Ind}_a\, \gamma_0 = \mathrm{Ind}_a\, \gamma_1$.*

Proof The implication (1) $\Rightarrow$ (2) follows from the fourth version of Cauchy's theorem (Theorem 6.18) and the integral definition of the index of a curve: if γ_0 and γ_1 are homotopic, then

$$\mathrm{Ind}_a\, \gamma_0 = \frac{1}{2\pi i} \int_{\gamma_0} \frac{dz}{z-a} = \frac{1}{2\pi i} \int_{\gamma_1} \frac{dz}{z-a} = \mathrm{Ind}_a\, \gamma_1.$$

To prove the implication (2) $\Rightarrow$ (1), set

$$\tilde{U} = \{z \colon \log r < \mathrm{Re}\, z < \log R\}$$

(we assume that $\log 0 = -\infty$, $\log(+\infty) = +\infty$.) The function $z \mapsto a + e^z$ maps $\tilde{U}$ onto the annulus U.

By Lemma 4.18, there exist piecewise smooth functions $\varphi_0, \varphi_1 \colon [A; B] \to \mathbb{R}$ such that

$$\gamma_0(t) = a + |\gamma_0(t) - a|e^{i\varphi_0(t)}, \quad \gamma_1(t) = a + |\gamma_1(t) - a|e^{i\varphi_1(t)}.$$

Now set

$$\tilde{\gamma}_0(t) = \log|\gamma_0(t) - a| + i\varphi_0(t), \quad \tilde{\gamma}_1(t) = \log|\gamma_1(t) - a| + i\varphi_1(t).$$

We can construct a linear homotopy $\tilde{F} \colon [A; B] \times [0; 1] \to \tilde{U}$ between the paths $\tilde{\gamma}_0$ and $\tilde{\gamma}_1$ in the convex open set U as follows:

$$\tilde{F} \colon (t, s) \mapsto (1-s)\tilde{\gamma}_0(t) + s\tilde{\gamma}_1(t).$$

Setting $F(t, s) = a + e^{\tilde{F}(t,s)}$, we obtain a homotopy between paths in U for which $F(t, 0) = \gamma_0(t)$, $F(t, 1) = \gamma_1(t)$. This construction can be applied to quite arbitrary paths γ_0 and γ_1 in U, but if γ_0 and γ_1 are closed and their indices coincide, then F turns out to be a homotopy of *closed* paths. Indeed, if $\mathrm{Ind}_a\, \gamma_0 = \mathrm{Ind}_a\, \gamma_1 = n$, then $\varphi_0(B) - \varphi_0(A) = \varphi_1(B) - \varphi_1(A) = 2\pi n$ by Proposition 4.19, so for every $s \in [0; 1]$ we have

$$\tilde{F}(1, s) - \tilde{F}(0, s) = (1-s)(\tilde{\gamma}_0(1) - \tilde{\gamma}_0(0)) + s(\tilde{\gamma}_1(1) - \tilde{\gamma}_1(0))$$
$$= (1-s)2\pi i n + s \cdot 2\pi i n = 2\pi i n,$$

whence $F(1, s) = F(0, s)$. Therefore, F defines a homotopy between γ_0 and γ_1 as closed paths, and we are done. $\square$

For a disk punctured at a point that is not necessarily its center, the homotopy class of a closed path is also completely determined by its index.

Proposition 6.21 *Let $D \subset \mathbb{C}$ be an open disk and $a \in D$. If γ_0 and γ_1 are two closed piecewise smooth paths in $D \setminus \{a\}$, then they are homotopic as closed paths if and only if $\operatorname{Ind}_a \gamma_0 = \operatorname{Ind}_a \gamma_1$.*

Proof The fact that the indices of homotopic paths coincide still follows from our last version of Cauchy's theorem (Theorem 6.18). Conversely, let $\operatorname{Ind}_a \gamma_0 = \operatorname{Ind}_a \gamma_1$; we will show that these closed paths are homotopic. To this end, pick a closed disk $D' \subset D$ centered at a. Applying to γ_0 and γ_1 a homothety centered at a with an appropriate coefficient $k \in (0; 1)$, we obtain closed paths γ_0' and γ_1' lying in $D' \setminus \{a\}$. It is easy to see that γ_0 and γ_0', and also γ_1 and γ_1', are homotopic as closed paths: for example, a homotopy between γ_0 and γ_0' can be defined by

$$(t, s) \mapsto a + (1 - s + sk)(\gamma_0(t) - a)$$

(all intermediate paths are homothetic to γ_0, the coefficient decreases uniformly from 1 to k). Thus,

$$\operatorname{Ind}_a \gamma_0' = \operatorname{Ind}_a \gamma_0 = \operatorname{Ind}_a \gamma_1 = \operatorname{Ind}_a \gamma_1',$$

therefore, by Proposition 6.20, the paths γ_0' and γ_1' are homotopic in $D' \setminus \{a\}$, and a fortiori in $D \setminus \{a\}$. Since the relation of being homotopic is transitive (see Exercise 6.5), γ_0 and γ_1 are also homotopic in $D \setminus \{a\}$, as required. $\square$

Remark 6.22 In Proposition 6.21, the disk D can be replaced by an arbitrary simply connected open set.

However, an attempt to generalize this proposition in another direction fails: if from a disk D we remove points $a_1, \ldots, a_k$ where $k > 1$, then the homotopy class of a closed path γ in $D \setminus \{a_1, \ldots, a_k\}$ *is not determined* by the indices of γ around $a_1, \ldots, a_k$: we can always find closed paths γ_0 and γ_1 such that they are not homotopic but $\operatorname{Ind}_{a_j} \gamma_0 = \operatorname{Ind}_{a_j} \gamma_1$ for $1 \leq j \leq k$. See an example in Fig. 6.2.

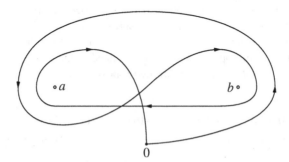

Fig. 6.2 The closed path starting at the origin has index 0 around the point a and index 0 around the point b, like a "constant path" which reduces to a single point. Nevertheless, in $\mathbb{C} \setminus \{a, b\}$ this path cannot be continuously deformed into a point.

Exercises

6.1. Show that the following germs of holomorphic functions cannot be analytically
continued along the interval $[-1; 1]$ directed from 1 to -1:
(a) the germ of the function $1/z$ at the point 1;
(b) the germ of the function $\sqrt{z}$ at the point 1, $\sqrt{1} = 1$.

6.2. In a neighborhood of the point $z = 4$, take the single-valued branch of the function $f(z) = \sqrt{z^2 + 9}$ for which $f(4) = 5$ and consider its analytic continuation
to the point $z = -4$ along the following paths:
(a) the semicircle $|z| = 4$, $\operatorname{Im} z > 0$;
(b) the semicircle $|z| = 4$, $\operatorname{Im} z < 0$;
(c) the interval $[-4; 4]$.
In each of these cases, find $f(-4)$.

6.3. On the open set U shown in Fig. 6.3, consider the single-valued branch of log
for which $\log 1 = 0$. Find $\log(3i)$.

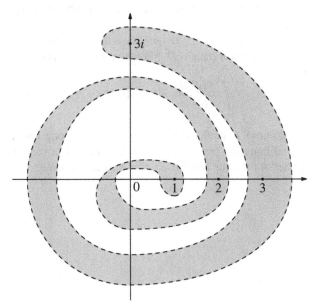

Fig. 6.3 $\log 1 = 0$; $\log(3i) = ?$

6.4. Consider the germ of the function $f(z) = \sqrt{(z-1)(z-2)}$ at 0 for which
$f(0) = 0$.
(a) Describe the analytic continuation of this germ along the closed curve γ in
Fig. 6.4.
(b) Find the value of the integral $\int_{\gamma} \sqrt{(z-1)(z-2)}\, dz$.

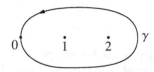

Fig. 6.4

The latter integral is to be understood as a "Riemann integral" from Exercise 4.8: the process of analytic continuation of f along γ fixes the value of f at each point of γ. Later you will learn to calculate such integrals in a fast and easy way, but at the moment you have to do some work "by hand."

6.5. Show that being homotopic is an equivalence relation on the set of all paths in U that connect given points p and q.

6.6. Show that every continuous path in an open set $U \subset \mathbb{C}$ is homotopic (with fixed endpoints) to a piecewise smooth path and, moreover, to a piecewise linear path.

6.7. Show that Theorem 6.15 ceases to be true if f cannot be analytically continued along a single intermediate path γ_s.

6.8. Let $D \subset \mathbb{C}$ be an open disk, $f : D \to \mathbb{C}$ be a holomorphic function, and γ be a closed piecewise linear path in D. Show that for every point $a \in D$ that does not lie on γ,

$$\int_\gamma \frac{f(\zeta)\, d\zeta}{\zeta - a} = 2\pi i \cdot \operatorname{Ind}_a \gamma \cdot f(a).$$

6.9. (M. A. Evdokimov) A math professor took a string long enough, tied it to a picture, and hung the picture with two nails in such a way that it will fall down if either of the nails is removed. Can you do the same?

Chapter 7
Laurent Series and Isolated Singularities

7.1 The Multiplicity of a Zero

Let f be a function holomorphic in a neighborhood of a point $a \in \mathbb{C}$ with $f(a) = 0$. It follows from the principle of analytic continuation in its simplest form (Corollary 5.20) that either f is identically zero in a neighborhood of a, or, on the contrary, in some neighborhood of a it does not vanish except at the point a itself. In the first case, there's not much to talk about, so we will assume that the second case holds.

Definition 7.1 We say that a holomorphic function f has an *isolated zero* at a point $a \in \mathbb{C}$ if f is holomorphic in a neighborhood of a, $f(a) = 0$, and there exists a neighborhood of a in which f does not vanish except at the point a itself.

"Isolated zero" is often abbreviated to just "zero."

Proposition 7.2 *Let f be a holomorphic function with an isolated zero at $a \in \mathbb{C}$ and k be a positive integer. Then the following conditions are equivalent.*

(1) There exists a holomorphic function g in a neighborhood of a such that $g(a) \neq 0$ and $f(z) = (z-a)^k g(z)$.

(2) The Taylor series of f at the point a has the form

$$f(z) = c_k(z-a)^k + c_{k+1}(z-a)^{k+1} + \dots \qquad (7.1)$$

with $c_k \neq 0$.

(3) $f(a) = f'(a) = f''(a) = \dots = f^{(k-1)}(a) = 0$, $f^{(k)}(a) \neq 0$.

A positive integer k satisfying these properties always exists.

Proof $(1) \Leftrightarrow (2)$. If the Taylor series of g has the form

$$g(z) = b_0 + b_1(z-a) + b_2(z-a)^2 + \dots,$$

then

$$f(z) = (z-a)^k g(z) = b_0(z-a)^k + b_1(z-a)^{k+1} + \dots,$$

© Springer Nature Switzerland AG 2020
S. Lvovski, *Principles of Complex Analysis*, Moscow Lectures 6,
https://doi.org/10.1007/978-3-030-59365-0_7

and $b_0 = g(a) \neq 0$.

Conversely, if the Taylor series of f has the form (7.1), then the series

$$c_k + c_{k+1}(z-a) + c_{c+2}(z-a)^2 + \ldots$$

has the same radius of convergence as (7.1), and hence converges to a holomorphic function g such that $g(0) = c_k \neq 0$.

(2) $\Leftrightarrow$ (3). Follows immediately from the fact that the coefficient of $(z-a)^n$ in the power series expansion of f is equal to $f^{(n)}(a)/n!$.

Finally, if there is no positive integer k satisfying the equivalent conditions (1)–(3), then the Taylor series of f is identically zero, and hence f is identically zero in a neighborhood of a, contradicting the assumption.   □

Definition 7.3 The number k satisfying the equivalent conditions (1)–(3) from Proposition 7.2 is called the *multiplicity*, or *order*, of the zero of f at a. Notation: $\mathrm{ord}_a f$. If $f(a) \neq 0$, we set $\mathrm{ord}_a f = 0$.

Definition 7.4 If f has an isolated zero of multiplicity 1 at a point a, we say that f has a *simple zero* at a.

Proposition 7.5 *If f_1 and f_2 are holomorphic functions in a neighborhood of a point $a \in \mathbb{C}$ that are not identically zero in any neighborhood of a, then $\mathrm{ord}_a(f_1 f_2) = \mathrm{ord}_a(f_1) + \mathrm{ord}_a(f_2)$.*

Proof This follows immediately from condition (1) of Proposition 7.2: if $f_1(z) = (z-a)^{k_1} g_1(z)$, $f_2(z) = (z-a)^{k_2} g_2(z)$ with $g_1(a) \neq 0$, $g_2(a) \neq 0$, then $f_1 f_2 = (z-a)^{k_1+k_2} g_1 g_2$ and the function $g_1 g_2$ does not vanish at a.   □

7.2 Laurent Series

As we know, a function holomorphic in an open disk centered at a can be uniquely expanded in a series in nonnegative powers of $z-a$. We need a slight generalization of this result: it turns out that a function holomorphic in an annulus centered at a (see Definition 6.19) can be expanded in a series in positive and negative powers of $z-a$.

Proposition 7.6 *Let $a \in \mathbb{C}$, $0 \leq r < R \leq +\infty$, and let f be a holomorphic function in the annulus $U = \{z : r < |z-a| < R\}$. Then*

$$f(z) = \sum_{n=-\infty}^{\infty} c_n (z-a)^n \quad \text{for all} \quad z \in U, \qquad (7.2)$$

the series in the right-hand side converges absolutely and uniformly on every compact subset $K \subset U$, and its coefficients are given by

$$c_n = \frac{1}{2\pi i} \int\limits_{\gamma} f(z)(z-a)^{-n-1}\, dz, \qquad (7.3)$$

where γ is any positively oriented circle in U centered at a (or, if you prefer, any closed path in U that has index 1 with respect to a).

An expansion (7.2) which converges uniformly on compact subsets in U is unique.

A series of the form (7.2) which represents a holomorphic function in an annulus is called a *Laurent series*.

Proof We begin with uniqueness. If γ is a closed path in U that has index 1 with respect to a, then $\int_{\gamma} \frac{dz}{z-a} = 2\pi i$; while $\int_{\gamma} (z-a)^k\, dz = 0$ for $k \neq -1$, because the function $(z-a)^k$ with such a k has an antiderivative. For given $k \in \mathbb{Z}$, multiply both sides of (7.2) by $(z-a)^{-k-1}$; since the series converges uniformly on every compact subset in U and, in particular, on the curve corresponding to the path γ, the expansion (7.2) multiplied by $(z-a)^{-k-1}$ can be integrated term by term; it follows from the above that this integration yields $2\pi i c_k = \int_{\gamma} f(z)(z-a)^{-k-1}\, dz$, which proves the uniqueness part.

Now we are going to prove the existence part. First of all, note that, by Proposition 6.20, all closed paths in U that have index 1 with respect to a are homotopic, so, by Theorem 6.18, the left-hand side of (7.3) does not indeed depend on the choice of γ. Now let $z \in U$; choose numbers r' and R' such that $r < r' < |z-a| < R' < R$, denote by U' the annulus $\{z: r' < |z-a| < R'\}$, and denote by γ_1 and γ_2 its boundary circles $z: |z-a| = r'$ and $z: |z-a| = R'$, respectively. Orient γ_1 and γ_2 in the positive direction (counterclockwise). If $\varepsilon > 0$ is so small that the circle of radius ε centered at z is contained in U', then it follows from Cauchy's formula applied to this circle and Example 5.7 that

$$f(z) = \frac{1}{2\pi i} \int\limits_{\gamma_2} \frac{f(\zeta)\, d\zeta}{\zeta - z} - \frac{1}{2\pi i} \int\limits_{\gamma_1} \frac{f(\zeta)\, d\zeta}{\zeta - z}. \qquad (7.4)$$

We now transform both integrals in the right-hand side. Considering the integral over γ_2, we proceed in exactly the same way as in the proof of Proposition 5.10: expand $\frac{1}{\zeta - z}$ in a geometric series, multiply by $f(\zeta)\, d\zeta$, and integrate term by term over γ_2. We write down the answer without repeating the calculations:

$$\frac{1}{2\pi i} \int\limits_{\gamma_2} \frac{f(\zeta)\, d\zeta}{\zeta - z} = \sum_{n=0}^{\infty} c_n(z-a)^n, \qquad (7.5)$$

where

$$c_n = \frac{1}{2\pi i} \int\limits_{\gamma_2} \frac{f(\zeta)\, d\zeta}{(\zeta - a)^{n+1}}. \qquad (7.6)$$

To transform the integral over γ_1, note that if $\zeta \in \gamma_1$ and $z \in U'$, then $|\zeta - a| < |z-a|$. Therefore, the fraction $\frac{1}{\zeta - z}$ can be expanded in a geometric series as follows:

$$\frac{1}{\zeta - z} = \frac{1}{(\zeta - a) - (z - a)} = -\frac{1}{z - a} \cdot \frac{1}{1 - \frac{\zeta - a}{z - a}}$$

$$= -\left(\frac{1}{z - a} + \frac{\zeta - a}{(z - a)^2} + \ldots + \frac{(\zeta - a)^k}{(z - a)^{k+1}} + \ldots \right).$$

If $\zeta \in \gamma_1$, then

$$\left| \frac{(\zeta - a)^k}{(z - a)^{k+1}} \right| = \frac{1}{|z - a|} \cdot \left(\frac{r'}{|z - a|} \right)^k \le \frac{1}{r'} \left(\frac{r'}{|z - a|} \right)^k,$$

so the geometric series in the brackets converges on γ_1 uniformly with respect to ζ. Multiplying by the bounded function $f(\zeta)$ preserves uniform convergence. Multiplying by $f(\zeta)\, d\zeta$ and integrating term by term yields

$$-\frac{1}{2\pi i} \int_{\gamma_1} \frac{f(\zeta)\, d\zeta}{\zeta - z} = \sum_{k=0}^{\infty} b_k (z - a)^{-k-1}, \tag{7.7}$$

where

$$b_k = \frac{1}{2\pi i} \int_{\gamma_1} f(\zeta)(\zeta - a)^k\, d\zeta. \tag{7.8}$$

Comparing (7.5)–(7.8) with the expression (7.4) for f, we see that $f(z)$ satisfies (7.2). It remains to verify that the series (7.2) converges uniformly on compact subsets in U. To this end, note that the series (7.5) converges uniformly even on all compact subsets in the disk $\{z\colon |z - a| < R'\}$, since this power series converges at every point of this disk. As to the series (7.7), the change of variable $1/(z - a) = t$ shows that the power series $\sum_{k=0}^{\infty} b_k t^k$ converges at every point of the disk $\{t\colon |t| < 1/r'\}$, so it converges uniformly on compact subsets of this disk; therefore, the original series (7.7) converges uniformly on every compact subset in $\{z\colon |z - a| > r'\}$, and a fortiori on every compact subset in U'. It remains to observe that every compact subset $K \subset U$ is contained in an open annulus $\{z\colon r' < |z - a| < R'\}$, $r < r' < R' < R$, for appropriate r' and R'. $\square$

We are especially interested in the case where the inner radius of the annulus is zero, i.e., the annulus is a punctured disk: $U = \{z\colon 0 < |z - a| < R\}$. In this case, the following terminology is used.

Definition 7.7 If a holomorphic function $f\colon U \to \mathbb{C}$ has a Laurent expansion

$$f(z) = \sum_{n=-\infty}^{\infty} c_n (z - a)^n,$$

then the part of this expansion containing nonnegative powers of $z - a$, i.e., the series $\sum_{n=0}^{\infty} c_n (z - a)^n$, is called the *regular part* of (the Laurent series of) f, while the series

$\sum\limits_{k<0} c_k (z-a)^k$ containing negative powers of $z - a$ is called the *principal part* of
(the Laurent series of) f.

The word "principal" is chosen for a good reason: as we will see in the next section, it is the principal part of the Laurent series that contains the most interesting information on the behavior of the function f.

Note that the principal and regular parts of a Laurent series converge separately (which follows from the absolute convergence or directly from the proof of Proposition 7.6). Thus, the radius of convergence of the regular part of the Laurent series (for a function holomorphic in a punctured disk of radius R) is at least R, and the regular part converges to a function that is holomorphic in the whole disk, including the removed point. On the other hand, the principal part $\sum\limits_{n=1}^{\infty} c_{-n}(z-a)^{-n}$ (for the Laurent series of a function holomorphic in a punctured disk) converges for z arbitrarily close to a. Denoting $t = (z-a)^{-1}$, we see that the series $\sum\limits_{n=1}^{\infty} c_{-n} t^n$ converges for arbitrarily large $|t|$; hence, it has infinite radius of convergence, and thus converges to a function holomorphic on all of $\mathbb{C}$.

Definition 7.8 A function holomorphic on all of $\mathbb{C}$ is said to be *entire*.

Our discussion can be summarized as follows.

Corollary 7.9 *Let $D = \{z: |z - a| < R\}$ be an open disk in the complex plane. Every holomorphic function $f: D \setminus \{a\} \to \mathbb{C}$ can be represented in the form $f(z) = \varphi(z) + \psi(1/(z - a))$ where φ is a function holomorphic on D and ψ is an entire function.*

7.3 Isolated Singularities

Let $f(z)$ be a function holomorphic in a punctured neighborhood of a point a; in classical texts, a is then called an *isolated singularity* of f. In this section, we study the behavior of such a function as z approaches a. As we will see, this behavior is subject to certain limitations.

Isolated singularities can be classified according to the form of the principal part of the corresponding Laurent series. We begin with the case where this principal part is identically zero. It can be characterized as follows.

Proposition 7.10 (Riemann removable singularity theorem) *Let*

$$D = \{z: |z - a| < r\},$$

and let $f: D \setminus \{a\} \to \mathbb{C}$ be a holomorphic function. Then the following three conditions are equivalent.

(1) *The function f is bounded in a punctured neighborhood of a (which may be smaller than $D \setminus \{a\}$).*

(2) *The function f extends to a holomorphic function on the whole disk D.*

(3) *The principal part of the Laurent series of f in a punctured neighborhood of a is identically zero (in other words, the coefficients of all negative powers of $z - a$ in the Laurent series are zero).*

Definition 7.11 If f is holomorphic on $D \setminus \{a\}$ and the principal part of its Laurent series at a is identically zero, then f is said to have a *removable singularity* at a.

According to Proposition 7.10, if f has a removable singularity, then it satisfies all equivalent conditions (1)–(3) from this proposition.

Proof of Proposition 7.10 (1) $\Rightarrow$ (3). Let $|f(z)| \leq C$ for $0 < |z - a| < \varepsilon$, $\varepsilon > 0$. If the Laurent expansion of f in $D \setminus \{a\}$ has the form

$$f(z) = \sum_{k=-\infty}^{\infty} c_k (z - a)^k,$$

then, by (7.3), for every positive integer $n > 0$ we have

$$c_{-n} = \frac{1}{2\pi i} \int_{\gamma} f(z)(z - a)^{n-1} \, dz,$$

where γ is an arbitrary circle centered at a. Take γ to be a circle of radius $\delta < \varepsilon$. Then, estimating the integral over γ by Proposition 4.15 and taking into account that for $z \in \gamma$ we have $|f(z)| \leq C$ and $|z - a| = \delta$, we obtain

$$|c_{-n}| \leq \frac{1}{2\pi} \cdot C \cdot \delta^{n-1} \cdot 2\pi\delta = C\delta^n.$$

Letting $\delta \to 0$, we see that $c_{-n} = 0$ for $n > 0$, i.e., the Laurent series does not contain terms with negative powers of $z - a$.

(3) $\Rightarrow$ (2). We have already discussed this implication at the end of the previous section: if a Laurent series has no principal part, then it is an ordinary power series which converges in the whole disk D and, consequently, defines a holomorphic function in D which coincides with f on $D \setminus \{a\}$.

(2) $\Rightarrow$ (1). By assumption, f extends to a function holomorphic at a; now, a holomorphic function, being continuous, is bounded in a neighborhood of every point of its domain of definition.   □

The second case is when the principal part of is not zero, but contains only finitely many terms. How this affects the behavior of the function is described in the next proposition.

Proposition 7.12 *Let $D = \{z : |z-a| < r\}$, and let $f : D \setminus \{a\} \to \mathbb{C}$ be a holomorphic function. Then the following four conditions are equivalent.*

(1) *The singularity at a is not removable, but there exist constants $N > 0$ and $C > 0$ such that $|f(z)| \le C|z - a|^{-N}$ for all z from a punctured neighborhood of a (which may be smaller than $D \setminus \{a\}$). In short, this condition can be stated as follows: $|f(z)| = O(|z - a|^{-N})$ as $z \to a$.*

(2) *There exist a holomorphic function $g: D \to \mathbb{C}$, $g(a) \ne 0$, and a positive integer $n > 0$ such that $f(z) = g(z)/(z - a)^n$ everywhere on $D \setminus \{a\}$.*

(3) *The principal part of the Laurent series of f in a punctured neighborhood of a is not identically zero but contains only finitely many nonzero terms.*

(4) $\lim_{z \to a} f(z) = \infty.$

Recall that the notation $\lim_{z \to a} f(z) = \infty$ means that $\lim_{z \to a} |f(z)| = +\infty$.

Definition 7.13 If f is holomorphic on $D \setminus \{a\}$ and the principal part of its Laurent series at a is not identically zero but contains only finitely many nonzero terms, then f is said to have a *pole* at a.

According to Proposition 7.12, if f has a pole, then it satisfies all equivalent conditions (1)–(4) from this proposition.

Partial proof of Proposition 7.12 For the time being, we prove that conditions (1)–(3) are equivalent and that they imply condition (4). The remaining implication (4) $\Rightarrow$ (3) will be established a little later.

(1) $\Rightarrow$ (2). Assume that condition (1) is satisfied; choose an integer $k \ge N$ and set $h(z) = (z - a)^k f(z)$. Then the function h is holomorphic on $D \setminus \{a\}$ and bounded in a punctured neighborhood of a. Thus, by the Riemann removable singularity theorem (Proposition 7.10), h extends to a function holomorphic on the whole disk D. Set $m = \operatorname{ord}_a h \ge 0$; then $h(z) = (z-a)^m g(z)$ where g is holomorphic on D and $g(a) \ne 0$. If $m \ge k$, then the function $f(z) = (z - a)^{m-k} g(z)$ extends to a holomorphic function on the whole of D, contradicting the assumption that the singularity is not removable. Therefore, $k > m$ and $f(z) = g(z)/(z - a)^{k-m}$ where g is holomorphic at a and $k - m > 0$. Thus, condition (2) is satisfied.

(2) $\Rightarrow$ (1). Obviously follows from the fact that the function g, being holomorphic at the point a, is bounded in some neighborhood of this point: $|g(z)| \le C$ for $|z - a| < \varepsilon$.

(2) $\Rightarrow$ (3). If the power series expansion of g has the form $g(z) = \sum_{k=1}^{\infty} c_m(z-a)^m$, then $c_0 = g(a) \ne 0$. Therefore, the Laurent series of f has the form

$$f(z) = \frac{g(z)}{(z - a)^n} = \frac{c_0}{(z - a)^n} + \frac{c_1}{(z - a)^{n-1}} + \ldots + \frac{c_{n-1}}{z - a} + \text{(regular part)},$$

as required.

(3) $\Rightarrow$ (2). The argument is similar: if c_{-n}, where $n > 0$, is a nonzero coefficient of the Laurent series of f with the smallest index, then the Laurent expansion of the function $g(z) = (z - a)^n f(z)$ has no terms with negative powers of $z - a$, hence this function is holomorphic on the whole disk D, while $g(0) = c_{-n} \ne 0$.

(2) $\Rightarrow$ (4). Obvious. $\qquad\square$

Pole is a commonly encountered isolated singularity, so let us study poles in slightly more detail.

Definition 7.14 Let f be a function defined in a punctured neighborhood of a point a that has the form $f(z) = g(z)/(z-a)^n$ where $n > 0$ is a positive integer and g is a holomorphic function in a neighborhood of a such that $g(a) \neq 0$. Then f is said to have a *pole of order n* at a. Sometimes, in this situation f is said to have a "pole of order $-n$" (which is more correct!). Also, the term "multiplicity" is sometimes used instead of "order."

Notation 7.15 If f is holomorphic in a punctured neighborhood of a and has the form $f(z) = (z-a)^m g(z)$ where $m \in \mathbb{Z}$ and g is a holomorphic function in a neighborhood of a such that $g(a) \neq 0$, then we write $\mathrm{ord}_a(f) = m$.

If f has a removable singularity at a, then, obviously, $\mathrm{ord}_a f$ in the sense of this notation coincides with $\mathrm{ord}_a f$ in the sense of Definition 7.3.

Proposition 7.16 *If f_1 and f_2 are holomorphic functions in a punctured neighborhood of a point a that have a removable singularity or a pole at a, and none of these functions is identically zero in a neighborhood of a, then the quotient f_1/f_2 also has a removable singularity or a pole at a, and*

$$\mathrm{ord}_a(f_1/f_2) = \mathrm{ord}_a(f_1) - \mathrm{ord}_a(f_2).$$

Proof If $f_1(z) = (z-a)^{n_1} g_1(z)$, $f_2(z) = (z-a)^{n_2} g_2(z)$ where g_1 and g_1 are holomorphic functions that do not vanish at a, then $f_1(z)/f_2(z) = (z-a)^{n_1-n_2}(g_1(z)/g_2(z))$; for $n_1 \geq n_2$, this function has a removable singularity; for $n_1 < n_2$, it has a pole; in any case, $\mathrm{ord}_a(f_1/f_2) = n_1 - n_2 = \mathrm{ord}_a(f_1) - \mathrm{ord}_a(f_2)$.   □

Finally, we introduce another useful term.

Definition 7.17 If a function f has a pole of order 1 at a point a (equivalently: if $\mathrm{ord}_a f = -1$), then f is said to have a *simple pole* at a.

Besides the cases of zero or finite principal part, the possibility exists that it is infinite. That's how the function behaves in this situation.

Proposition 7.18 *Let $D = \{z: |z-a| < r\}$, and let $f: D\backslash\{a\} \to \mathbb{C}$ be a holomorphic function. Then the following two conditions are equivalent.*

   (1) The principal part of the Laurent series of f in a punctured neighborhood of a contains infinitely many nonzero terms.
   (2) For every $c \in \mathbb{C} \cup \{\infty\}$ there exists a sequence of points $z_n \in D \setminus \{a\}$ such that $\lim_{n\to\infty} z_n = a$ and $\lim_{n\to\infty} f(z_n) = c$.

Definition 7.19 If f is holomorphic on $D \setminus \{a\}$ and the principal part of its Laurent series at a contains infinitely many nonzero terms, then f is said to have an *essential singularity* at a.

According to Proposition 7.18, if f has an essential singularity, then it satisfies condition (2) from this proposition.

Proof of Proposition 7.18 (1) $\Rightarrow$ (2). For $c = \infty$, this can be proved as follows. If there is no sequence $z_n \to a$ for which $f(z_n) \to \infty$, then f is bounded for $0 < |z-a| < \varepsilon$ with some $\varepsilon > 0$, so the singularity is removable by Proposition 7.10, and the principal part of the Laurent series is zero, contradicting condition (1). Thus, the limit ∞ can be obtained.

If $c \in \mathbb{C}$ is finite and there is no sequence $z_n \to a$ for which $f(z_n) \to c$, then there exist $\varepsilon > 0$ and $\delta > 0$ such that $|f(z) - c| \geq \delta$ for $0 < |z - a| < \varepsilon$. Thus, the absolute value of the function $g(z) = \frac{1}{f(z)-c}$ is bounded by $1/\delta$ for $0 < |z - a| < \varepsilon$, and hence g has a removable singularity at a. Since $f(z) = c + \frac{1}{g(z)}$, Proposition 7.16 implies that f has at a a removable singularity or a pole, a contradiction.

(2) $\Rightarrow$ (1). Arguing by contradiction, if f has a removable singularity or a pole, then, by Proposition 7.10 and the part of Proposition 7.12 already proved, $f(z)$ has a finite or infinite limit as $z \to a$, contradicting condition (2). $\qquad\square$

Since the principal part of a Laurent series can be either identically zero, or nonzero but finite, or infinite, no other case is possible besides those of a removable singularity, pole, or essential singularity, and these three cases themselves are mutually exclusive.

We also emphasize the sharp contrast between an essential singularity and a removable singularity or a pole: in the latter two cases there exists a (finite or infinite) limit $\lim_{z \to a} f(z)$, while for an essential singularity this limit does not exist in the strongest possible sense.

Now we can easily complete the proof of Proposition 7.12.

Completing the proof of Proposition 7.12 It remains to verify that if condition (4) of this proposition is satisfied, then f has a pole at a. Indeed, a cannot be a removable singularity, since the condition $\lim_{z \to a} f(z) = \infty$ implies that f is not bounded in any neighborhood of a; and it cannot be an essential singularity, since otherwise f would have no limit at all as $z \to a$. $\qquad\square$

We have already mentioned that Proposition 7.10 is called the Riemann removable singularity theorem. Proposition 7.12 has no special name, but, by way of compensation, Proposition 7.18 has too many names. In Germany, it is known as the Weierstrass theorem; in Italy, the Casorati or Casorati–Weierstrass theorem; the latter name is also quite popular in France and English-speaking countries; while in Russia, this proposition is called Sokhotski's theorem.

In conclusion, note that the Casorati–Weierstrass theorem can be considerably strengthened: it is not only true that as z approaches an essential singularity, $f(z)$ may approach any limit: the so-called Picard's great theorem says that in any punctured neighborhood of an essential singularity, a holomorphic function assumes all complex values except possibly one! We will prove this theorem in Chap. 12.

7.4 The Point ∞ as an Isolated Singularity

Of course, the point ∞ of the Riemann sphere $\overline{\mathbb{C}}$ is not a complex number, but it makes perfect sense to define its punctured neighborhoods and, by analogy with the previous section, study the behavior of holomorphic functions in these punctured neighborhoods.

We now introduce a notion which has been already discussed a little in Sec. 1.6.

Definition 7.20 A *punctured neighborhood of infinity* is a set of the form

$$\{z \in \mathbb{C} : |z| > R\}, \qquad R \in \mathbb{R}.$$

When working with functions of a complex variable z in a punctured neighborhood of infinity, one applies the change of variable $t = 1/z$.

Definition 7.21 Let f be a holomorphic function in a punctured neighborhood of infinity. It is said to have a *removable singularity* (*pole, essential singularity*) at *infinity* if the function $t \mapsto f(1/t)$ has the same type of singularity at 0.

According to the results of the previous section, the latter type is determined by the Laurent series of the function $t \mapsto f(1/t)$ at the origin. Switching back from the variable t to the variable z, we obtain the following definitions and results.

Definition 7.22 The *regular part* of the Laurent series of a holomorphic function in a punctured neighborhood of infinity is the sum of its terms with *nonpositive* powers of the variable. The *principal part* of the Laurent series of a holomorphic function in a punctured neighborhood of infinity is the sum of its terms with *positive* powers of the variable.

Proposition 7.23 *A function f holomorphic in a punctured neighborhood of infinity has a removable singularity at infinity if and only if it is bounded in some punctured neighborhood of infinity (which may be smaller than the original one). Equivalent conditions: there exists a finite limit* $\lim\limits_{|z| \to \infty} f(z)$*; the principal part of the Laurent series of f at infinity is zero.*

A function f holomorphic in a punctured neighborhood of infinity has a pole at infinity if and only if f has no removable singularity at infinity and $|f(z)| = O(|z|^N)$ *as* $|z| \to \infty$ *for some* $N > 0$*. Equivalent conditions:* $\lim\limits_{|z| \to \infty} f(z) = \infty$*; the principal part of the Laurent series of f at infinity has finitely many nonzero terms but is not identically zero.*

A function f holomorphic in a punctured neighborhood of infinity has an essential singularity at infinity if and only if the principal part of the Laurent series of f at infinity has infinitely many nonzero terms.

There is no need to prove this proposition: its first and second paragraphs follow from Propositions 7.10 and 7.12 by the change of variable $t = 1/z$, and the third paragraph is a tautology; we leave it to the reader to reformulate the Casorati–Weierstrass theorem for this case.

We now deduce some corollaries of the results already proved in the context of singularities at infinity.

Recall that an entire function is a function holomorphic on all of $\mathbb{C}$.

Proposition 7.24 (Liouville's theorem) *If f is an entire function and $|f(z)| = O(|z|^N)$ as $|z| \to \infty$ for some $N > 0$, then f is a polynomial of degree at most N.*

Recall that $|f(z)| = O(|z|^N)$ means that there exist $C > 0$ and $M > 0$ such that $|f(z)| \le C|z|^N$ for $|z| \ge M$.

Proof It follows from the assumptions of the theorem and Proposition 7.12 (as well as its proof) that the principal part of the Laurent series of the function $t \mapsto f(1/t)$ contains only terms of degree at least $-N$ in t, so the power series of the function $z \mapsto f(z)$ contains only terms of degree at most N in z. But this means exactly that f is a polynomial of degree at most N. □

Corollary 7.25 *A bounded entire function is constant.*

This follows from Proposition 7.24 by setting $N = 0$. The reader who views this proof as cheating is encouraged to apply the Riemann removable singularity theorem to the function $t \mapsto f(1/t)$.

Sometimes, by Liouville's theorem one means not Proposition 7.24, but Corollary 7.25, which is logically weaker.

Functions whose all isolated singularities are "not worse" than poles deserve a special term.

Definition 7.26 Let $U \subset \mathbb{C}$ be an open set. A *meromorphic function* on U is a holomorphic function $f \colon U \setminus S \to \mathbb{C}$ where $S \subset U$ is a subset with no accumulation points in U satisfying the following property: at each point $s \in S$, the function f has a pole or a removable singularity.

Proposition 7.27 *If $U \subset \mathbb{C}$ is a connected open set and f and g are holomorphic functions on U that are not identically zero, then the quotient f/g is a meromorphic function on U.*

Proof Let $S = \{z \in U \colon g(z) = 0\}$. Since g is not identically zero, S has no accumulation points in U. At points not in S, the function f/g is holomorphic; at each point $s \in S$, it has an isolated singularity by Proposition 7.16. □

Actually, the converse is also true: every meromorphic function on an open set $U \subset \mathbb{C}$ is the quotient of two holomorphic functions. However, this result is not so easy to prove (this will be done in Chap. 11); moreover, it is not always true in more general situations (for Riemann surfaces).

Definition 7.28 A function is said to be *meromorphic at infinity* if it is holomorphic in a punctured neighborhood of infinity and has a pole at infinity (in the sense of Definition 7.21).

To conclude this chapter, we describe the functions meromorphic on the whole Riemann sphere.

Proposition 7.29 *The functions meromorphic on the whole Riemann sphere $\overline{\mathbb{C}}$ are exactly the rational functions.*

Some clarifications are in order. First, a rational function is nothing else than the quotient of two polynomials. Second, it is understood that the subset $S \subset \overline{\mathbb{C}}$ from the definition of a meromorphic function has no accumulation points in the whole Riemann sphere, and this is equivalent to its being finite (if S contains complex numbers with arbitrarily large absolute values, then ∞ is an accumulation point of S; and if all the complex numbers from S have bounded absolute values, then S is necessarily finite, otherwise there would exist an accumulation point in the "finite" part of $\overline{\mathbb{C}}$).

Proof It is easy to see that every rational function f is meromorphic on $\overline{\mathbb{C}}$: this is obvious for finite points, and if $f = g/h$ where g and h are polynomials of degree m and n, respectively, then $|f(z)| = O(|z|^{m-n})$ as $|z| \to \infty$, so at infinity f also has a pole or a removable singularity.

Conversely, let f be a meromorphic function on $\overline{\mathbb{C}}$. For each point $a \in S$, denote by f_a the principal part of f at a. Since all singularities of f are poles or removable singularities, for each finite a (i.e., $a \in \mathbb{C}$) this principal part is either identically zero or a finite sum of the form $c_1/(z-a) + \cdots + c_m/(z-a)^m$, i.e., a rational function. For the point ∞, the principal part is also a finite sum of the form $b_1 z + \ldots + b_k z^k$ (or identically zero), i.e., a polynomial. Since S is finite, we can construct the sum of all principal parts at the points from S: this is a rational function, which we denote by φ. The difference $f - \varphi$ has no poles or essential singularities neither on $\mathbb{C}$, nor at infinity. Therefore, it is an entire function, and, since it has a removable singularity at infinity, it is bounded in some punctured neighborhood of infinity $\{z: |z| > R\}$. Since on the disk $\{z: |z| \le R\}$ this difference is also bounded by continuity, the entire function $f - \varphi$ is bounded on the whole plane and, consequently, is constant. Thus, $f(z) = \varphi(z) + \text{const}$. Since φ is a rational function of z, the same is true for f. $\square$

One can easily see that the decomposition, obtained in the proof, of a rational function f into the sum of a constant and finitely many principal parts is nothing else than the partial fraction decomposition of f.

Exercises

7.1. Find all zeros (and their multiplicities) of the following functions:
   (a) $f(z) = \sin z$;
   (b) $f(z) = z \sin z$;
   (c) $f(z) = \sin^3 z$;
   (d) $f(z) = \sin(z^3)$.
7.2. For each of the following functions, find all its isolated singularities and determine their types (removable singularity, pole, essential singularity). For poles, find the corresponding orders.

(a) $f(z) = \frac{1}{z(z^2-4)^2}$;

(b) $f(z) = \frac{z}{\sin(z^2)}$;

(c) $f(z) = e^{\tan z}$.

7.3. Show that every germ of the function $f(z) = \frac{\sin \sqrt{z}}{\sqrt{z}}$ (at a point other than the origin) extends to an entire function, and find all zeros of this function and their multiplicities.

7.4. Does there exist a function f holomorphic in the unit disk $D = \{z : |z| < 1\}$ and satisfying the condition $|z|^{5/3} \le |f(z)| \le |z|^{4/3}$ for all $z \in D$?

7.5. Find the Laurent series expansion of the function $f(z) = \frac{1}{z(z-1)}$
(a) in the annulus $\{z : 0 < |z| < 1\}$;
(b) in the annulus $\{z : |z| > 1\}$.

7.6. Find the Laurent series expansion of the function $f(z) = \frac{1}{(z-1)(z-2)}$
(a) in the disk $\{z : |z| < 1\}$;
(b) in the annulus $\{z : 1 < |z| < 2\}$;
(c) in the annulus $\{z : |z| > 2\}$.

7.7. Let a and b be two distinct nonzero complex numbers.
(a) Show that the function $f(z) = \sqrt{(z-a)(z-b)}$ has a single-valued branch in the infinite annulus $\{z : |z| > \max(|a|, |b|)\}$; find the coefficients of z^2 and z^{-2} in the Laurent series expansion of f in this annulus.
(b) The same questions for the function $f(z) = \log \frac{z-a}{z-b}$.

7.8. In light of part (a) of the previous exercise, try to find a more economical solution of Exercise 6.4 (b) (you may need some other coefficients of the Laurent series).

7.9. Let f be a holomorphic function in the punctured disk $D^* = \{z : 0 < |z| < 1\}$ that satisfies the condition $|f(z)| \le 1/|z|^{\sqrt{7}}$ for all $z \in D^*$. What can be said about the isolated singularity of f at the origin?

7.10. Let f be a holomorphic function in the punctured disk $D^* = \{z : 0 < |z| < 1\}$ that satisfies the condition $|f(z)| \le 1 + \log(1/|z|)$ for all $z \in D^*$. What can be said about the isolated singularity of f at the origin?

7.11. Let f be a holomorphic function in the punctured disk $D^* = \{z : 0 < |z| < 1\}$ such that $\operatorname{Re} f(z) > 0$ for all $z \in D^*$.
(a) Can f have an essential singularity at 0?
(b) Can f have a pole at 0?

7.12. Let f be a holomorphic function in the punctured disk $D^* = \{z : 0 < |z| < 1\}$ such that $f(z) \notin [0; +\infty) \subset \mathbb{R}$ for all $z \in D^*$. What can be said about the isolated singularity of f at the origin?

Chapter 8
Residues

8.1 Basic Definitions

In this chapter, we draw further consequences from Cauchy's theorem. We begin with an innocent-looking but extremely important definition.

Definition 8.1 Let f be a holomorphic function in a punctured neighborhood $U = \{z : 0 < |z - a| < \varepsilon\}$ of a point $a \in \mathbb{C}$. The *residue* of f at a is the number

$$\operatorname{Res}_a f(z) = \frac{1}{2\pi i} \int_\gamma f(z)\, dz$$

where γ is an arbitrary closed piecewise smooth curve in U that has index 1 with respect to a.

The residue does not depend on the choice of γ by Proposition 6.21.

It is clear from definition that if f has a removable singularity at a, then the residue of f at a is zero.

It is convenient to calculate residues using the following fact.

Proposition 8.2 *If f is a holomorphic function in a punctured neighborhood of a point a, then the residue $\operatorname{Res}_a f(z)$ is equal to the coefficient of $(z - a)^{-1}$ in the Laurent series expansion of f at a.*

Proof Consider the Laurent series

$$f(z) = \sum_{n=-\infty}^{\infty} c_n (z - a)^n$$

and integrate it term by term over a closed curve γ with $\operatorname{Ind}_a \gamma = 1$; then the integrals of all terms except $c_{-1}(z - a)^{-1}$ vanish, and the integral of the latter term is equal to $2\pi i c_{-1}$. □

© Springer Nature Switzerland AG 2020
S. Lvovski, *Principles of Complex Analysis*, Moscow Lectures 6,
https://doi.org/10.1007/978-3-030-59365-0_8

Now we prove a simple but useful proposition.

Proposition 8.3 *Let $\bar{U} \subset \mathbb{C}$ be a closed subset of the complex plane bounded by a closed non-self-crossing piecewise smooth curve γ (the curve γ belongs to $\bar{U}$); set $\mathrm{Int}(\bar{U}) = U$. Assume that $a_1, \ldots, a_n \in U$ and f is a function that is continuous on $\bar{U} \setminus \{a_1, \ldots, a_n\}$ and holomorphic on the open set $U \setminus \{a_1, \ldots, a_n\}$. Then*

$$\int_\gamma f(z)\, dz = 2\pi i \sum_{k=1}^{n} \mathrm{Res}_{a_k}\, f(z),$$

where it is understood that γ is positively oriented.

Here, exactly the same reservations apply as for Theorem 5.4: we have no general definition of the notion "the part of the plane bounded by a curve," so, if one wishes to be completely rigorous, Proposition 8.3 should be regarded as a recipe for constructing rigorous statements about each individual curve (a circle, the boundary of a semi-disk, a rectangle, etc.).

Proof As in Example 5.6, draw a circle of radius $\varepsilon_j > 0$ around each point a_j in such a way that all these circles lie in U and are pairwise disjoint. Then, by (5.5), the integral $\int_\gamma f(z)\, dz$ is equal to the sum of the integrals over the (positively oriented) circles of radii ε_k centered at a_k; by definition, the integral over the kth circle is equal to $2\pi i\, \mathrm{Res}_{a_k}\, f(z)$.   □

The remaining part of this chapter is devoted to applications of Proposition 8.3.

8.2 The Argument Principle

Here is the first example of calculating a residue.

Lemma 8.4 (logarithmic residue theorem) *Let f be a function that is either holomorphic in a neighborhood of a point $a \in \mathbb{C}$ and is not identically zero in any neighborhood of a, or is holomorphic in a punctured neighborhood of a point a and has a pole at a. Then at the point a the function f'/f has either a removable singularity or a pole, and $\mathrm{Res}_a(f'(z)/f(z)) = \mathrm{ord}_a(f)$.*

Proof If $\mathrm{ord}_a(f) = k$, then in some (punctured) neighborhood of a we have $f(z) = (z-a)^k g(z)$ where g is holomorphic in a neighborhood of a and $g(a) \neq 0$. A direct calculation shows that

$$\frac{f'(z)}{f(z)} = \frac{k(z-a)^{k-1}g(z) + (z-a)^k g'(z)}{(z-a)^k g(z)} = \frac{k}{z-a} + \frac{g'(z)}{g(z)}.$$

The second term in the right-hand side is a function that is holomorphic at a and hence does not affect the residue, while the residue of the first term is equal to k by Proposition 8.2.   □

Remark 8.5 The quotient f'/f is called the *logarithmic derivative* of f, because in a neighborhood of a point where f is holomorphic and does not vanish, the function $\log f$ is well defined and f'/f coincides with its derivative. However, the logarithmic derivative is well defined also in domains where $\log f$ is not defined as a single-valued holomorphic function: analytic continuation along a closed path changes $\log f$ at most by an additive constant ($2\pi i k$ with $k \in \mathbb{Z}$), so the ambiguity disappears after differentiation.

A residue of the logarithmic derivative of f is called a logarithmic residue of f.

The simple Lemma 8.4 combined with the even simpler Proposition 8.3 implies a beautiful result called the argument principle. Informally, it says the following: if f is meromorphic on a domain bounded by a closed non-self-crossing curve and continuous on its boundary, then the number of turns made by the image of the boundary around the origin is equal to the number of zeros of f in this domain minus the number of poles (counting multiplicities). Expressed in more precise terms, the statement necessarily becomes longer.

Theorem 8.6 (argument principle) *Let $\bar{U} \subset \mathbb{C}$ be a closed subset of the complex plane bounded by a closed non-self-crossing piecewise smooth curve γ (the curve γ belongs to $\bar{U}$); set $\mathrm{Int}(\bar{U}) = U$. Assume that we are given a finite subset $S \subset U$ and a function $f : \bar{U} \setminus S \to \mathbb{C}$ with the following properties:*

(1) f extends to a holomorphic function on $U' \setminus S$ where $U' \supset U$ is an open set (informally, this idea is sometimes expressed as follows: f is holomorphic on $U \setminus S$ and analytic on the boundary of U);

(2) f has no essential singularities at points of S;

(3) f does not vanish on γ.

If we now equip γ with a bijective parametrization with respect to which it is positively oriented, then

$$\sum_{a \in U} \mathrm{ord}_a(f) = \mathrm{Ind}_0(f \circ \gamma). \tag{8.1}$$

The left-hand side of (8.1) involves a sum over all points of U, but actually it is a finite sum of integers, because $\mathrm{ord}_z(f) = 0$ unless z is a zero or a pole of f; the number of poles is finite by assumption, and the assumption implies that the number of zeros is also finite (see below). Note also that it follows from condition (1) that if γ is a piecewise smooth curve, then so is $f \circ \gamma$.

Proof Note that f has at most finitely many zeros in U. Indeed, if there are infinitely many of them, then they have an accumulation point somewhere in $\bar{U}$. If this accumulation point lies in U, then f is identically zero on $U \setminus S$ and hence, by continuity, on γ, contradicting condition (3); if, however, the accumulation point lies on γ, then f vanishes at this point, which also contradicts condition (3).

Now the proof follows by a direct calculation. Namely, regarding γ not as a subset in $\mathbb{C}$, but as a parametrization $\gamma : [A; B] \to \mathbb{C}$ which is bijective onto its image and with respect to which the boundary of U is positively oriented, we have

$$\mathrm{Ind}_0(f \circ \gamma) = \frac{1}{2\pi i} \int\limits_{f \circ \gamma} \frac{dw}{w} = \frac{1}{2\pi i} \int\limits_A^B \frac{\frac{d}{dt}(f(\gamma(t)))}{f(\gamma(t))}\, dt$$

$$= \frac{1}{2\pi i} \int\limits_A^B \frac{f'(\gamma(t))}{f(\gamma(t))} \gamma'(t)\, dt = \frac{1}{2\pi i} \int\limits_\gamma \frac{f'(z)}{f(z)}\, dz$$

$$= \frac{1}{2\pi i} \cdot 2\pi i \sum_{a \in U} \mathrm{Res}_a\, \frac{f'(z)}{f(z)} = \sum_{a \in U} \mathrm{ord}_a(f), \quad (8.2)$$

as required. □

The theorem we have just proved is not the strongest possible version of the argument principle. Here is a more precise result.

Theorem 8.7 (strengthened argument principle) *Theorem 8.6 remains valid if condition* (1) *is replaced by the following weaker condition:*
  (1′) *the function f is continuous on $\bar{U} \setminus S$ and holomorphic on $U \setminus S$.*

Remark 8.8 Some comments are necessary concerning the statement of this theorem. First, we give no proof of the strengthened argument principle. Second, things are not that simple with its statement too. Namely, if we require f to be only continuous on the boundary, then we cannot guarantee that the closed path $f \circ \gamma$ is piecewise smooth: the assumptions imply only its continuity. So, the right-hand side of (8.1) may raise a question, because indices were defined only for piecewise smooth curves.

All these difficulties are not insuperable. Indices can be defined for arbitrary continuous closed paths by replacing, in the part of Proposition 4.19 related to the definition, "piecewise smooth" with "continuous." Formula (4.2), which we used to verify that this definition is well defined, no longer makes sense for an arbitrary continuous path, but this can also be verified by more elementary means: one can directly check that, in terms of Lemma 4.18, the difference $\varphi(B) - \varphi(A)$ does not depend on the choice of φ.

Having defined indices for arbitrary continuous paths, we can derive the strengthened argument principle (using a larger dose of topology than in the main part of our course) from local properties of holomorphic (and meromorphic) functions which will be discussed in the next chapter.

One can also reduce Theorem 8.7 to the rigorously proved Theorem 8.6 as follows. Note that the number of zeros and poles of f in U is finite; this can be established in the same way as in the proof of Theorem 8.6, observing additionally that if a sequence of poles converged to a point of the boundary, then it would not be possible to extend f to this point by continuity. Therefore, there exists a closed non-self-crossing piecewise smooth curve $\gamma_1 \subset U$ that is close to γ and surrounds all zeros and poles of f. Now apply Theorem 8.6 to the curve γ_1 and observe that since γ_1 is sufficiently close to γ, it follows that these curves are homotopic (this can be proved by an argument similar to that used in the homotopy enhancement lemma (Lemma 6.4). Thus, the compositions $f \circ \gamma_1$ and $f \circ \gamma$ are also homotopic, and hence their indices with respect to the origin coincide.

In what follows, I will state corollaries of the argument principle which rely on its strengthened version; however, in the cases where we derive something important, the rigorously proved weaker version (Theorem 8.6) will usually suffice. Still, at least once we will need precisely the strengthened version of the argument principle.

Along with a geometric (or, more exactly, topological) statement of the argument principle given in Theorem 8.6, it is useful to formulate also its analytic statement in terms of integrals.

Corollary 8.9 *Under the assumptions of Theorem 8.6,*

$$\sum_{z \in U} \mathrm{ord}_z(f) = \frac{1}{2\pi i} \int_\gamma \frac{f'(z)}{f(z)}\, dz.$$

Proof This assertion is contained in the chain of equations (8.2).   □

Especially interesting corollaries of the argument principle arise when the function f has no poles.

Corollary 8.10 *Let $\bar{U} \subset \mathbb{C}$ be a closed subset of the complex plane bounded by a closed non-self-crossing piecewise smooth curve γ (the curve γ belongs to $\bar{U}$); set $\mathrm{Int}(\bar{U}) = U$. Assume that a function $f : \bar{U} \to \mathbb{C}$ extends to a holomorphic function on an open set $U' \supset \bar{U}$ (a stronger version: f is continuous on $\bar{U}$ and holomorphic on U) and is not constant.*

Equip γ with a bijective parametrization with respect to which it is positively oriented.

If now a point $a \in \mathbb{C}$ does not belong to $f(\gamma)$, then it lies in $f(U)$ if and only if $\mathrm{Ind}_a(f \circ \gamma) \neq 0$; moreover, the index $\mathrm{Ind}_a(f \circ \gamma)$ is equal to the sum of the orders of all zeros of the function $f(z) - a$ (in U).

Proof It suffices to apply the argument principle to the function $z \mapsto f(z) - a$.   □

This, in turn, implies that in order to establish that a holomorphic function is a conformal map, it suffices to know only its behavior on the boundary of the domain being mapped. Here is the precise statement (with the same reservations as apply to Proposition 8.3).

Proposition 8.11 *Let $\bar{U}$ and $\bar{U}'$ be closed bounded subsets in $\mathbb{C}$ bounded by closed non-self-crossing piecewise smooth curves γ and γ'. Denote the interiors of the sets $\bar{U}$ and $\bar{U}'$ by U and U', respectively.*

Now let $f : \bar{U} \to \mathbb{C}$ be a continuous map with the following properties:

(1) f is holomorphic on U;

(2) f induces a bijection between γ and γ', and if γ is equipped with a bijective piecewise smooth parametrization with respect to which it is positively oriented, then f induces a parametrization of γ' with respect to which it is also positively oriented.

Then $f(U)$ is contained in U', and the restriction $f|_U : U \to U'$ is a conformal map.

Proof If a is a point outside $\bar{U}'$, then the closed curve γ' has index 0 with respect to it (for the notions of "outside" and "inside," see the discussion at p. 53). Therefore, by Corollary 8.10, the point a does not lie in $f(U)$; we have shown that $f(U) \subset \bar{U}'$.

Now let a be a point on γ'; we will show that in this case, again, $a \notin f(U)$. Indeed, arguing by contradiction, assume that $a = f(b)$, $b \in U$. Since f is continuous and $f(\gamma) = \gamma'$, it is clear that f cannot be constant on U, so the principle of analytic continuation shows that there exists a closed disk $\Delta \subset U$ centered at b such that f does not assume the value a on the boundary of Δ. Denote by μ the boundary circle of the disk Δ (positively oriented and parametrized appropriately); then Corollary 8.10 shows that $\mathrm{Ind}_a(f \circ \mu) > 0$. Since the set $f(\mu)$ is compact and hence closed in $\mathbb{C}$, the point a has a neighborhood $V \ni a$ contained in the same connected component of $\mathbb{C} \setminus f(\mu)$ as a; by Proposition 4.20, for every point $a' \in V$ we have $\mathrm{Ind}_{a'}(f \circ \mu) = \mathrm{Ind}_a(f \circ \mu) > 0$, so $a' \in f(U)$ by Corollary 8.10. However, this cannot be the case: since a lies on the boundary of U', any its neighborhood contains points lying outside γ', and such points, as we have established above, cannot lie in $f(U)$.

Hence, $f(U) \subset U'$. Let now $a \in U'$ be an arbitrary point. Since f induces a bijection between γ and γ', and the curve γ has been equipped with a bijective parametrization, we have $\mathrm{Ind}_a(f \circ \gamma) = 1$. Therefore, Corollary 8.10 implies that the equation $f(z) = a$ has exactly one solution in U (even counting multiplicities). Thus, the function f, being holomorphic on U, induces a bijection between U and U', as required. $\square$

Corollary 8.10 can be used to find the number of zeros of a function in a given domain. The following trick can sometimes simplify the task.

Proposition 8.12 (Rouché's theorem) *Let* $\gamma \colon [A; B] \to \mathbb{C}$ *be a closed continuous path, and denote by* $|\gamma|$ *the set* $\gamma([A; B]) \subset \mathbb{C}$. *Let* $f, g \colon |\gamma| \to \mathbb{C}$ *be continuous maps such that* $f(z) \neq 0$ *and* $|f(z)| > |g(z)|$ *for every* $z \in |\gamma|$. *Then* $\mathrm{Ind}_0(f \circ \gamma) = \mathrm{Ind}_0((f + g) \circ \gamma)$.

Here $(f + g) \circ \gamma$ stands for the path given by the formula $t \mapsto f(\gamma(t)) + g(\gamma(t))$.

Proof Since $|f(z)| > |g(z)|$ for every $z \in |\gamma|$, we see that $f(z) + g(z) \neq 0$ for such z. Thus, not only the path $f \circ \gamma$ does not pass through the origin, which is stated explicitly in the assumptions, but the path $(f + g) \circ \gamma$ too. Now we define a homotopy $F \colon [A; B] \times [0; 1] \to \mathbb{C}$ by the formula $F(t, s) = f(\gamma(t)) + sg(\gamma(t))$. Since

$$|sg(\gamma(t))| = s|g(\gamma(t))| \leq |g(\gamma(t))| < |f(\gamma(t))|$$

for any $t \in [A; B]$ and $s \in [0; 1]$, we see that $f(\gamma(t)) + sg(\gamma(t)) \neq 0$. Thus, F is a homotopy of closed paths in $\mathbb{C} \setminus \{0\}$. Since $F(t, 0) = F(\gamma(t))$ and $F(t, 1) = f(\gamma(t)) + g(\gamma(t))$, it follows that $f \circ \gamma$ and $(f + g) \circ \gamma$ are homotopic as closed paths in $\mathbb{C} \setminus \{0\}$; therefore, their indices coincide. $\square$

A good illustration of how Rouché's theorem works is the following proof of the "fundamental theorem of algebra."

Proposition 8.13 *A polynomial of degree n in one variable with complex coefficients has exactly n roots counting multiplicities.*

Proof We may assume without loss of generality that the leading coefficient of the polynomial is equal to 1. Let the polynomial have the form

$$P(z) = z^n + a_{n-1}z^{n-1} + \ldots + a_1 z + a_0.$$

Clearly, $\lim_{|z| \to \infty} (P(z)/z^n) = 1$, hence there exists $R > 0$ such that the quotient $|P(z) - z^n|/|z^n|$ does not exceed, say, $1/2$ whenever $|z| \geq R$. In particular, for $|z| \geq R$ we have $|P(z) - z^n| < |z^n|$. Since $P(z) = z^n + (P(z) - z^n)$, it follows that $|P(z)| > 0$ for $|z| \geq R$, so P has no roots outside the disk $\{z : |z| < R\}$.

It remains to count the roots of P inside this disk. To this end, set $f(z) = z^n$, $g(z) = P(z) - z^n$, and denote by γ the circle of radius R centered at 0 with the standard parametrization (a single traversal in the positive direction). Since $|f(z)| > |g(z)|$ for $z \in \gamma$, Rouché's theorem implies that

$$\text{Ind}_0(P \circ \gamma) = \text{Ind}_0((f + g) \circ \gamma) = \text{Ind}_0(f \circ \gamma).$$

Since $f(z) = z^n$, the index in the right-hand side is equal to n. Applying now Corollary 8.10 to P and γ, we see that inside the disk of radius R the polynomial P has n roots counting multiplicities, which completes the proof. ☐

8.3 Computing Integrals

We have already seen how residues can be used to find indices of curves. Another, somewhat unexpected, application of residues is the evaluation of (improper and other) integrals. There is no general theory here: in this section, you will see a series of artificial tricks, sometimes very beautiful.

To begin with, we practice the skill of finding residues.

Example 8.14 Let us find $\text{Res}_1 \frac{e^z}{z^3 - 1}$. Since the function has a simple pole at 1, we can proceed as follows. Considering its Laurent series expansion

$$\frac{e^z}{z^3 - 1} = \frac{c_{-1}}{z - 1} + c_0 + \ldots,$$

we see that

$$\text{Res}_1 \frac{e^z}{z^3 - 1} = c_{-1} = \lim_{z \to 1} (z - 1) \frac{e^z}{z^3 - 1} = \left. \frac{e^z}{z^2 + z + 1} \right|_{z=1} = \frac{e}{3}.$$

Let us extract the technique used in this calculation into a separate proposition.

Proposition 8.15 *If f and g are holomorphic functions in a neighborhood of a point a and g has a simple zero at a, then*

$$\operatorname{Res}_a \frac{f(z)}{g(z)} = \frac{f(a)}{g'(a)}. \tag{8.3}$$

Proof It follows from the assumptions that the function f/g has a simple pole (or even a removable singularity) at a. Hence we may proceed as in Example 8.14:

$$\operatorname{Res}_a \frac{f(z)}{g(z)} = \lim_{z \to a} (z-a) \frac{f(z)}{g(z)} = \lim_{z \to a} \frac{f(z)(z-a)}{g(z) - g(a)}$$

$$= \lim_{z \to a} \left(f(z) \bigg/ \frac{g(z) - g(a)}{z - a} \right) = \frac{f(a)}{g'(a)}.$$

For a pole of order greater than 1, it is also not difficult to find the residue.

Example 8.16 Let us find $\operatorname{Res}_1 \frac{1}{(z^2-1)^2}$. In this case, the pole is of order 2. Consider the Laurent series expansion at the point 1 and multiply both sides by $(z-1)^2$:

$$\frac{1}{(z^2 - 1)^2} = \frac{c_{-2}}{(z-1)^2} + \frac{c_{-1}}{z-1} + c_0 + \dots,$$

$$\frac{(z-1)^2}{(z^2 - 1)^2} = c_{-2} + c_{-1}(z-1) + c_0(z-1)^2 + \dots,$$

$$\frac{1}{(z+1)^2} = c_{-2} + c_{-1}(z-1) + c_0(z-1)^2 + \dots.$$

Thus, the residue in question, i.e., c_{-1}, is equal to the derivative of the function $1/(z+1)^2$ at the point 1:

$$\operatorname{Res}_1 \frac{1}{(z^2 - 1)^2} = \frac{d}{dz} \frac{1}{(z+1)^2} \bigg|_{z=1} = -\frac{1}{4}.$$

In the case of a pole of order greater than 1, we can also obtain a formula similar to (8.3), but it turns out to be more involved; the reader is encouraged to work it out in Exercise 8.3. Those reluctant to memorize this formula may follow the pattern outlined in the above example each time it is needed.

We finally turn to integrals.

Example 8.17 Let us evaluate the integral

$$\int_{-\infty}^{\infty} \frac{e^{itx} \, dx}{x^2 + 1}$$

where t is real.

First, assume that $t > 0$. Consider the closed curve that is the boundary of the semi-disk of a large radius R centered at the origin that lies in the upper half-plane (it is required that $R > 1$, and later we will let R tend to infinity; see Fig. 8.1a). Denoting this boundary, positively oriented, by Γ_R, we evaluate the integral $\int_{\Gamma_R} \frac{e^{itz} \, dz}{z^2 + 1}$.

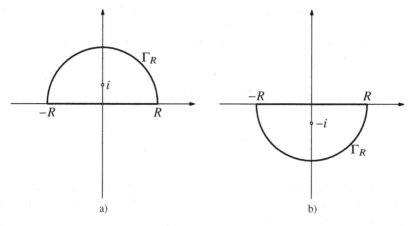

Fig. 8.1

The function $e^{itz}/(z^2+1)$ is holomorphic inside the contour Γ_R except for the point i, at which it has a simple pole. It follows from (8.3) that

$$\operatorname{Res}_i \frac{e^{itz}}{z^2+1} = \frac{e^{it\cdot i}}{2i} = \frac{-ie^{-t}}{2}.$$

Therefore, by Proposition 8.3, we have

$$\int_{\Gamma_R} \frac{e^{itz}\,dz}{z^2+1} = 2\pi i \frac{-ie^{-t}}{2} = \frac{\pi}{e^t}.$$

Now we split the integral over the closed contour Γ_R into two integrals: the integral over the line segment $[-R; R]$ (directed from $-R$ to R) and the integral over the semicircle of radius R centered at the origin (directed from R to $-R$), which we denote by γ_R. Bearing in mind that the integral of $\frac{e^{itz}}{z^2+1}$ over $[-R; R]$ is nothing else than $\int_{-R}^{R} (e^{itx}/(x^2+1))\,dx$, we can write

$$\int_{-R}^{R} \frac{e^{itx}\,dx}{x^2+1} + \int_{\gamma_R} \frac{e^{itz}\,dz}{z^2+1} = \frac{\pi}{e^t}. \tag{8.4}$$

Now let R tend to infinity. The first term in the left-hand side of (8.4) tends to the integral in question $\int_{-\infty}^{\infty} \frac{e^{itx}\,dx}{x^2+1}$. We will show that the second term tends to zero. Indeed, using the estimate (4.1), we obtain

$$\left| \int_{\gamma_R} \frac{e^{itz}\,dz}{z^2+1} \right| \leq \pi R \sup_{z \in \gamma_R} \frac{|e^{itz}|}{|z^2+1|}. \tag{8.5}$$

If $z \in \gamma_R$, i.e., $z = R(\cos \varphi + i \sin \varphi)$, $0 \le \varphi \le \pi$, then $|e^{itz}| = e^{-Rt \sin \varphi} \le 1$ (here we have used the condition $t \ge 0$) and $|z^2 + 1| \ge |z^2| - 1 = R^2 - 1$. Therefore, the right-hand side of (8.5) does not exceed $\pi R/(R^2 - 1)$, which tends to zero as $R \to \infty$. Thus,

$$\lim_{R \to +\infty} \int_{\gamma_R} \frac{e^{itz} \, dz}{z^2 + 1} = 0,$$

whence

$$\int_{-\infty}^{\infty} \frac{e^{itx} \, dx}{x^2 + 1} = \frac{\pi}{e^t} \quad \text{for } t \ge 0.$$

Note, by the way, that though the convergence of the improper integral in this case is obvious, we could, in principle, derive it from (8.4), by moving the second term to the right-hand side and letting R tend to infinity.

It remains to consider the case where $t < 0$. Here, the integration over Γ_R does not help, since for z lying "high" in the upper half-plane, $|e^{itz}|$ is exponentially large and the second term in (8.4) does not tend to zero. For this reason, we consider the contour Γ_R' that is the positively oriented boundary of the semicircle of radius R centered at the origin and lying in the lower half-plane (Fig. 8.1 b). It is important not to get confused with signs: if Γ_R' is positively oriented, then the integral over the part of this curve lying on the real axis is the integral from R to $-R$. Evaluating the integral of $e^{itz} \, dz/(z^2 + 1)$ over Γ_R' by residues, we find that it is equal to $-\pi e^t$. Denoting by γ_R' the part of the circle of radius R centered at the origin that lies in the lower half-plane, we obtain

$$-\int_{-R}^{R} \frac{e^{itx} \, dx}{x^2 + 1} + \int_{\gamma_R'} \frac{e^{itz} \, dz}{z^2 + 1} = -\pi e^t. \tag{8.6}$$

Since $t < 0$, we have $|e^{itz}| \le 1$ for z from the lower half-plane, so the absolute value of the second term in (8.6) again does not exceed $1/(R^2 - 1)$ and thus tends to zero as $R \to \infty$; taking the limit, we obtain

$$\int_{-\infty}^{\infty} \frac{e^{itx} \, dx}{x^2 + 1} = \pi e^t \quad \text{for } t < 0.$$

We can also combine these two formulas into one:

$$\int_{-\infty}^{\infty} \frac{e^{itx} \, dx}{x^2 + 1} = \frac{\pi}{e^{|t|}} \quad \text{for } t \in \mathbb{R}.$$

Example 8.18 Now, let us evaluate the integral

$$\int_{-\infty}^{\infty} \frac{\cos(tx)\,dx}{x^2+1}$$

where t is again real.

Most of the work has already been done in the previous example: since $\cos(tx) = \operatorname{Re} e^{itx}$, we have

$$\int_{-\infty}^{\infty} \frac{\cos(tx)\,dx}{x^2+1} = \operatorname{Re}\int_{-\infty}^{\infty} \frac{e^{itx}\,dx}{x^2+1} = \operatorname{Re}\left(\frac{\pi}{e^{|t|}}\right) = \frac{\pi}{e^{|t|}}.$$

If we tried, as in the previous example, to directly evaluate this integral by integrating over the boundaries of semi-circles and using residues, we would fail. Indeed, $\cos(tz) = (e^{itz} + e^{-itz})/2$; the second term prevents us from estimating the integral over a semi-circle lying in the upper half-plane, while the first term in the same way prevents us from estimating the integral over a semi-circle lying in the lower half-plane!

The moral is that sometimes, in order to evaluate an integral by residues, one should not take the analytic continuation of the integrand to the complex plane right away, as we have done in Example 8.17, but first represent it as the real or imaginary part of an appropriate holomorphic function.

Let us look at Example 8.17 once again. The trick we have used in this example works for any integrals of the form

$$\int_{-\infty}^{\infty} e^{itx} \cdot \frac{P(x)}{Q(x)}\,dx$$

where $t \in \mathbb{R}$ and P, Q are polynomials, provided that Q has no real roots and, most importantly, $\deg Q - \deg P \geq 2$: otherwise, the method does not allow one to prove that the integral over an arc of radius R tends to zero as $R \to \infty$. However, if $\deg Q - \deg P = 1$, something can nevertheless be done (if $\deg Q \leq \deg P$, then, for an obvious reason, there is no hope). It is convenient to extract the main trick into a separate lemma. See Definition 4.16 for the arc length integral involved in the statement.

Lemma 8.19 (Jordan) *Denote by $\gamma_R = \{z : |z| = R,\ \operatorname{Im} z \geq 0\}$ the semicircle of radius R centered at the origin that lies in the upper half-plane. Then for every $a > 0$ there exists a constant C depending only on a such that*

$$\int_{\gamma_R} |e^{iaz}|\,|dz| \leq C$$

for all $R > 0$.

Proof Consider the standard parametrization of the semicircle γ_R, i.e., $z = Re^{i\varphi}$, $\varphi \in [0; \pi]$. Then $dz = iRe^{i\varphi} \, d\varphi$, $|dz| = R \, d\varphi$, whence

$$\int_{\gamma_R} |e^{iaz}| \, |dz| = \int_0^\pi R|e^{ia(R\cos\varphi + iR\sin\varphi)}| \, d\varphi = \int_0^\pi Re^{-aR\sin\varphi} \, d\varphi$$

$$= 2R \int_0^{\pi/2} e^{-aR\sin\varphi} \, d\varphi \quad (8.7)$$

(the last equality follows from the symmetry $\sin(\pi - \varphi) = \sin\varphi$).

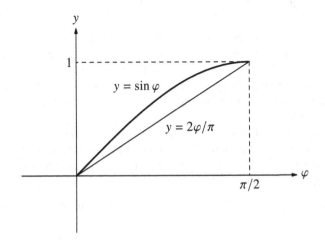

Fig. 8.2

Observe that, since the sine function is convex upward on the interval $[0; \pi/2]$, for $\varphi \in [0; \pi/2]$ we have $\sin\varphi \geq 2\varphi/\pi$ (Fig. 8.2). Hence, the chain of equalities (8.7) can be continued as follows:

$$\int_{\gamma_R} |e^{iaz}| \, |dz| = 2R \int_0^{\pi/2} e^{-aR\sin\varphi} \, d\varphi \leq 2R \int_0^{\pi/2} e^{-2aR\varphi/\pi} \, d\varphi = \frac{\pi}{a}(1 - e^{-aR}) \leq \frac{\pi}{a}.$$

It remains to set $C = \pi/a$.   □

Example 8.20 Now, armed with Jordan's lemma, we consider an example of using it. Let us evaluate the integral $\int_{-\infty}^{\infty} \frac{x \sin x \, dx}{x^2 + 1}$. The first step is the same as in Example 8.18: since for $x \in \mathbb{R}$ we have

$$\frac{x \sin x}{x^2 + 1} = \text{Im} \, \frac{xe^{ix}}{x^2 + 1},$$

we will evaluate the integral $\displaystyle\int\limits_{-\infty}^{\infty} \frac{xe^{ix}\,dx}{x^2+1}$ and then take its imaginary part. Consider

the same contour Γ_R, which is the positively oriented boundary of the semi-disk $\{z: |z| \le R,\ \mathrm{Im}\,z \ge 0\}$ (with $R > 1$). From Proposition 8.3 we have

$$\int\limits_{\Gamma_R} \frac{ze^{iz}\,dz}{z^2+1} = 2\pi i\,\mathrm{Res}_i\left(\frac{ze^{iz}}{z^2+1}\right) = \frac{\pi i}{e}.$$

Denoting, as in Jordan's lemma, the semicircle $\{z: |z| = R,\ \mathrm{Im}\,z \ge 0\}$ by γ_R, we have

$$\int\limits_{-R}^{R} \frac{xe^{ix}\,dx}{x^2+1} + \int\limits_{\gamma_R} \frac{ze^{iz}\,dz}{z^2+1} = \frac{\pi i}{R}, \tag{8.8}$$

and the integral in question will be evaluated as soon as we establish that the integral over γ_R tends to zero as $R \to \infty$. This time, the estimate from Proposition 4.15 is inadequate, but we can use the sharper estimate from Proposition 4.17. Namely, this proposition implies that

$$\left|\int\limits_{\gamma_R} \frac{ze^{iz}\,dz}{z^2+1}\right| \le \int\limits_{\gamma_R} \left|\frac{ze^{iz}\,dz}{z^2+1}\right| |dz| \le \sup_{z\in\Gamma_R}\left|\frac{z}{z^2+1}\right| \cdot \int\limits_{\gamma_R} |e^{iz}|\,|dz|.$$

In the right-hand side, the first factor does not exceed $R/(R^2-1)$, while the second one, by Jordan's lemma, does not exceed some constant C not depending on R. Therefore, we have (for all $R > 1$)

$$\left|\int\limits_{\gamma_R} \frac{ze^{iz}\,dz}{z^2+1}\right| \le \frac{CR}{R^2-1};$$

since the right-hand side tends to zero as $R \to \infty$, the same is true for the left-hand side, so we have established that the second term in the left-hand side of (8.8) tends to zero. Therefore,

$$\int\limits_{-\infty}^{\infty} \frac{xe^{ix}\,dx}{x^2+1} = \lim_{R\to+\infty}\int\limits_{-R}^{R} \frac{xe^{ix}\,dx}{x^2+1} = \frac{\pi i}{e},$$

whence, taking the imaginary part, we obtain

$$\int\limits_{-\infty}^{\infty} \frac{x\sin x\,dx}{x^2+1} = \frac{\pi}{e}.$$

There is a subtlety to be mentioned in connection with this calculation. As you know, by definition,

$$\int\limits_{-\infty}^{\infty} f(x)\, dx := \lim_{\substack{A \to -\infty \\ B \to \infty}} \int\limits_{A}^{B} f(x)\, dx$$

where A and B tend to $-\infty$ and ∞ independently of each other. However, in all examples so far, we established the existence (and found the values) of limits of the form $\lim\limits_{R \to +\infty} \int_{-R}^{R} f(x)\, dx$. In the first two examples, where the absolute value of the integrand decreased quadratically, this did not matter, since the integral converged absolutely; however, in the present example there is no absolute convergence, so the convergence of the integral $\int\limits_{-\infty}^{\infty} \dfrac{x \sin x\, dx}{x^2 + 1}$ must be verified separately. This integral does indeed converge, which can be seen, for example, from the Abel–Dirichlet test (see [5, Chap. VI, Sec. 5]; in this case, we can establish the convergence directly by integrating by parts).

The limit $\lim\limits_{R \to +\infty} \int_{-R}^{R} f(x)\, dx$, if it exists, is called the *principal value* of the improper integral. In general, if f is a function defined, say, on the whole of $\mathbb{R}$ except for a point a, then the principal value integral of f from $-\infty$ to ∞ is the following limit (if it exists):

$$\text{V. p.} \int\limits_{-\infty}^{\infty} f(x)\, dx := \lim_{\substack{R \to +\infty \\ \varepsilon \to 0}} \left(\int\limits_{-R}^{a - \varepsilon} f(x)\, dx + \int\limits_{a + \varepsilon}^{R} f(x)\, dx \right).$$

In other words, neighborhoods of all points where the function is not defined, infinity included, need to be symmetric. Along the way, we have introduced notation: V. p. before the integral sign (abbreviation for the French *valeur principale*) means that the integral is to be understood as a principal value integral. Sometimes, instead of V. p. one writes P. v. (abbreviation for the English *principal value*).

If an integral converges in the ordinary sense, then it certainly converges (to the same number) as a principal value integral. The converse is totally wrong: for example, V. p. $\int\limits_{-\infty}^{\infty} \frac{dx}{x} = 0$. Residues are especially well adapted to evaluating principal value integrals.

Example 8.21 An estimate necessary for evaluating the integral $\int\limits_{-\infty}^{\infty} \frac{x e^{ix}\, dx}{x^2 + 1}$ from Example 8.20 can be obtained without Jordan's lemma. Namely, to prove that

$$\lim_{R \to +\infty} \int\limits_{\gamma_R} \frac{z e^{iz}\, dz}{z^2 + 1} = 0,$$

where γ_R is still the semicircle $\{z : |z| = R, \operatorname{Im} z \geq 0\}$, split γ_R into two parts:

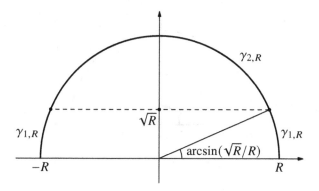

Fig. 8.3

$$\gamma_{1,R} = \{z : |z| = R, 0 \le \operatorname{Im} z \le \sqrt{R}\},$$
$$\gamma_{2,R} = \{z : |z| = R, \operatorname{Im} z \ge \sqrt{R}\}.$$

We have

$$\int_{\gamma_R} \frac{ze^{iz}\, dz}{z^2 + 1} = \int_{\gamma_{1,R}} \frac{ze^{iz}\, dz}{z^2 + 1} + \int_{\gamma_{2,R}} \frac{ze^{iz}\, dz}{z^2 + 1}. \qquad (8.9)$$

Denote the terms in the right-hand side by I_1 and I_2, respectively (thus, I_1 is the sum of integrals over two arcs). We will estimate I_1 and I_2 separately.

As we have already observed, if $z \in \gamma_R$, then $\left|\frac{z}{z^2+1}\right| \le \frac{R}{R^2-1}$; besides, for all z from the upper half-plane we have $|e^{iz}| \le 1$. Now we estimate the integral $I_1(R)$. Since $\gamma_{1,R}$ is the union of two arcs of angular measure $\arcsin(\sqrt{R}/R)$ each (Fig. 8.3), the total length of these arcs is $2R \arcsin(1/\sqrt{R})$. Hence, the elementary estimate (8.9) implies that

$$|I_1| \le \frac{R}{R^2 - 1} \cdot 2R \arcsin\left(\frac{1}{\sqrt{R}}\right);$$

clearly, the right-hand side tends to zero as $R \to +\infty$, so $I_1(R)$ tends to zero too.

To estimate $I_2(R)$, observe that if $\operatorname{Im} z \ge \sqrt{R}$, then $|e^{iz}| \le e^{-\sqrt{R}}$. Again applying (8.9), we obtain

$$|I_2(R)| \le e^{-\sqrt{R}} \cdot \frac{R}{R^2 - 1} \cdot \operatorname{length}(\gamma_{2,R}) \le \frac{Re^{-\sqrt{R}}}{R^2 - 1} \cdot \pi R,$$

and the right-hand side also tends to zero as $R \to +\infty$. Hence, the left-hand side of (8.9) tends to zero too, as required.

In this argument, $\sqrt{R}$ could be replaced by R^α for every $\alpha \in (0; 1)$. Also, instead of the boundary of a semi-disk we could integrate, say, over the boundary of the rectangle with vertices $-R$, R, $R + iR$, and $-R + iR$.

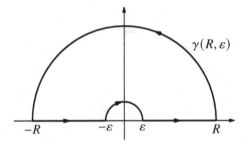

Fig. 8.4

Example 8.22 Now let us evaluate the integral $\displaystyle\int_{-\infty}^{\infty} \frac{\sin x \, dx}{x}$. The function $\sin z / z$

itself is entire, so it makes no sense to consider its residues. However, observing that

$\frac{\sin x}{x} = \mathrm{Im}\, \frac{e^{ix}}{x}$ for $x \in \mathbb{R}$, we can try to evaluate the integral $\displaystyle\int \frac{e^{ix} \, dx}{x}$. Unfortunately,

the function e^{iz}/z has a pole right on the real axis, so we cannot use the semicircle

contour directly. Instead, we will integrate $\frac{e^{iz} \, dz}{z}$ over the contour $\gamma_{R,\varepsilon}$ shown in

Fig. 8.4. We need the following lemma.

Lemma 8.23 (fractional residue theorem) *Let f be a holomorphic function that
has a simple pole at a point a, and let ℓ_1 and ℓ_2 be fixed rays starting at a. For every
sufficiently small $\varepsilon > 0$, denote by γ_ε the arc of the circle of radius ε centered at a
between the rays ℓ_1 and ℓ_2 oriented counterclockwise. Then*

$$\lim_{\varepsilon \to 0} \int_{\gamma_\varepsilon} f(z) \, dz = \alpha i \cdot \mathrm{Res}_a(f(z))$$

where α is the angle between ℓ_1 and ℓ_2 (expressed in radians).

Proof Let $\mathrm{Res}_a(f(z)) = c$. Since the pole at a is simple, we have

$$f(z) = \frac{c}{z - a} + g(z)$$

where the function g is holomorphic (and hence bounded) in a neighborhood of a.

A direct calculation shows that $\displaystyle\int_{\gamma_\varepsilon} \frac{c \, dz}{z - a} = \alpha i \cdot c$ (indepentently of ε); while the

integral of $g(z) \, dz$ tends to zero: if $|g(z)| \leq M$ in a neighborhood of a, then

$$\left| \int_{\gamma_\varepsilon} g(z) \, dz \right| \leq \sup_{z \in \gamma_\varepsilon} |g(z)| \cdot \mathrm{length}(\gamma_\varepsilon) \leq M \alpha \varepsilon \to 0.$$

The lemma is proved.   □

Now we return to the integral of $e^{ix}\,dx/x$ and the contour $\gamma_{R,\varepsilon}$ (Fig. 8.4). Since the function e^{iz}/z has no singularities inside this contour, we have $\displaystyle\int_{\gamma_{R,\varepsilon}} \frac{e^{iz}\,dz}{z} = 0.$

Therefore, denoting by γ_R the arc of the semicircle of radius R directed from R to $-R$, and by γ_ε the arc of the semicircle of radius ε directed from $-\varepsilon$ to ε, we have

$$0 = \int_{-R}^{-\varepsilon} \frac{e^{ix}\,dx}{x} + \int_{\gamma_\varepsilon} \frac{e^{iz}\,dz}{z} + \int_{\varepsilon}^{R} \frac{e^{ix}\,dx}{x} + \int_{\gamma_R} \frac{e^{iz}\,dz}{z}.$$

Now let ε tend to zero and R tend to infinity. The integral over γ_R in the right-hand side tends to zero; this can be deduced from Jordan's lemma (Lemma 8.19) in the same way as in Example 8.20. The integral over γ_ε tends to $-\pi i \cdot \mathrm{Res}_0 \frac{e^{iz}}{z} = -\pi i$ by the fractional residue theorem (Lemma 8.23); the minus sign appears because the arc γ_ε is negatively oriented. Taking the limit yields

$$\mathrm{V.\,p.} \int_{-\infty}^{\infty} \frac{e^{ix}\,dx}{x} - \pi i = 0,$$

and taking the imaginary part yields

$$\mathrm{V.\,p.} \int_{-\infty}^{\infty} \frac{\sin x}{x}\,dx = \pi. \tag{8.10}$$

It remains to observe that the integral in the left-hand side of (8.10) converges (again, this can be deduced from the Abel–Dirichlet test or simply by integration by parts), so the original ordinary integral also equals π.

Example 8.24 Residues are useful not only for evaluating improper integrals. For a change, let us evaluate the integral

$$\int_0^{2\pi} \frac{dx}{2 + \cos x}. \tag{8.11}$$

Every calculus textbook explains how the indefinite integral of such a function (a rational function of sines and cosines) can be expressed in terms of elementary functions. However, residues allow one to obtain the result much faster and simpler[1].

To evaluate our integral, we represent it as the integral of a holomorphic function over the unit circle γ parametrized in the standard way: $z = e^{ix}$, $x \in [0; 2\pi]$. Then we have

---

[1] Of course, there is no question of competing with programs such as Mathematica or Maple.

$$dz = ie^{ix}\,dx = iz\,dx, \quad dx = \frac{dz}{iz}, \quad \cos x = \frac{e^{ix} + e^{-ix}}{2} = \frac{z + z^{-1}}{2};$$

substituting these expressions into (8.11), we obtain

$$\int_0^{2\pi} \frac{dx}{2 + \cos x} = \int_\gamma \frac{dz/iz}{2 + (z + z^{-1})/2} = -2i \int_\gamma \frac{dz}{z^2 + 4z + 1}.$$

Inside the unit disk, the function $1/(z^2 + 4z + 1)$ has a unique simple pole at the point $-2 + \sqrt{3}$; by (8.3), its residue at this point is equal to

$$\left.\frac{1}{2z + 4}\right|_{z=-2+\sqrt{3}} = \frac{\sqrt{3}}{6};$$

therefore, the integral in question is equal to $-2i \cdot 2\pi i(\sqrt{3}/6) = 2\pi\sqrt{3}/3$.

Example 8.25 Sometimes, before calculating the sum of the residues, it is useful to replace z with $1/z$. For example, let us try to evaluate the integral $\displaystyle\int_{|z|=2} \frac{z^6\,dz}{z^7 + z + 1}$ (the circle is positively oriented).

To begin with, note that for $|z| \geq 2$ we have $|z|^7 > |z + 1|$, so all roots of the denominator (and thus all poles of the integrand) lie inside the disk $\{z : |z| < 2\}$. Since it does not seem possible to find the poles explicitly, we make the change of variable $z = 1/w$. Thus,

$$dz = -\frac{dw}{w^2},$$

$$\frac{z^6}{z^7 + z + 1} = \frac{w}{1 + w^6 + w^7},$$

$$\frac{z^6\,dz}{z^7 + z + 1} = -\frac{dw}{w(1 + w^6 + w^7)}.$$

As z describes the circle $\{z : |z| = 2\}$ counterclockwise, the variable $w = 1/z$ describes the circle $\{w : |w| = 1/2\}$ clockwise! Hence, we finally obtain

$$\int_{|z|=2} \frac{z^6\,dz}{z^7 + z + 1} = -\int_{|w|=\frac{1}{2}} \left(-\frac{dw}{w(1 + w^6 + w^7)}\right) = \int_{|w|=\frac{1}{2}} \frac{dw}{w(1 + w^6 + w^7)},$$

where the circle of radius $1/2$ is positively oriented. Now the integrand has no poles inside the disk of radius $1/2$ centered at the origin, except for a simple pole at 0. It is easy to see that the residue at this pole is equal to 1, so the integral in question is equal to $2\pi i$.

Now we return to improper integrals and consider two examples involving roots and logarithms, functions that in the complex plane should be handled with care.

Example 8.26 Let us evaluate the integral $\displaystyle\int_0^\infty \frac{\log x \, dx}{x^2+4}$ (here log stands for the "ordi-

nary" logarithm from the calculus course). It is easy to see that this integral converges absolutely; anyway, this will be established in our calculation.

To evaluate the integral, consider the contour shown in Fig. 8.5 (denote it by $\Gamma_{R,\varepsilon,\delta}$), where both horizontal segments are at distance $\delta > 0$ from the real axis, and integrate over $\Gamma_{R,\varepsilon,\delta}$ the function $\frac{\log z}{z^2+4}$, where $\log z$ stands for the branch of log on the set $\mathbb{C} \setminus [0;+\infty)$ for which $\log(re^{i\varphi}) = \log r + i\varphi,\ 0 < \varphi < 2\pi$. As δ decreases, the integral $\displaystyle\int_{\Gamma_{R,\varepsilon,\delta}} \frac{\log z \, dz}{z^2+4}$ does not change; if we let δ tend to zero, both upper and lower horizontal segments approach the interval $[\varepsilon; R]$ of the real axis; the integral over the upper horizontal segment tends to $\int_\varepsilon^R \frac{\log x \, dx}{x^2+4}$, while the integral over the lower one tends to $-\displaystyle\int_\varepsilon^R \frac{\log x + 2\pi i}{x^2+4}\, dx$: as z approaches the real axis from the lower half-plane, the imaginary part of $\log z$ (for the chosen branch of the logarithm) tends to $2\pi i$, and the minus sign before the integral appears because the lower horizontal segment is traversed in the "negative" direction, from larger values of the real part to smaller ones[2].

Finally, taking the limit as $\delta \to 0$, we obtain the following equality:

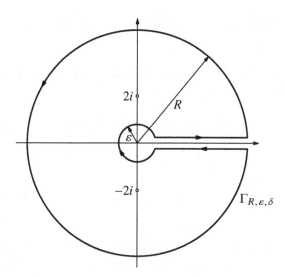

Fig. 8.5

---

[2] The classical authors did not bother with such tiresome explanations. In the first half of the 20th century, one would write the following: we cut the plane along the ray $[0;+\infty)$, choose such and such a branch of the logarithm on the resulting cut plane, and observe that on the upper *edge of the*

$$\int_{\Gamma_{R,\varepsilon,\delta}} \frac{\log z \, dz}{z^2 + 4} = \int_{\varepsilon}^{R} \frac{\log x \, dx}{x^2 + 4} + \int_{|z|=R} \frac{\log z \, dz}{z^2 + 4}$$

$$- \int_{\varepsilon}^{R} \frac{\log x + 2\pi i}{x^2 + 4} \, dx + \int_{|z|=\varepsilon} \frac{\log z \, dz}{z^2 + 4}. \quad (8.12)$$

The integral in the left-hand side of (8.12) can be evaluated by residues: the function $\log z / (z^2 + 4)$ has simple poles at the points $\pm 2i$, so, using (8.3) to find the residues, we obtain

$$\int_{\Gamma_{R,\varepsilon,\delta}} \frac{\log z \, dz}{z^2 + 4} = 2\pi i \left(\frac{\log(2i)}{2 \cdot 2i} + \frac{\log(-2i)}{2 \cdot (-2i)} \right) = -\frac{\pi^2 i}{2}.$$

Now let ε tend to zero and R tend to infinity. Then both integrals over circles in the right-hand side of (8.12) tend to zero. Indeed, for $|z| = R$ we have $|\log z| = |\log R + i \arg z| \le \log R + 2\pi$ and $|z^2 + 4| \ge R^2 - 4$, so

$$\left| \int_{|z|=R} \frac{\log z \, dz}{z^2 + 4} \right| \le \frac{\log R + 2\pi}{R^2 - 4} \cdot 2\pi R \to 0,$$

and for $|z| = \varepsilon$ we have $|\log z| \le |\log \varepsilon| + 2\pi$, $|1/(z^2 + 4)| \le \text{const}$, so

$$\left| \int_{|z|=\varepsilon} \frac{\log z \, dz}{z^2 + 4} \right| \le \text{const}(|\log \varepsilon| + 2\pi) \cdot 2\pi \varepsilon \to 0.$$

Thus, taking the limit in (8.12) and substituting the value of the left-hand side found by residues, we have

$$-\frac{\pi^2 i}{2} = \int_{0}^{\infty} \left(\frac{\log x}{x^2 + 4} - \frac{\log x + 2\pi i}{x^2 + 4} \right) dx,$$

which we simplify to

$$\int_{0}^{\infty} \frac{dx}{x^2 + 4} = \frac{\pi}{4}.$$

That's all true, but not what we need! To obtain what we need, integrate over the same contour the function $\frac{(\log z)^2}{z^2 + 4}$, instead of the obvious $\frac{\log z}{z^2 + 4}$ (the branch of the logarithm is the same as above). The integrals over the circles of radii ε and R still tend to zero, which can be deduced using similar estimates, so we have

---

cut the logarithm coincides with $\log x$, while on the lower edge it is equal to $\log x + 2\pi i$. Such an exposition is less rigorous, but perhaps more clear.

$$\int_{\Gamma_{R,\varepsilon,\delta}} \frac{(\log z)^2}{z^2 + 4} \, dz = \int_0^\infty \left(\frac{(\log x)^2}{x^2 + 4} - \frac{(\log x + 2\pi i)^2}{x^2 + 4} \right) dx. \qquad (8.13)$$

The left-hand side can be again evaluated by residues:

$$\int_{\Gamma_{R,\varepsilon,\delta}} \frac{(\log z)^2}{z^2 + 4} \, dz = 2\pi i \left(\frac{(\log z)^2}{2z} \bigg|_{z=2i} + \frac{(\log z)^2}{2z} \bigg|_{z=-2i} \right) = \pi^2 (1 - i \log 2).$$

Now, simplifying the right-hand side of (8.13), we obtain

$$\pi^2 (1 - i \log 2) = 4\pi i \int_0^\infty \frac{\log x \, dx}{x^2 + 4} + 4\pi^2 \int_0^\infty \frac{dx}{x^2 + 4}.$$

Substituting the value $\int_0^\infty dx/(x^2 + 4)$ found along the way (or simply taking the imaginary part), we finally obtain

$$\int_0^\infty \frac{\log x \, dx}{x^2 + 4} = -\frac{\pi \log 2}{4}.$$

Example 8.27 Let us find $\int_0^1 \sqrt{x(1-x)} \, dx$. For this, we need the function $\sqrt{z(1-z)}$.
First, observe that the germ of this function at a point not on the interval $[0; 1]$ of the real axis extends to a holomorphic function on the whole of $\mathbb{C} \setminus [0; 1]$. Indeed, a closed path in $\mathbb{C} \setminus [0; 1]$ that has index 1 with respect to 0 has the same index also with respect to 1, as follows from Proposition 4.20. Hence, if we analytically continue the germs of the functions $\sqrt{z}$ and $\sqrt{1-z}$ along such a path, then each of them gets multiplied by -1, while their product remains the same. Since every closed path in $\mathbb{C} \setminus [0; 1]$ is homotopic to a circle around the interval $[0; 1]$ traversed several times (this is easy to prove rigorously, and even easier to believe), we see that the germ of $\sqrt{z(1-z)}$ extends to a holomorphic function on $\mathbb{C} \setminus [0; 1]$. Choose the branch of this function for which the values on the upper edge of the cut along the interval $[0; 1]$ are positive (more formally: the limit of the function as z approaches a point of $(0; 1)$ from the upper half-plane along a vertical line is positive). Now consider the contour shown in Fig. 8.6, which we denote by $\gamma_{\varepsilon,\delta}$ where δ is the distance from its horizontal segments to the real axis.

As in the previous example, the integral of $\sqrt{z(1-z)}$ over this contour does not change as $\delta \to 0$. At the upper edge of the cut, $\sqrt{z(1-z)}$ is equal to the positive value of the square root; at the lower edge, to the negative one. Bearing in mind that, besides, the lower edge is traversed in the direction from larger numbers to smaller ones, we obtain

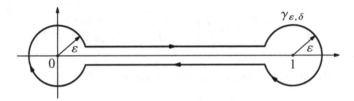

Fig. 8.6

$$\int_{\gamma_{\varepsilon,\delta}} \sqrt{z(1-z)}\,dz = 2\int_{\varepsilon}^{1-\varepsilon} \sqrt{x(1-x)}\,dx$$

$$+ \int_{|z|=\varepsilon} \sqrt{z(1-z)}\,dz + \int_{|z-1|=\varepsilon} \sqrt{z(1-z)}\,dz. \quad (8.14)$$

Obviously, the second and third terms in the right-hand side tend to zero as $\varepsilon \to 0$. Thus, taking in (8.14) the limit as $\varepsilon \to 0$, we have

$$\int_{0}^{1} \sqrt{x(1-x)}\,dx = \frac{1}{2}\int_{\gamma_{\varepsilon,\delta}} \sqrt{z(1-z)}\,dz.$$

It remains to evaluate the integral in the right-hand side. It is clear that the path $\gamma_{\varepsilon,\delta}$ is homotopic in $\mathbb{C}\setminus[0;1]$ to the negatively oriented circle of radius $R > 1$ centered at the origin. Hence, integrating the Laurent series expansion

$$\sqrt{z(1-z)} = \sum_{n=-\infty}^{\infty} c_n z^n$$

term by term in a punctured neighborhood of infinity, we obtain

$$\int_{\gamma_{\varepsilon,\delta}} \sqrt{z(1-z)}\,dz = -2\pi i c_{-1}.$$

To find c_{-1}, write

$$\sqrt{z(1-z)} = \pm iz \cdot \left(1 - \frac{1}{z}\right)^{1/2}$$

(the plus or minus sign depends on the choice of a branch). For $|z| > 1$, the right-hand side can be expanded using the binomial theorem, whence

$$\sqrt{z(1-z)} = \pm iz\left(-\frac{1}{2}\cdot\frac{1}{z} - \frac{1}{8}\cdot\frac{1}{z^2} + \dots\right),$$

so $c_{-1} = \pm i/8$ and

$$\int\limits_0^1 \sqrt{x(1-x)}\, dx = \frac{1}{2}\cdot(-2\pi i c_{-1}) = \pm\frac{\pi}{8}.$$

Since the function $\sqrt{x(1-x)}$ is nonnegative on $[0;1]$, the sign is not minus but plus, so we finally see that the integral in question is equal to $\pi/8$.

Our last example is amusing because it actually does not involve residues: the function we are going to integrate has no singularities at all.

Example 8.28 As is well known (see, for instance, [6, Chap. XVII, Sec. 2, Example 17]),

$$\int\limits_{-\infty}^\infty e^{-x^2}\, dx = \sqrt{\pi}.$$

Using this formula, let us evaluate the integral

$$\int\limits_{-\infty}^\infty e^{-x^2+itx}\, dx$$

where $t \in \mathbb{R}$.

To begin with, complete the square in the exponent:

$$-x^2 + itx = -\left(x - \frac{it}{2}\right)^2 - \frac{t^2}{4}.$$

Now we have

$$\int\limits_{-\infty}^\infty e^{-x^2+itx}\, dx = e^{-t^2/4}\int\limits_{-\infty}^\infty e^{-\left(x-\frac{it}{2}\right)^2}\, dx = e^{-t^2/4}\lim_{R\to\infty}\int\limits_{-R-\frac{it}{2}}^{R-\frac{it}{2}} e^{-z^2}\, dz$$

(in the last integral, we mean integration over the horizontal line segment connecting the integration "limits").

By Cauchy's theorem, the integral of e^{-z^2} over the boundary of the rectangle with vertices R, $R - \frac{it}{2}$, $-R - \frac{it}{2}$, and $-R$ is zero, so

$$\int\limits_{-R}^R e^{-z^2}\, dz - \int\limits_{-R-\frac{it}{2}}^{R-\frac{it}{2}} e^{-z^2}\, dz = \int\limits_{-R}^{-R-\frac{it}{2}} e^{-z^2}\, dz - \int\limits_R^{R-\frac{it}{2}} e^{-z^2}\, dz,$$

whence

$$\left| \int_{-R}^{R} e^{-z^2} dz - \int_{-R-\frac{it}{2}}^{R-\frac{it}{2}} e^{-z^2} dz \right| \le \left| \int_{R}^{R-\frac{it}{2}} e^{-z^2} dz \right| + \left| \int_{-R}^{-R-\frac{it}{2}} e^{-z^2} dz \right| \tag{8.15}$$

(in the right-hand side, we mean integration over vertical segments). Since we have $|e^{-(\pm R+iy)^2}| = e^{-R^2+y^2}$ and $|y| \le |t|/2$, the right-hand side of (8.15) does not exceed $|t| \cdot \frac{|t|^2}{4} \cdot e^{-R^2}$, and thus tends to zero as $R \to \infty$. Therefore,

$$\int_{-\infty}^{\infty} e^{-\left(x-\frac{it}{2}\right)^2} dx = \int_{-\infty}^{\infty} e^{-x^2} dx = \sqrt{\pi},$$

whence

$$\int_{-\infty}^{\infty} e^{-x^2+itx} dx = \sqrt{\pi} e^{-t^2/4}.$$

The right-hand side of this equality, regarded as a function of t, is called the Fourier transform of the function e^{-x^2} (and the integral we have found in Example 8.17 is called the Fourier transform of the function $\frac{1}{1+x^2}$).

Exercise

8.1. For each of the following functions, find all its residues at all isolated singularities:
(a) $\tan z$; (b) $\frac{1}{z^3+z}$; (c) $\frac{e^z}{z^2-4}$; (d) $z^3 \cos \frac{1}{z+1}$; (e) $\sin z \sin \frac{1}{z}$; (f) $\sin \frac{z}{z+1}$.
8.2. And now, find the residues (again, at all isolated singularities) of the following functions:
(a) $\frac{1}{z^3+z^2}$; (b) $\frac{z^2}{(z^2-1)^2}$; (c) $\frac{e^z}{z^2(z^2+1)}$; (d) $\frac{1}{z \sin^2 z}$; (e) $\frac{1}{\sin^{2016} z}$; (f) $\tan^3 z$.
8.3. Let f and g be holomorphic at a point a, with $\operatorname{ord}_a g = n$. Show that

$$\operatorname{Res}_a \frac{f(z)}{g(z)} = (n-1)! \lim_{z \to a} \frac{d^{n-1}}{dz^{n-1}} \frac{(z-a)^n f(z)}{g(z)}.$$

8.4. Evaluate the integral

$$\int_{|z|=1} \frac{\sin \pi z \, dz}{3z^2 - 10z + 3}$$

(the unit circle is positively oriented).
8.5. A function f is continuous on the unit circle $\{z: |z| \le 1\}$ and holomorphic on its interior. Show that f cannot map the positively oriented unit circle to a negatively oriented ellipse.
8.6. (a) A function f is continuous on the closed unit disk $\bar{U} = \{z: |z| \le 1\}$ and holomorphic on its interior. Can it map the unit circle to the "figure eight"

shown in Fig. 8.7 (the arrows indicate the orientation induced by traversing the unit circle in the positive direction)?

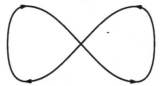

Fig. 8.7

(b) The same question as in part (a), but this time the orientation (and the number of traversals) of the "figure eight" is not specified: we know only that, as a set, it is the image of the boundary circle.

8.7. A function f is continuous on the unit disk $\bar{U} = \{z: |z| \leq 1\}$ and holomorphic on its interior. Set $\partial D = \{z: |z| = 1\}$. Show that if $f(\partial D) \subset \partial D$ and f is not a constant function, then $f(\partial D) = \partial D$.

8.8. How many roots (counting multiplicities) has the polynomial $z^7 + 3z + 1$ inside the unit disk $\{z: |z| < 1\}$?

8.9. How many roots (counting multiplities) has the function $f(z) = 2z^3 + e^z + 1$ inside the unit disk $\{z: |z| < 1\}$?

Evaluate the following integrals.

8.10. $\displaystyle\int_{-\infty}^{\infty} \frac{dx}{x^4 + 1}$.

8.11. $\displaystyle\int_{-\infty}^{\infty} \frac{dx}{(x^2 + 1)^2}$.

8.12. $\displaystyle\int_{-\infty}^{\infty} \frac{dx}{x^{100} + 1}$. (*Hint.* You can integrate over the boundary of a semi-disk and then add up 50 residues, but perhaps you will prefer the contour shown in Fig. 8.8; at least, there is only one pole inside it!)

8.13. $\displaystyle\int_{0}^{2\pi} \frac{1 + \cos x}{2 - \sin x}\, dx$.

8.14. $\displaystyle\int_{-\infty}^{\infty} \frac{\cos tx\, dx}{1 + x^4}$ $(t \in \mathbb{R})$.

8.15. $\displaystyle\int_{-\infty}^{\infty} \frac{x \cos x\, dx}{x^2 + 2x + 2}$.

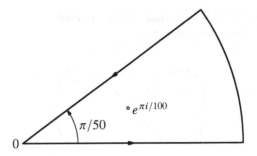

Fig. 8.8

8.16. $\displaystyle\int_{-\pi}^{\pi}\frac{\sin^2\!\left(n+\frac{1}{2}\right)dx}{\sin^2\dfrac{x}{2}}$ for an arbitrary positive integer n.

8.17. V. p. $\displaystyle\int_{-\infty}^{\infty}\frac{\sin x\,dx}{x^2-x}.$

8.18. $\displaystyle\int_{|z|=2}\frac{dz}{(z-3)(z^5-1)}$ (the circle is positively oriented).

8.19. $\displaystyle\int_{0}^{1}\frac{x^2\,dx}{\sqrt{x(1-x)}}.$

8.20. $\displaystyle\int_{0}^{\infty}\frac{dx}{\sqrt[3]{x}\cdot(1+x^2)}.$

8.21. $\displaystyle\int_{0}^{\infty}\frac{x^\lambda\,dx}{1+x^2}$ where $|\operatorname{Re}\lambda|<1$ and t^λ stands for $e^{\lambda\log t}$, $\log t\in\mathbb{R}$. (*Hint*. It is convenient to integrate first from $-\infty$ to ∞.)

8.22. Find all positive integers n such that $\displaystyle\int_{|z|=2}\frac{z^n\,dz}{z^{10}-1}\neq 0$.

8.23. Let f be a function holomorphic on $\mathbb{C}$ except for finitely many isolated singularities. Show that
$$\sum_{a\in\mathbb{C}}\operatorname{Res}_a(f(z))=c_{-1},$$
where c_{-1} is the coefficient of z^{-1} in the Laurent series of f in a punctured neighborhood of infinity.

The number $-c_{-1}$ (note the sign) is called the *residue at infinity* of $f(z)$. In this terminology, the result of this exercise can be stated as follows: if f has only finitely many isolated singularities, then the sum of all residues of $f(z)$, including the residue at infinity, is zero. We will encounter residue at infinity again in Chap. 13.

8.24. Let P be a polynomial of degree $n > 1$ without multiple roots; denote its roots by $a_1, \ldots, a_n$. Show that if k is an integer, $0 \le k \le n - 2$, then

$$\frac{a_1^k}{P'(a_1)} + \ldots + \frac{a_n^k}{P'(a_n)} = 0.$$

8.25. Let $F(x) = P(x)/Q(x)$ be a rational function, where P and Q are polynomials with real coefficients, $\deg P < \deg Q$, and all roots of Q are real and distinct. Show that

$$\text{V. p.} \int_{-\infty}^{\infty} F(x)\, dx = 0.$$

8.26. Let D be an open disk, $a_1, \ldots, a_n \in U$, let f be a holomorphic function on $D \setminus \{a_1, \ldots, a_n\}$, and let γ be a closed piecewise smooth curve in $D \setminus \{a_1, \ldots, a_n\}$. Show that

$$\int_{\gamma} f(z)\, dz = 2\pi i \sum_{k=1}^{n} \text{Ind}_{a_k}\, \gamma \cdot \text{Res}_{a_k}\, f(z).$$

Chapter 9
Local Properties of Holomorphic Functions

9.1 The Open Mapping Theorem

The open mapping theorem is the following fact.

Theorem 9.1 *Let $U \subset \mathbb{C}$ be a connected open set and $f: U \to \mathbb{C}$ be a nonconstant holomorphic function. Then for every open subset $V \subset U$, its image $f(V) \subset \mathbb{C}$ is open too.*

In topology, a map that takes open sets to open sets is said to be open. Thus, the open mapping theorem says that a nonconstant holomorphic map defined on a connected open set is open.

For functions of a real variable, an analog of the open mapping theorem does not hold, as can be seen from very simple examples. For instance, a "best possible" function $f: x \mapsto x^2$, regarded as a function from $\mathbb{R}$ to $\mathbb{R}$, takes the open interval $(-1; 1) \subset \mathbb{R}$ to the interval $[0; 1) \subset \mathbb{R}$, which is not open.

Proof The argument from this proof was already used in Chap. 8. So, let $a \in V \subset U$ where $V \subset U$ is open, and let $b = f(a)$. We must prove that some δ-neighborhood of b is contained in $f(V)$. Note that the set $f^{-1}(b) \subset U$ has no accumulation points in U: otherwise, it would follow from the principle of analytic continuation that f is identically equal to b, while this function is nonconstant by assumption. Therefore, there exists $\varepsilon > 0$ such that the closed disk $\bar{D} = \{z: |z - a| \leq \varepsilon\}$ is contained in $V \subset U$ and $f^{-1}(b) \cap \bar{D} = \{a\}$. Denote by $\gamma = \{z: |z - a| = r\}$ the boundary of this disk. We equip it with the standard parametrization $t \mapsto a + re^{it}, t \in [0; 2\pi]$, which orients γ positively. Below, by the same letter γ we denote both the curve as a subset in $\mathbb{C}$ and the closed path oriented as described above, so, for example, writing $f(\gamma)$ we mean the image of the set, and writing $f \circ \gamma$ we mean the closed path.

Since $f(\gamma) \not\ni b$ but $f(a) = b$, the argument principle (more exactly, Corollary 8.10) shows that $\mathrm{Ind}_b(f \circ \gamma) > 0$. Since $\gamma \subset \mathbb{C}$ is a compact subset, its image $f(\gamma)$ is closed in $\mathbb{C}$. Since $b \notin f(\gamma)$, there exists $\delta > 0$ such that the δ-neighborhood of b (denote it by V_δ) is disjoint with $f(\gamma)$. By Proposition 4.20, for every point $b' \in V_\delta$ we have

© Springer Nature Switzerland AG 2020

S. Lvovski, *Principles of Complex Analysis*, Moscow Lectures 6,
https://doi.org/10.1007/978-3-030-59365-0_9

$$\text{Ind}_{b'}(f \circ \gamma) = \text{Ind}_b(f \circ \gamma) > 0;$$

applying the argument principle once again, we see that $b' = f(a')$ for some $a' \in \bar{D}$. Therefore, $V_\delta \subset f(V)$, and we are done. $\qquad\qquad\qquad\qquad\qquad\qquad\qquad\square$

Looking more closely at the proof of Theorem 9.1, we can extract important additional information.

Proposition 9.2 (inverse function theorem) *Let f be a holomorphic function on an open set $U \subset \mathbb{C}$, and let $f'(a) \neq 0$ at a point $a \in \mathbb{C}$. Then there exist open sets $U_1 \ni a$ and $V_1 \ni f(a)$ such that $f(U_1) = V_1$ and f induces a bijection between U_1 and V_1; the inverse map $f^{-1}: V_1 \rightarrow U_1$ is also holomorphic.*

Proof Set $f(a) = b$. As in the proof of Theorem 9.1, we conclude that there exists a closed disk $\bar{D} = \{z: |z - a| \le \varepsilon\}$, $\bar{D} \subset U$, on which f does not assume the value b except at the point a (since $f'(a) \neq 0$, the function f is not constant in any neighborhood of a, so the existence of such a disk $\bar{D}$ follows from Corollary 5.20). Denote the boundary of $\bar{D}$ by γ; we assume that γ is parametrized in the same way as in the proof of Theorem 9.1.

Since $f'(a) \neq 0$, the function $z \mapsto f(z) - b$ has a simple zero at a; by construction, it has no other zeros in $\bar{D}$. Therefore, it follows from Corollary 8.10 that $\text{Ind}_b(f \circ \gamma) = 1$. Let $V' \ni b$ be a δ-neighborhood of b (for some $\delta > 0$) disjoint with $f(\gamma)$. As above, for every $b' \in V'$ we have $\text{Ind}_{b'}(f \circ \gamma) = \text{Ind}_b(f \circ \gamma) = 1$. Hence, the same Corollary 8.10 shows that there exists a unique point a' with $|a' - a| < \varepsilon$ for which $f(a') = b'$. Now set

$$U' = f^{-1}(V') \cap \{z: |z - a| < \varepsilon\}.$$

Obviously, U' is open and f maps U' bijectively onto V_1. Further, let $U_1 \subset U'$, $U_1 \ni a$, be an open set on which f' does not vanish, and let $V_1 = f(U_1)$. It follows from what we have already proved that f maps U_1 bijectively onto V_1, and $f'(z) \neq 0$ for all $z \in U_1$ by construction; furthermore, the inverse map $f^{-1}: V_1 \rightarrow U_1$ is continuous: to establish this, we must verify that if $W \subset U_1$ is an open subset, then the set $(f^{-1})^{-1}(W) \subset V_1$ is also open, but $(f^{-1})^{-1}(W) = f(W)$, and $f(W)$ is open by the open mapping theorem. Hence, the restriction of f to U_1, which maps U_1 bijectively onto V_1, satisfies all assumptions of Proposition 2.10, so the inverse map $f^{-1}: V_1 \rightarrow U_1$ is holomorphic, as required. $\qquad\qquad\qquad\square$

Remark 9.3 Proposition 9.2 immediately follows from the "inverse function theorem" of real analysis (see [5, Chap. VIII, Sec. 6]): indeed, holomorphic functions are automatically continuously differentiable, and the derivative of f at a point a (see Sec. 1.5) is the multiplication by a nonzero complex number, which is an invertible linear operator. However, since I tried to keep to a minimum the prerequisites from multivariable analysis needed for reading the book, an independent proof is given above.

The key point of this proof is the application of the argument principle, i.e., a topological argument. Similar topological considerations allow one to prove the inverse function theorem for maps from (an open subset of) $\mathbb{R}^n$ to $\mathbb{R}^n$ for arbitrary n, but you will not find such an approach in textbooks: notions necessary for the topological proof in dimension greater than 2 are studied later than multivariable analysis.

9.2 Ramification

We have explored the behavior of a holomorphic function near a point at which its derivative does not vanish. The case where the derivative does vanish is also amenable to comprehensive analysis.

Definition 9.4 Let f be a holomorphic function in a neighborhood of a point a that is not constant in any neighborhood of a. Set $f(a) = b$. Then the *ramification index* of f at a is the order of the zero of the function $z \mapsto f(z) - b$ at a.

Proposition 9.5 *Let f be a holomorphic function in a neighborhood of a point a that is not constant in any neighborhood of a. Set $f(a) = b$, and let*

$$f(z) = c_0 + c_1(z - a) + \ldots + c_n(z - a)^n + \ldots$$

be the power series expansion of f in a neighborhood of a. Then the following three numbers coincide:
(1) *the ramification index of f at a;*
(2) $\min\{k > 0 : c_k \neq 0\}$;
(3) $\min\{k > 0 : f^{(k)}(a) \neq 0\}$.

This follows immediately from Proposition 7.2 applied to the function $z \mapsto f(z) - b$.

It is clear from Proposition 9.5 that ramification index is a generalization of the notion of "order of a zero."

Corollary 9.6 *If a holomorphic function f has an isolated zero at a point a, then its ramification index at a is equal to $\mathrm{ord}_a(f)$.*

Obviously, the ramification index of f at a is equal to 1 if and only if $f'(a) \neq 0$. If the ramification index of f at a is greater than 1, then a is called a *ramification point* of f. Sometimes, the term "critical point" is used instead of "ramification point."

The notion of ramification index allows one to describe the behavior of a holomorphic function at a point where its derivative vanishes.

Proposition 9.7 *Let f be a holomorphic function in a neighborhood of a point a that is not constant in any neighborhood of a. Set $f(a) = b$, and assume that the ramification index of f at a is equal to k. Then there exist neighborhoods $U \ni a$ and $V \ni b$ such that $f(U) = V$, every point $b' \in V$ except b has exactly k preimages in U (with respect to f), and the point b has only one preimage in U, namely, a.*

Proposition 9.7 is, in turn, a consequence of the following more precise result. Informally, it says that in a neighborhood of a ramification point of order k, a holomorphic function looks like the map $z \mapsto z^k$ in a neighborhood of the origin.

Proposition 9.8 *Let f be a holomorphic function in a neighborhood of a point a that is not constant in any neighborhood of a. Set $f(a) = b$, and assume that the ramification index of f at a is equal to k. Then there exist neighborhoods $U \ni a$ and $V \ni b$ for which $f(U) = V$, and conformal isomorphisms $\alpha : U \to U'$, $\beta : V \to V'$, where V and V' are open disks centered at the origin, such that $\alpha(a) = 0$, $\beta(b) = 0$, and $\beta(f(z)) = (\alpha(z))^k$ for all $z \in U$.*

The main part of Proposition 9.8 can be summarized in the following commutative square:

$$
\begin{array}{ccc}
U & \xrightarrow{\ \alpha\ } & U' \\
{\scriptstyle f}\downarrow & & \downarrow{\scriptstyle w \mapsto w^k} \\
V & \xrightarrow{\ \beta\ } & V'.
\end{array}
$$

Proof Since the function $z \mapsto f(z) - b$ has a zero of order k at a, we can write $f(z) = b + (z-a)^k g(z)$ where g is holomorphic in a neighborhood of a and $g(a) \neq 0$. Since $g(a) \neq 0$, in some neighborhood of $g(a)$ there is a well-defined branch of the function $z \mapsto \sqrt[k]{z}$; thus, in some neighborhood $U_1 \ni a$ we have a well-defined function $h(z) = \sqrt[k]{g(z)}$ for which $(h(z))^k = g(z)$. Set $\alpha(z) = (z-a)h(z)$; then the function α is defined on the same neighborhood U_1 and $f(z) = b + \alpha(z)^k$. Since $\alpha'(a) \neq 0$, Proposition 9.2 shows that α is a conformal map from some neighborhood of a onto its image.

Also, set $\beta(w) = w - b$: it is quite obvious that β is a conformal map from an (arbitrary) neighborhood of b onto its image. The identity $\beta(f(z)) = (\alpha(z))^k$ is clear by construction. Now choose $\varepsilon > 0$ and set $U_\varepsilon = \{z : |z| < \varepsilon\}$, $U = \alpha^{-1}(U_\varepsilon)$. If ε is sufficiently small, then it follows from the above that $\alpha : U \to U_\varepsilon$ is a holomorphic bijection. Setting, in obvious notation, $V = \beta^{-1}(U_{\varepsilon^k})$, $U' = U_\varepsilon$, $V' = U_{\varepsilon^k}$, all the required objects are constructed. $\square$

Now we are ready to fulfill the promise made in Chap. 2 (Remark 2.11).

Proposition 9.9 *Let $f : U \to V$ be a holomorphic and bijective map between two open subsets of the complex plane. Then the derivative of f does not vanish on U and the inverse map $f^{-1} : V \to U$ is also holomorphic.*

Proof Proposition 9.7 implies that if f' vanishes even at one point, then f cannot be one-to-one. Hence, the derivative of f does not vanish. Now, Proposition 9.2 implies that for every point $a \in U$ there exist neighborhoods $U_1 \ni a$, $V_1 \ni f(a)$ such that f induces a bijection from U_1 onto V_1 and the inverse map f^{-1} is holomorphic on V_1. Thus, f^{-1} is holomorphic in a neighborhood of every point of V, i.e., holomorphic on V. $\square$

It follows from what we have proved, among other things, that the map inverse to a conformal map is also holomorphic, and that open subsets $U, V \subset \mathbb{C}$ are conformally isomorphic if and only if there are mutually inverse holomorphic maps $f: U \to V$ and $g: V \to U$ between them. Note also that, in light of Proposition 9.9, we can restate the definition of a conformal isomorphism between open sets $U, V \subset \mathbb{C}$ (Definition 3.1) as follows: U and V are conformally isomorphic if and only if there exist holomorphic maps $f: U \to V$ and $g: V \to U$ such that both compositions $f \circ g$ and $g \circ f$ are identity maps. This agrees with the interpretation of the term "isomorphism" adopted in other mathematical theories.

We remember from Chap. 2 that a holomorphic map preserves angles between curves at a point where its derivative does not vanish. Now we are able to find out what happens with these angles at points where the derivative does vanish. It turns out that they get multiplied by the ramification index.

Proposition 9.10 *Let f be a holomorphic function at a point a that has ramification index k at this point. If γ_1 and γ_2 are smooth curves passing through a and making angle φ, then the angle between the curves $f(\gamma_1)$ and $f(\gamma_2)$ at the point $f(a)$ is equal to $k\varphi$.*

Proof Apply Proposition 9.8 to the function f and the point a. Since α and β, being conformal maps, do preserve angles, it remains to prove that the map $z \mapsto z^k$ multiplies angles between curves passing through the origin by k.

So, let $f: z \to z^k$. It suffices to show that if the tangent at 0 to a smooth curve γ makes an angle φ with the real axis (more exactly, with its positive direction), then the tangent to the curve $f \circ \gamma$, where $f: z \to z^k$, makes the angle $k\varphi$ with the real axis. For this, observe that the tangent to γ at 0 makes an angle φ with the real axis if and only if $\lim_{t \to 0}(\gamma(t)/|\gamma(t)|) = e^{i\varphi}$ (the tangent is the limiting position of the secant: see Fig. 9.1). Since $(f \circ \gamma)(t) = (\gamma(t))^k$, we have

$$\lim_{t \to 0} \frac{(f \circ \gamma)(t)}{|(f \circ \gamma)(t)|} = \lim_{t \to 0}\left(\frac{\gamma(t)^k}{|\gamma(t)^k|} \right) = \left(\lim_{t \to 0} \frac{\gamma(t)}{|\gamma(t)|} \right)^k = e^{ik\varphi},$$

as required. $\square$

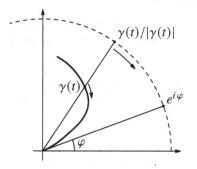

Fig. 9.1

9.3 The Maximum Modulus Principle and Its Corollaries

Consider a function $\varphi \colon U \to \mathbb{R}$ defined on an open set $U \subset \mathbb{C}$. It is said to have a *local maximum* at a point $a \in U$ if there exists an open set $V \ni a$, $V \subset U$, such that $\varphi(a) \geq \varphi(z)$ for all $z \in U$. A local minimum is defined in a similar way.

Note that, according to our definitions, local maxima and minima are non-strict extrema.

Proposition 9.11 (maximum modulus principle) *If f is a nonconstant holomorphic function on a connected open set $U \subset \mathbb{C}$, then the function $z \mapsto |f(z)|$ cannot have a local maximum at any point of U.*

Proof Arguing by contradiction, assume that the function $|f(z)|$ attains a local maximum at some point $a \in U$. Then there exists an open subset $V \subset U$, $V \ni a$, such that $|f(a)| \geq |f(z)|$ for all $z \in V$. By the open mapping theorem, the set $f(V)$ is open in $\mathbb{C}$, hence $f(V)$ contains an open disk D centered at $f(a)$. But every open disk in the plane contains a point that lies further from the origin than its center. Since $D \subset f(V)$, this point is $f(b)$ for some $b \in V$, hence $|f(b)| > |f(a)|$, a contradiction.   □

Here is another version of the maximum modulus principle. I take the liberty of omitting the standard remarks about the status of assertions involving the phrase "the part of the plane bounded by a curve."

Proposition 9.12 *Let $\bar{U} \subset \mathbb{C}$ be a closed bounded subset of the complex plane bounded by finitely many piecewise smooth curves, and let $f \colon U \to \mathbb{C}$ be a continuous function holomorphic on its interior. If the interior of U is connected and f is not a constant, then the maximum value of the function $z \mapsto |f(z)|$ is attained on the boundary of the set $\bar{U}$, but not in its interior.*

Proof Since $\bar{U}$ is closed and bounded and the function $z \mapsto |f(z)|$ is continuous, the maximum value is attained somewhere, but it cannot be attained at an interior point of $\bar{U}$ by Proposition 9.11.   □

Now we are going to use the maximum modulus principle to derive the so-called Schwarz lemma. While seemingly a rather special result, this lemma has a number of important corollaries.

Proposition 9.13 (Schwarz lemma) *Let $D = \{z \colon |z| < 1\}$ be the unit disk and $f \colon D \to D$ be a holomorphic map such that $f(0) = 0$. Then:*
   (1) $|f(z)| \leq |z|$ for all $z \in D$;
   (2) $|f'(0)| \leq 1$;
   (3) if inequality (1) (for at least one $z \neq 0$) or inequality (2) becomes an equality, then there exists $\theta \in \mathbb{R}$ such that $f(z) = e^{i\theta} z$ for all $z \in D$.

Of course, part (2) of this proposition follows from Cauchy's estimates (5.23), but proving the Schwarz lemma in its entirety requires more subtle considerations.

Proof Since $f(0) = 0$, the function $g(z) = f(z)/z$ extends to a holomorphic function on the whole disk D with $g(0) = f'(0)$. For every positive $r < 1$, consider the closed disk $\bar{D}_r = \{z : |z| \le r\} \subset D$. By Proposition 9.12, the maximum of the function $|g|$ on the disk $\bar{D}_r$ is attained on the boundary of $\bar{D}_r$. Since $|f(z)| < 1$ for all $z \in D$ and $|z| = r$ for all $z \in \partial \bar{D}_r$, we have $|g(z)| < 1/r$ for all $z \in \bar{D}_r$. Now, for every $z \in D$ take a radius r such that $|z| < r < 1$ and let r tend to 1; then we see that $|g(z)| \le 1$ for every $z \in D$. If $z \ne 0$, this means that $|f(z)| \le |z|$; and if $z = 0$, this means that $|f'(0)| \le 1$. Thus, parts (1) and (2) are proved. If $|f'(0)| = 1$ or $|f(z)| = |z|$ for some $z \in D \setminus \{0\}$, this means that $|g(z)| = 1$ for some $z \in D$. Therefore, $|g(z)|$ attains a maximum at z, so, by Proposition 9.11, the function g is constant on D, $g(z) = c$ for all z, and, moreover, $|c| = 1$, i.e., $c = e^{i\theta}$, $\theta \in \mathbb{R}$. This proves part (3).□

Here is the first of the promised important corollaries of the Schwarz lemma. Another (more serious) one will be considered in Chap. 12.

Proposition 9.14 *Every conformal automorphism of the disk $D = \{z : |z| < 1\}$ is linear fractional.*

Recall that all linear fractional automorphisms of the disk were described in Proposition 3.5.

Proof Let $f \colon D \to D$ be a conformal automorphism with $f(0) = a \in D$. As we know (see Proposition 3.5), the linear fractional transformation

$$g \colon z \mapsto \frac{z - a}{1 - \bar{a}z}$$

is a linear fractional automorphism of D; then the map $\varphi = g \circ f$ is also a conformal automorphism of D, and $\varphi(0) = 0$. If we show that φ is linear fractional, then it will follow that the map $f = g^{-1} \circ \varphi$ is also linear fractional, proving the proposition. We will even show that φ is a rotation about the center of the disk.

Indeed, the map $\varphi \colon D \to D$ satisfies the assumptions of the Schwarz lemma, so $|\varphi(z)| \le |z|$ for all z. The inverse map $\varphi^{-1} \colon D \to D$ also satisfies the assumptions of the Schwarz lemma; therefore, for every $z \in D$ we have

$$|z| = |\varphi^{-1}(\varphi(z))| \le |\varphi(z)|.$$

Comparing these two inequalities, we see that $|\varphi(z)| = |z|$ for all z; consequently, by the Schwarz lemma, we have $\varphi(z) = e^{i\theta}z$ for all z, with $\theta \in \mathbb{R}$. Hence, φ is indeed a rotation, and the proposition is proved.   □

Corollary 9.15 *Every conformal automorphism of the upper half-plane is linear fractional.*

As you remember, the linear fractional automorphisms of the upper half-plane were also described earlier (Proposition 3.4).

Proof Let $\varphi \colon H \to U$ be an arbitrary linear fractional isomorphism between the upper half-plane H and the unit disk U (see Chap. 3). If $f \colon H \to H$ is a conformal automorphism, then $g = \varphi \circ f \circ \varphi^{-1} : U \to U$ is a conformal automorphism of the unit disk U (see the diagram):

$$
\begin{array}{ccc}
H & \xrightarrow{\ \varphi\ } & U \\
{\scriptstyle f}\big\downarrow & & \big\downarrow{\scriptstyle g} \\
H & \xrightarrow{\ \varphi\ } & U.
\end{array}
$$

By Proposition 9.14, the automorphism g is linear fractional; therefore, the automorphism $f = \varphi^{-1} \circ g \circ \varphi$ is also linear fractional as the composition of three linear fractional transformations.

9.4 Bloch's Theorem

In this section, we prove a "quantitative version" of the open mapping theorem.

Namely, the open mapping theorem (Theorem 9.1) says, in particular, that the image of a connected open set under a nonconstant holomorphic map necessarily contains open disks. Paying due attention to quantitative issues when proving this theorem, one can obtain information on the radii of such disks. We begin with the following result.

Proposition 9.16 *Let $U \subset \mathbb{C}$ be an open disk of radius R centered at a point z_0, and let $f : U \to \mathbb{C}$ be a holomorphic function. Assume that $|f'(z_0)| = a \neq 0$ and $\sup_{z \in U} |f(z) - f(z_0)| \leq M < +\infty$. Then the set $f(U)$ contains an open disk of radius $a^2 R^2 / 9M$ centered at $f(z_0)$. Moreover, f is a bijection from some neighborhood of z_0 onto this disk.*

The essence of this proposition is not the factor 9 in the denominator (as we will see, it is certainly not the optimal one), but the fact that, first, $f(U)$ necessarily contains a disk of radius not less than a constant depending only on $f'(z_0)$, $\sup |f(z)|$, and the radius R of the disk U; and, second, that this constant is proportional to the square of $|f'(z_0)|$ and the square of R and inversely proportional to M.

Proof We may (and will) assume without loss of generality that $z_0 = 0$ and $f(z_0) = 0$. The plan of the proof is as follows. We are going to find a number ρ such that $0 < \rho < R$ and $|f(z)| \geq c$ for $|z| = \rho$, where $c > 0$ is a constant depending on a, R, and M. Then it follows from the argument principle that $f(U) \supset \{z : |z| < c\}$: if γ_ρ is the positively oriented circle of radius ρ, then $\mathrm{Ind}_0(f \circ \gamma_\rho) > 0$ because $f(0) > 0$, and $\mathrm{Ind}_w(f \circ \gamma_\rho) = \mathrm{Ind}_0(f \circ \gamma_\rho)$ as soon as $|w| < c$ by Proposiion 4.20, since the disk of radius c centered at the origin is disjoint with $f(\gamma_\rho)$.

Let us put this plan into action. If $f(z) = \sum\limits_{n=1}^{\infty} a_n z^n$ is the power series expansion of f in the disk U, then $|a_1| = a$ by assumption and $|a_n| \leq M/R^n$ for all n by Cauchy's estimates (5.23). If now $|z| = r$, where $0 < r < R$, then we have

$$|f(z)| \geq |a_1 z| - |a_2 z^2 + a_3 z^3 + \ldots| = ar - |a_2 z^2 + a_3 z^3 + \ldots|$$

$$\geq ar - \|a_2|r^2 + |a_3|r^3 + \ldots| \geq ar - \left(\frac{M}{R^2} r^2 + \frac{M}{R^3} r^3 + \ldots\right) = ar - \frac{Mr^2/R^2}{1 - (r/R)}.$$

A standard calculation shows that the maximum value of the function

$$r \mapsto ar - \frac{Mr^2/R^2}{1 - (r/R)}$$

on the interval $(0; R)$ is attained at the point $r = R\left(1 - \sqrt{\frac{M}{aR+M}}\right)$ and equal to $c = (\sqrt{M + aR} - \sqrt{M})^2$. It is convenient to make c smaller (we are after a simple formula, not a sharp estimate), using Cauchy's estimates to see that $aR \leq M$:

$$(\sqrt{M + aR} - \sqrt{M})^2 = \frac{a^2 R^2}{(\sqrt{M + aR} + \sqrt{M})^2} \geq \frac{a^2 R^2}{(\sqrt{2M} + \sqrt{M})^2} > \frac{a^2 R^2}{9M}.$$

So, for $|z| = R\left(1 - \sqrt{\frac{M}{aR+M}}\right)$ we have $|f(z)| > a^2 R^2/(9M)$, hence the disk of radius $a^2 R^2/(9M)$ centered at the origin is contained in $f(U)$, as required. □

In Proposition 9.16, it is important that the lower bound on the radius of a disk contained in $f(U)$ depends on the maximum of $|f|$. It turns out that we can get rid of this dependence. The corresponding result is called Bloch's theorem.

Theorem 9.17 (Bloch theorem) *Let $U \subset \mathbb{C}$ be an open disk of radius R centered at a point z_0. If $f: U \to \mathbb{C}$ is a holomorphic function for which $|f'(z_0)| = a > 0$, then the set $f(U) \subset \mathbb{C}$ contains an open disk of radius $aR/40$. Moreover, f is a bijection from some open subset in U onto this disk.*

It still does not matter that the numerical factor is equal to $1/40$, and this estimate is still not the optimal one (by the way, the optimal estimate is not known to this day). The important things are, first, the fact that the estimate depends only on the radius of the disk and on the value of the derivative at its center but not on the function f; and, second, the form of its dependence on $f'(z_0)$ and R.

Proof We are going to apply Proposition 9.16 to the restriction of f to an appropriate open disk $U_{b,\rho} = \{z: |z - b| < \rho\}$ contained in U, trying to make the product $|f'(b)| \cdot \rho$ as large as possible and the supremum $\sup_{z \in U_{b,\rho}} |f(z) - f(b)|$ as small as possible. First, consider the case where the derivative of f extends to a continuous function on the closure of U. If for every $r \in [0; R]$ we set

$$h(r) = (R - r) \sup_{|z - z_0| \leq r} |f'(z)|,$$

then the function h is, obviously, continuous on the interval $[0; R]$, with $h(0) = aR$ and $h(R) = 0$. Set $r_0 = \max\{r: h(r) = aR\}$; clearly, such a number $r_0 < R$ does exist and $h(r) < aR$ for $r > r_0$. Also, let b be a point in the disk of radius r_0

centered at a at which $|f'(z)|$ attains the maximum value; by the maximum modulus principle, we may assume that b lies on the boundary of this disk, i.e., $|b - z_0| = r_0$; since $h(r_0) = aR$, we obtain $|f'(b)| = aR/(R - r_0)$. Now set $\rho = (R - r_0)/2$. Then the disk $U_{b,\rho}$ is contained in the disk of radius $(R + r_0)/2$; therefore, for $\zeta \in U_{b,\rho}$ we have

$$|f'(\zeta)| \leq \sup_{|z-z_0| \leq (R+r_0)/2} |f'(z)| = \frac{h(R + r_0/2)}{(R - r_0)/2} \leq \frac{aR}{(R - r_0)/2}, \tag{9.1}$$

since $h(r) < aR$ for $r > r_0$. Now for every $z \in U_{b,\rho}$ we have (below $[z_0; z]$ stands for the line segment between z_0 and z)

$$|f(z) - f(z_0)| = \left| \int_{[z_0;z]} f'(\zeta)\, d\zeta \right| \leq |z - z_0| \cdot \sup_{\zeta \in U_{b,\rho}} |f'(\zeta)|$$

$$\leq |z - z_0| \cdot \frac{aR}{(R - r_0)/2} \leq \frac{R - r_0}{2} \cdot \frac{aR}{(R - r_0)/2} = aR$$

(we have used inequality (9.1)). Applying now Proposition 9.16 to the restriction of f to the disk $U_{b,\rho}$, we see that $f(U_{b,\rho})$ contains the disk of radius

$$\frac{|f'(b)| \cdot \rho}{9 \sup\limits_{z \in U_{b,\rho}} |f(z) - f(z_0)|} \geq \frac{(aR/(R - r_0))^2 ((R - r_0)/2)^2}{9aR} \geq \frac{aR}{36}$$

centered at $f(b)$ (and, moreover, some neighborhood of b is mapped bijectively onto this disk). Since $aR/36 > aR/40$, we have proved Bloch's theorem for the case where f' extends by continuity to the closure of the disk of radius R centered at z_0. In the general case, consider the restriction of f to the disk of radius $R(1 - \varepsilon)$ centered at z_0, where $\varepsilon > 0$ is sufficiently small. Since f' extends by continuity to the closure of this smaller disk, we see that $f(U)$ contains a disk of radius $aR(1 - \varepsilon)/36$, which is greater than $aR/40$ for sufficiently small ε. $\square$

Exercises

9.1. (a) Let f be a function continuous on the disk $\bar{D} = \{z : |z| \leq 1\}$ and holomorphic on its interior. Can it map the unit circle $\bar{D} = \{z : |z| = 1\}$ to the set shown in Fig. 9.2 (a) by bold lines (the union of a circle and a line segment)?
   (b) The same question for Fig. 9.2 (b).

9.2. Show that every conformal automorphism of the complex plane has the form $z \mapsto az + b$, $a \neq 0$. (*Hint.* Let $f : \mathbb{C} \to \mathbb{C}$ be such an automorphism. To begin with, show that f has no essential singularity at infinity.)

9.3. Let f be a function holomorphic and one-to-one in a punctured neighborhood of a point a.

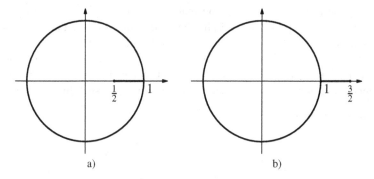

Fig. 9.2

(a) Show that f cannot have an essential singularity at a.

(b) If f has a pole at a, what can be said about the order of this pole?

9.4. Let f be a nonconstant holomorphic function on a connected open set $U \subset \mathbb{C}$. Show that if f has no zeros in U, then the function $z \mapsto |f(z)|$ has no local minima in U.

9.5. Let f be an entire function that is not constant. Set $M(r) = \max_{|z|=r} |f(z)|$. Show that $\lim_{r \to +\infty} M(r) = +\infty$.

9.6. Let $P_1, \ldots, P_n$ be points outside a disk centered at a point O. Show that on the boundary of the disk there is a point for which the product of the distances to $P_1, \ldots, P_n$ is greater than $OP_1 \cdot \ldots \cdot OP_n$, and also a point for which this product is less than $OP_1 \cdot \ldots \cdot OP_n$.

9.7. Show that there is no conformal isomorphism between the open sets $\mathbb{C} \setminus \{0, 1, 2\}$ and $\mathbb{C} \setminus \{0, 1, 2017\}$. (*Hint.* See Exercise 9.3.)

9.8. Let P be a polynomial of degree n in one variable with complex coefficients. Set $M(r) = \max_{|z|=r} |P(z)|$. Show that the function $r \mapsto M(r)/r^n$ decreases on $(0; +\infty)$.

9.9. Let $D = \{z : |z| < 1\}$ be the unit disk and $f : D \to D$ be a holomorphic map such that $f(0) = f'(0) = 0$. Show that $|f(z)| \le |z|^2$ for all $z \in D$.

9.10. Let $D = \{z : |z| < 1\}$ be the unit disk. Are the open sets $D \setminus \{0, 1/2\}$ and $D \setminus \{0, 1/3\}$ isomorphic?

9.11. Show that the punctured disk $\Delta^* = \{z : 0 < |z| < 1\}$ is not isomorphic to the annulus $A = \{z : a < |z| < b\}$ $(0 < a < b < \infty)$.

9.12. Let $H = \{z : \text{Im}(z) > 0\}$, and let $f : H \to H$ be a holomorphic function such that $f(i) = i$. Show that $|f'(i)| \le 1$. (*Hint.* Map H onto the unit disk and use the Schwarz lemma.)

9.13. Let $D = \{z : D \to D\}$ be the unit disk and $f : D \to D$ be a holomorphic map such that $f(1/2) = 2/3$. What is the greatest possible value of $|f'(1/2)|$? (*Hint.* Again, this problem can be reduced to the Schwarz lemma.)

9.14. Let $D = \{z : |z| < 1\}$ be the unit disk. Does there exist a holomorphic map $f : D \to D$ such that $f(0) = 1/2$ and $f(1/2) = 7/8$?

9.15. Let f be a nonconstant function that is continuous on the closed unit disk
$\bar{D} = \{z : |z| \le 1\}$ and holomorphic on its interior. Assume that $|f(z)| = 1$ as
soon as $|z| = 1$. Show that f has a zero in the interior of $\bar{D}$.

9.16. Let $D = \{z : |z| < 1\}$ be the unit disk and $f : D \to \mathbb{C}$ be a bounded holomorphic
function such that $f'(0) = 0$ and $f''(0) \ne 0$. Show that there exists a constant c
depending only on $a = |f''(0)|$ and $M = \sup_{z \in U} |f(z)|$ such that $f(U)$ contains
a disk of radius c.

Chapter 10
Conformal Maps. Part 1

10.1 Holomorphic Functions on Subsets of the Riemann Sphere

Getting down to a more serious study of conformal maps, we must learn to work with functions defined on open subsets of the Riemann sphere as confidently as we do with functions defined on open subsets of $\mathbb{C}$.

We begin with an official definition of what is meant by an open subset in $\overline{\mathbb{C}}$ (some indications have already been given in the previous chapters, but let us dot all the i's).

It is unnecessary to remind the reader what is meant by an ε-neighborhood of a point in $\mathbb{C}$; every such ε-neighborhood can (and will) be regarded also as a subset in $\overline{\mathbb{C}}$. Now let $R > 0$; then the *R-neighborhood of* ∞ is the set

$$\{z \in \mathbb{C} \colon |z| > R\} \cup \{\infty\} \subset \overline{\mathbb{C}}.$$

Removing the point ∞ from this set, we obtain a punctured neighborhood of infinity, familiar to us from Chap. 7.

(I wish to emphasize that while the ε-neighborhood of a point $a \in \mathbb{C}$ is defined as the set of points at distance less than ε from a, there is no talk of "distances to infinity," and we will not measure distances on the Riemann sphere.)

Further, let $X \subset \overline{\mathbb{C}}$; a point $p \in X$ is said to be an *interior* point of X if there exists a set $V \subset X$ that is an ε-neighborhood of p if $p \neq \infty$, or an R-neighborhood of p if $p = \infty$. A subset $X \subset \overline{\mathbb{C}}$ is said to be *open* if every point $p \in X$ is an interior point.

If $X \subset \mathbb{C} \subset \overline{\mathbb{C}}$, then this definition of an open set agrees with that given in Chap. 1.

Now we can define holomorphic functions on open subsets of the Riemann sphere.

Definition 10.1 Let $X \subset \overline{\mathbb{C}}$ be an open subset containing ∞. A function $f \colon X \to \mathbb{C}$ is said to be *holomorphic in a neighborhood of infinity* if the function $g \colon t \mapsto f(1/t)$ extends to a function holomorphic in a neighborhood of zero.

Clearly, this condition is equivalent to f having a removable singularity at infinity in the sense of Definition 7.21. In texts dealing with Riemann surfaces, one says that the function $t = 1/z$ is a "local coordinate" in a neighborhood of ∞ (see Chap. 13).

© Springer Nature Switzerland AG 2020
S. Lvovski, *Principles of Complex Analysis*, Moscow Lectures 6,
https://doi.org/10.1007/978-3-030-59365-0_10

Definition 10.2 Let $X \subset \overline{\mathbb{C}}$ be an open set. A function $f : X \to \mathbb{C}$ is said to be *holomorphic on X* if it is holomorphic in some (ε- or R-) neighborhood of every point $p \in X$.

Of course, if $X \not\ni \infty$, then this definition is equivalent to the usual one.

Now let us discuss what part of what we know about holomorphic functions extends to holomorphic functions in the sense of Definition 10.2. First of all, given a function f holomorphic in a neighborhood of infinity, we *do not* define its derivative at ∞. However, many other things allow for a natural generalization. For example, the whole theory of isolated singularities does, as discussed in Sec. 7.4. We only add that a function has a zero of order $k > 0$ at infinity if and only if its Laurent series expansion in a punctured neighborhood of infinity has the form

$$f(z) = \frac{c_k}{z^k} + \frac{c_{k+1}}{z^{k+1}} + \cdots$$

where $c_k \neq 0$: it suffices to look at what we obtain after the change of variable $z = 1/t$. The notion of a ramification point and Proposition 9.7 still make sense exactly in the same way.

Once we have defined what a pole is for functions on arbitrary open subsets in $\overline{\mathbb{C}}$, we obtain the definition of meromorphic functions for this case. Remarkably, every meromorphic function f can also be regarded as a holomorphic map with values in $\overline{\mathbb{C}}$. Namely, outside the poles, f is even a holomorphic map with values in $\mathbb{C} \subset \overline{\mathbb{C}}$. If now f has a pole at a point $a \in \overline{\mathbb{C}}$, then it is holomorphic in some punctured neighborhood of a and $\lim_{z \to a} f(z) = \infty$. Set $f(a) = \infty$. Considering instead of the function $w = f(z)$, in a neighborhood of a, the function $t = 1/w = 1/f(z)$, we see that the "local coordinate" at infinity, i.e., $1/w$, is a holomorphic function of z (since $1/f$ has a removable singularity at a with value 0). This means exactly that the meromorphic function redefined at a as described above is a holomorphic map with values in $\overline{\mathbb{C}}$.

In particular, every linear fractional transformation and, more generally, every rational function is a holomorphic map from $\overline{\mathbb{C}}$ to $\overline{\mathbb{C}}$.

If f has not a pole but an essential singularity at a, then we cannot redefine it as a holomorphic (or even continuous) map into $\overline{\mathbb{C}}$: this is forbidden by the Casorati–Weierstrass theorem.

For holomorphic functions on open subsets in $\overline{\mathbb{C}}$, we still have the open mapping theorem (Theorem 9.1) and the principle of analytic continuation (Proposition 5.21). In order to extend these results to subsets in $\overline{\mathbb{C}}$, it suffices to consider, for the open mapping theorem, the behavior of the function in a neighborhood of infinity, and for the principle of analytic continuation, the case where ∞ is an accumulation point of zeros. In both cases, the matter is resolved by the change of variable $z = 1/t$; the details are left to the reader as an exercise. Also, it is still true that a one-to-one holomorphic map is a conformal isomorphism (another exercise). The notion of analytic continuation along a path and the monodromy theorem can be extended to $\overline{\mathbb{C}}$ in the same way.

However, the case of the argument principle is not that easy. The reason is that there is no meaningful way to define the index of a closed path in $\overline{\mathbb{C}}$ with respect to a point: if γ is a closed path and $a \in \overline{\mathbb{C}}$ is a point not on γ, then γ can always be contracted to a point without touching a (see the plan of the proof in Exercise 10.4); since the index of a curve must remain unchanged under deformations of this curve, we see that the index of every curve with respect to every point is zero — a lot of good such an "invariant" will do us! When needing to apply the argument principle to subsets of $\overline{\mathbb{C}}$ containing ∞, we will use linear fractional transformations to reduce the problem to the case where everything lives inside $\mathbb{C}$. In particular, such an argument will be used below in Sec. 10.3.

We conclude this section by proving just one result about complex analysis on $\overline{\mathbb{C}}$ (see several more examples in the exercises).

Proposition 10.3 *Every function holomorphic on the whole Riemann sphere is a constant.*

Proof I will give a proof using as few topological notions as possible. So, let $f \colon \overline{\mathbb{C}} \to \mathbb{C}$ be a holomorphic function. Since f is continuous at ∞, it is bounded in some neighborhood of ∞ and, consequently, on some set of the form $\{z \in \mathbb{C} \colon |z| > R\}$. For every $R_1 > R$, the closed disk $\{z \in \mathbb{C} \colon |z| \le R_1\}$ is compact, hence the continuous function $z \mapsto |f(z)|$ is bounded on it. Therefore, the function f is bounded on all of $\mathbb{C}$ and, being holomorphic, is a constant on $\mathbb{C}$ by Liouville's theorem (Corollary 7.25); by continuity, $f(\infty)$ is equal to the same constant.   □

10.2 The Reflection Principle

The reflection principle is a geometric construction that makes it possible in some situations to predict the existence of an analytic continuation and, moreover, to describe it. We begin with notation.

In Sec. 1.6, we introduced the notion of reflection with respect to a circle or a line on the Riemann sphere (Definition 1.32). If $C \subset \overline{\mathbb{C}}$ is such a "generalized circle," then the reflection with respect to it will be denoted by $S_C \colon \overline{\mathbb{C}} \to \overline{\mathbb{C}}$.

Here is the first and weakest version of the reflection principle, which, actually, does not yet involve analytic continuation.

Proposition 10.4 *Let $C \subset \overline{\mathbb{C}}$ be a generalized circle (a circle or a line), and let $U \subset \overline{\mathbb{C}}$ be an open set such that $U \cap C = \varnothing$ and $U \cap S_C(U) = \varnothing$.*

If $f \colon U \to \overline{\mathbb{C}}$ is a holomorphic map, then the map $\tilde{f} \colon S_C(U) \to \overline{\mathbb{C}}$ given by

$$\tilde{f}(z) = S_C(f(S_C(z)))$$

is also holomorphic.

Proof First, consider the case where C is the real axis (plus the point ∞, of course). Then $\tilde{f}(z) = \overline{f(\overline{z})}$. If f is a map with values in $\mathbb{C}$ (i.e., has no poles), then the function $\tilde{f}$ is also holomorphic: if, for example, in a neighborhood of a point $a \in U$ we have a power series expansion

$$f(z) = c_0 + c_1(z-a) + c_2(z-a)^2 + \dots,$$

then in a neighborhood of the point $\bar{a} = S_C(a)$ we have

$$\tilde{f}(z) = \overline{f(\bar{z})} = \bar{c}_0 + \bar{c}_1(z-\bar{a}) + \bar{c}_2(z-\bar{a})^2 + \dots,$$

i.e., the function $\tilde{f}$ can be represented as a power series in a neighborhood of every point of $S_C(U)$ and, consequently, is holomorphic. For a function with poles, it suffices to observe that if a is a pole for f, then, by construction, $\lim_{z \to \bar{a}} \tilde{f}(z) = \infty$, so $\bar{a}$ is a pole for $\tilde{f}$.

In the general case, let $\varphi \colon \overline{\mathbb{C}} \to \overline{\mathbb{C}}$ be a linear fractional transformation mapping C to the real axis. Set $V = \varphi^{-1}(U)$ and $g = \varphi^{-1} \circ f \circ \varphi$; this is also a holomorphic map from V to $\overline{\mathbb{C}}$. We know that linear fractional transformations preserve the symmetry of points with respect to a generalized circle, hence $\varphi(S_C(z)) = \overline{\varphi(z)}$ for all z; besides, denoting by V' the open set obtained from V by reflecting in the real axis, we have $V \cap V' = \varnothing$, and neither V nor V' intersects the real axis. As proved above, the function $\tilde{g}(z) = \overline{g(\bar{z})}$ (with values in $\overline{\mathbb{C}}$) is holomorphic on V'. Denoting the map $z \mapsto \bar{z}$ by Conj, we can rewrite the identity $\varphi(S_C(z)) = \overline{\varphi(z)}$ in the form $\varphi \circ S_C \circ \varphi^{-1} = \text{Conj}$, whence

$$\varphi \circ \tilde{g} \circ \varphi^{-1} = \varphi \circ (\text{Conj} \circ g \circ \text{Conj}) \circ \varphi^{-1} = S_C \circ f \circ S_C = \tilde{f};$$

since the linear fractional transformations φ and φ^{-1} are holomorphic, the map $\tilde{f}$ is holomorphic too. $\qquad\square$

Now we state and prove the main version of the reflection principle. I will not reiterate the warnings about statements of theorems involving piecewise smooth curves bounding domains, see the remarks after the proof of Theorem 5.4.

Proposition 10.5 (reflection principle) *Let $U \subset \overline{\mathbb{C}}$ be an open connected set and $f \colon U \to \overline{\mathbb{C}}$ be a holomorphic map. Assume that the following conditions are satisfied:*

(1) the boundary of U contains a subset γ that is an open arc or an open interval of a generalized circle (a circle or a line) $C \subset \overline{\mathbb{C}}$;

(2) the map f extends to a continuous map $F \colon U \cup \gamma \to \overline{\mathbb{C}}$;

(3) the set $F(\gamma)$ is contained in a generalized circle $C' \subset \overline{\mathbb{C}}$;

(4) $U \cap S_C(U) = \varnothing$.

Define a map $\tilde{f} \colon U \cup \gamma \cup S_C(U) \to \overline{\mathbb{C}}$ as follows:

$$\tilde{f}(z) = \begin{cases} f(z), & z \in U, \\ F(z), & z \in \gamma, \\ S_{C'}(f(S_C(z))), & z \in S_C(U). \end{cases}$$

Then $\tilde{f}$ is holomorphic.

Proof The continuity of $\tilde{f}$ is clear by construction. If $a \in S_C(U)$, then $\tilde{f}$ is holo-morphic in a neighborhood of a by Proposition 10.4. It remains to verify that $\tilde{f}$ is holomorphic in a neighborhood of every point $a \in \gamma$.

First, let both C and C' coincide with the real axis, and let $\tilde{f}$ be finite on γ (i.e., $f(\gamma) \subset \mathbb{C} \subset \overline{\mathbb{C}}$). Then, by Morera's theorem, in order to prove that $\tilde{f}$ is holomorphic in a neighborhood of a point $a \in C$, it suffices to verify that a has a neighborhood $V \subset U \cup \gamma \cup S_C(U)$ such that $\int_{\partial\Delta} f(z)\, dz = 0$ for every triangle $\Delta \subset V$. If Δ lies entirely inside U or $S_C(U)$, this follows from the above and Cauchy's theorem. If Δ intersects both U and $S_C(U)$, then the real axis C cuts it into a triangle and a quadrangle (or into two triangles, one lying in $U \cup \gamma$ and the other lying in $S_C(U) \cup \gamma$). The integral over the boundary of each of these polygons (oriented positively) vanishes by Cauchy's theorem, since $\tilde{f}$ is continuous on γ; but the sum of these two integrals is equal to the integral over the boundary of Δ, so the latter integral vanishes, too. Thus, we have proved the proposition in the case where C and C' coincide with the real axis and the function f has no poles.

If $f(z)$ tends to infinity as z from U tends to a point $a \in \gamma$, then observe that the function $1/f$ extended by continuity to γ is bounded in a neighborhood of a and, besides, since $\tilde{f}(z) = \overline{f(\bar{z})}$, we have $(\widetilde{1/f})(z) = \overline{1/f(\bar{z})}$; applying to $1/f$ the same Morera's theorem argument, we see that the function $\widetilde{1/f}$ is holomorphic in a neighborhood of a and has a zero at a. Thus, the function $\tilde{f}$ has a pole at a and is a holomorphic map to $\overline{\mathbb{C}}$ in a neighborhood of a. So, the proposition is proved for the case where $C = C'$ is the real axis.

The general case can be reduced to this one by linear fractional transformations, as in the proof of Proposition 10.4. Namely, if ℓ is the real axis and $\varphi, \psi : \overline{\mathbb{C}} \to \overline{\mathbb{C}}$ are linear fractional transformations such that $\varphi(\ell) = C$, $\psi(C') = \ell$, then the map $g = \psi \circ f \circ \varphi$ satisfies the assumptions of the proposition for the case $C = C' = \ell$; for the same reasons as in the proof of Proposition 10.4, we have $\tilde{g} = \psi \circ \tilde{f} \circ \varphi$; by the case already proved, the map $\tilde{g}$ is holomorphic, hence so is $\tilde{f} = \psi^{-1} \circ \tilde{g} \circ \varphi^{-1}$. $\square$

Corollary 10.6 *Let $V \subset \overline{\mathbb{C}}$ be an open connected set and $f : V \to V'$ be a conformal isomorphism. Assume that*

(1) the boundary of V contains a subset γ that is an open arc or an open interval of a generalized circle $C \subset \overline{\mathbb{C}}$, and the boundary of V' contains a subset γ' that is an open arc or an open interval of a generalized circle $C' \subset \overline{\mathbb{C}}$;

(2) f extends to a continuous and bijective map $F : V \cup \gamma \to V' \cup \gamma'$;

(3) $V \cap S_C(V) = \varnothing$.

Then the map $\tilde{f}$ defined by

$$\tilde{f}(z) = \begin{cases} f(z), & z \in V, \\ F(z), & z \in \gamma, \\ S_{C'}(f(S_C(z))), & z \in S_C(V), \end{cases}$$

is a conformal isomorphism between the domains $V \cup \gamma \cup S_C(V)$ and $V' \cup \gamma' \cup S_{C'}(V')$.

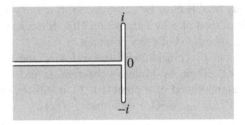

Fig. 10.1

Indeed, Proposition 10.5 implies all we need except the bijectivity of $\tilde{f}$, and the latter follows from the bijectivity of $\hat{f}$ which holds by assumption.

Note also that in most interesting cases, we need not spend much effort on checking condition (2): as we will see in Sec. 10.4, a conformal isomorphism between domains with "good" boundaries extends to a continuous bijection between their boundaries.

Corollary 10.6 helps in constructing conformal maps.

Example 10.7 Set $U = \mathbb{C} \setminus ((-\infty; 0] \cup [-i; i])$ (see Fig. 10.1). Let us find a conformal map from the domain U onto the upper half-plane.

In Example 3.9, we saw that the function $f: z \mapsto \sqrt{z^2 + 1}$ with $\sqrt{-1} = i$ is a conformal map from the open set $V = \{z : \operatorname{Im} z > 0\} \setminus [0; i]$ onto the upper half-plane $H = \{z : \operatorname{Im} z > 0\}$ (we have changed notation as compared with Chap. 3). It is easy to see that f extends by continuity to a bijection between $(0; +\infty)$ and $(1; +\infty)$. Setting $V' = H$, $C = C' = (-\infty; +\infty)$ (the real axis), $\gamma = (0; +\infty)$, $\gamma' = (1; +\infty)$ in Corollary 10.6, all assumptions of the corollary are satisfied. Since $V \cup \gamma \cup S_C(V) = U$ and $V' \cup \gamma' \cup S_{C'}(V') = \mathbb{C} \setminus (-\infty; 1]$, it shows that f extends to an isomorphism between the domain U and the set $\mathbb{C} \setminus (-\infty; 1]$. Since the function $w \mapsto i\sqrt{w - 1}$ with $\sqrt{1} = 1$ is a conformal map from $\mathbb{C} \setminus (-\infty; 1]$ onto the upper half-plane, we see, taking the composition, that the formula

$$z \mapsto i\sqrt{\sqrt{z^2 + 1} - 1}$$

defines a conformal map from the domain U onto the upper half-plane. Here, for the inner square root we choose the branch with $\sqrt{-1} = i$, and for the outer square root, the branch with $\sqrt{1} = 1$.

10.3 Mapping the Upper Half-Plane onto a Rectangle

This whole section is devoted to an analysis of one important example.

Let $a_1 < a_2 < a_3$ be real numbers; as usual, the upper half-plane will be denoted by $H = \{z : \operatorname{Im} z > 0\}$. Throughout the section, we fix a real number z_0 different from a_1, a_2, a_3 (in fact, the latter condition is not important), and define a function

$F: H \rightarrow \mathbb{C}$ by the formula

$$F(z) = \int_{z_0}^{z} \frac{dz}{\sqrt{(z - a_1)(z - a_2)(z - a_3)}}. \tag{10.1}$$

This formula is in need of clarification. First, by $\int_{z_0}^{z}$ we mean the integral over some path from z_0 to z in the upper half-plane H. Since H is convex, all such paths are homotopic to each other, so, by Cauchy's theorem, the integral does not depend on the choice of a path. Second, we must fix a branch of the square root. We choose it as follows: $\sqrt{(z - a_1)(z - a_2)(z - a_3)}$ is understood as the product of $\sqrt{z - a_j}$ over j from 1 to 3, and for each factor $\sqrt{z - a_j}$ we choose the branch such that $\sqrt{z - a_j} > 0$ for $z \in \mathbb{R}$ and $z > a_j$. With these refinements, the function F is, obviously, holomorphic on H: its derivative is the chosen branch of $\sqrt{\prod(z - a_j)}$.

Having defined the function F, let us study it. Denote by $\ell \subset \overline{\mathbb{C}}$ the generalized circle corresponding to the real axis. In other words, ℓ consists of all points of the real axis and the point ∞; the union $H \cup \ell$ is the closure of H in $\overline{\mathbb{C}}$; correspondingly, we denote it by $\overline{H}$.

Proposition 10.8 *The function F defined by* (10.1) *extends to a continuous function on $H \cup \ell$ with values in $\mathbb{C}$ (and not just in $\overline{\mathbb{C}}$).*

Proof To begin with, observe that the improper integral

$$\int_{-\infty}^{\infty} \frac{dx}{\sqrt{(x - a_1)(x - a_2)(x - a_3)}}$$

is convergent and, moreover, absolutely convergent; indeed, the integrand is $O(1/x^{3/2})$ as $|x| \rightarrow \infty$, and $O(|x - a_j|^{-1/2})$ as $x \rightarrow a_j$, so the integral converges by the comparison test. Therefore, the integral of $dx/\sqrt{(x - a_1)(x - a_2)(x - a_3)}$ over any interval of the real axis also converges; then it is clear that

$$\lim_{\substack{z \rightarrow x \\ z \in H}} F(z) = \int_{z_0}^{x} \frac{dx}{\sqrt{(x - a_1)(x - a_2)(x - a_3)}}$$

for every $x \in \mathbb{R}$, and we have obtained a continuous extension of F to the union of H and the real axis. Now we will show that there is also a continuous extension to the point ∞. Obviously, it suffices to establish that there exists a finite limit $\lim_{\substack{|z| \rightarrow \infty \\ \operatorname{Im} z \geq 0}} F(z)$.

In turn, by the Cauchy convergence criterion, for this it suffices to establish that for every $\varepsilon > 0$ there is $R > 0$ such that $|F(z_1) - F(z_2)| \leq \varepsilon$ whenever $|z_1|, |z_2| > R$. Indeed, let $|z_1| > R$, $|z_2| > R$, and $z_1, z_2 \in H$; we may assume without loss of generality that $|z_2| \geq |z_1|$. Note that the points z_1 and z_2 can be connected by a path in $\{z: |z| > R,\ \text{Im}\ z > 0\}$ consisting of two parts: a segment γ_1 of a ray starting at the origin and an arc γ_2 of the circle of radius $|z_2|$ (see Fig. 10.2).

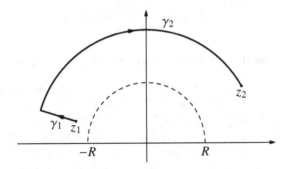

Fig. 10.2

Denoting by $\gamma_1 + \gamma_2$ the path "first γ_1, then γ_2," we have

$$|F(z_1) - F(z_2)| = \left| \int_{\gamma_1+\gamma_2} \frac{dz}{\sqrt{(z - a_1)(z - a_2)(z - a_3)}} \right|$$

$$\leq \left| \int_{\gamma_1} \frac{dz}{\sqrt{(z - a_1)(z - a_2)(z - a_3)}} \right| + \left| \int_{\gamma_2} \frac{dz}{\sqrt{(z - a_1)(z - a_2)(z - a_3)}} \right|. \quad (10.2)$$

Obviously, there exists a constant $C > 0$ such that

$$\left| \frac{1}{\sqrt{(z - a_1)(z - a_2)(z - a_3)}} \right| \leq \frac{C}{|z|^{3/2}} \quad (10.3)$$

for sufficiently large $|z|$. Now we estimate both terms in the right-hand side of (10.2). The first one (the absolute value of the integral over γ_1) can be estimated by Proposition 4.17: parametrizing γ_1 as $z = t z_1/|z_1|$, $|z_1| \leq t \leq |z_2|$, we have $|dz| = dt$ and, in view of (10.3),

$$\left| \int_{\gamma_1} \frac{dz}{\sqrt{(z - a_1)(z - a_2)(z - a_3)}} \right| \leq \int_{|z_1|}^{|z_2|} \frac{C}{t^{3/2}}\,dt \leq \frac{2C}{\sqrt{|z_1|}} \leq \frac{2C}{\sqrt{R}}.$$

As to the second term in (10.2), it suffices to estimate it using the coarser inequality from Proposition 4.15:

$$\left| \int_{\gamma_2} \frac{dz}{\sqrt{(z - a_1)(z - a_2)(z - a_3)}} \right| \leq \frac{C}{R^{3/2}} \cdot \text{length}(\gamma_2) \leq \frac{C \cdot \pi R}{R^{3/2}} = \frac{\pi C}{\sqrt{R}}.$$

Finally, we see that

$$|F(z_1) - F(z_2)| \leq \frac{2C}{\sqrt{R}} + \frac{\pi C}{\sqrt{R}}$$

for $|z_1| > R$, $|z_2| > R$; since the right-hand side tends to zero as $R \to +\infty$, the assumptions of the Cauchy convergence criterion are satisfied and there exists a finite limit $\lim\limits_{\substack{|z| \to \infty \\ \mathrm{Im}\, z \geq 0}} F(z)$. This is all we needed to prove the proposition. $\qquad\square$

From now on, we denote by F the continuous extension of the holomorphic function F to $\bar{H}$ whose existence has been established in Proposition 10.8; recall that $\bar{H} \ni \infty$.

Note that changing z_0 ("the lower limit of integration") in (10.1) alters the function F by an additive constant, which does not affect its qualitative behavior. We will assume that z_0 lies on the real axis to the left of a_1, the smallest root of the denominator in (10.1).

Proposition 10.9 *The function F takes the curve $\ell \subset \bar{\mathbb{C}}$ (the real axis with the point ∞ added) to the boundary of a rectangle with sides parallel to the coordinate axes. When ℓ is traversed once from left to right, the boundary of the rectangle is also traversed once in the positive direction.*

Proof Our choice of branches of the square root implies that for $t \in \mathbb{R}$ the integrand behaves as follows:

t	$(-\infty; a_1)$	$(a_1; a_2)$	$(a_2; a_3)$	$(a_3; +\infty)$
$\dfrac{1}{\sqrt{(t-a_1)(t-a_2)(t-a_3)}}$	$i\alpha,\ \alpha > 0$	< 0	$i\alpha,\ \alpha < 0$	> 0

Since $F(z_0) = 0$, the table shows that for $x \in (z_0; a_1]$ the value of the integral (i.e., of the function $F(x)$) is an imaginary number with positive imaginary part, and as x grows from z_0 to a_1, this imaginary part increases monotonously. Thus, the interval $[z_0; a_1]$ oriented from left to right is mapped onto a vertical interval traversed once in the bottom-up direction (and lying on the imaginary axis, but this does not really matter much).

However, what happens when x, continuing to move in the positive direction along the real axis, passes the point a_1? If $a_1 < t < a_2$, then, according to the table, the integrand $1/\sqrt{(t - a_1)(t - a_2)(t - a_3)}$ is real and negative in the upper half-plane, so for $x \in [a_1; a_2]$ we have

$$F(x) = F(a_1) - \int_{a_1}^{x} \frac{dt}{\sqrt{|(t - a_1)(t - a_2)(t - a_3)|}}.$$

It follows that as x grows from a_1 to a_2, the image of the real axis under the map F turns left at $90°$ and moves in the negative direction along a horizontal interval until it reaches $F(a_2)$.

Let's move on. If $a_2 < t < a_3$, then the above table states that the integrand is imaginary with negative imaginary part, so for $x \in [a_2; a_3]$ we have

$$F(x) = F(a_2) - i \cdot \int_{a_2}^{x} \frac{dt}{\sqrt{|(t - a_1)(t - a_2)(t - a_3)|}},$$

i.e., the image of this interval goes vertically again, but in the negative direction. The image of the real axis has once again turned left at $90°$.

In exactly the same way we can show that for $x > a_3$,

$$F(x) = F(a_3) + \int_{a_3}^{x} \frac{dt}{\sqrt{|(t - a_1)(t - a_2)(t - a_3)|}}.$$

In other words, our polygonal curve again turns left at $90°$ (and goes horizontally in the positive direction). Since the integral $\int_{-\infty}^{\infty} dt/\sqrt{(t - a_1)(t - a_2)(t - a_3)}$ is absolutely convergent, the image of the infinite interval $(a_3; +\infty)$ is a finite interval; the right endpoint of this interval is nothing else than $F(\infty)$. Recall that on the Riemann sphere one does not distinguish between $+\infty$ and $-\infty$, and that, by Proposition 10.8, $\lim_{t \to +\infty} F(t) = \lim_{t \to -\infty} F(t)$ (both these limits are equal to $F(\infty)$).

When the generalized circle ℓ is traversed in the direction in which we move, the point ∞ is succeeded by the interval $(-\infty; z_0]$. On this interval, the integrand is also imaginary with positive imaginary part, hence after $F(\infty)$ the image of the real axis again goes vertically up, since for $x \leq z_0 < a_1$ we have

$$F(x) = F(\infty) + i \cdot \int_{-\infty}^{x} \frac{dt}{\sqrt{|(t - a_1)(t - a_2)(t - a_3)|}}.$$

So, the image of the interval $(-\infty; a_1]$ is a vertical interval, and this vertical interval ends at the same point $F(z_0)$ from which we have started. Thus, we have shown that the image of the real axis with the point ∞ added is a rectangle with sides parallel to the coordinate axes, as required. See Fig. 10.3.   □

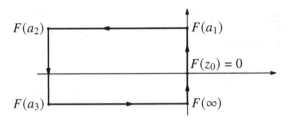

Fig. 10.3

Note that since the polygonal curve $F(\ell)$ has turned into a rectangle, it follows that the opposite sides of this rectangle are equal. The lengths of these sides are some integrals of $1/\sqrt{|(t - a_1)(t - a_2)(t - a_3)|}$ over (finite or infinite) intervals with endpoints at a_j and/or $\pm\infty$, so as a byproduct we have obtained two identities for definite integrals. Yet it is very easy to obtain them directly (see Exercise 10.8).

Proposition 10.10 *The function F defined above is a conformal map from the upper half-plane onto the interior of a rectangle with sides parallel to the coordinate axes. The map F extends to a continuous bijection between $\bar{H} = H \cup \ell$ and the closure of the rectangle; the points a_1, a_2, a_3, and ∞ are mapped to the vertices of the rectangle.*

Proof Consider a linear fractional transformation $\varphi\colon D \to H$ that sends the unit disk $D = \{z\colon |z| < 1\}$ onto the upper half-plane. This map extends by continuity to the boundaries, and Proposition 10.9 implies that $F \circ \varphi$ is a bijective map from the boundary of D traversed in the positive direction onto the boundary of a rectangle traversed in the positive direction. Now it follows from Proposition 8.11 that $F \circ \varphi\colon D \to F(\varphi(D))$ is a conformal map; hence, $F\colon H \to D$ is also a conformal map. We could not apply Proposition 8.11 right away, since H is not a bounded set with a piecewise smooth boundary. □

Note that the version of Proposition 8.11 we have used relies on the strengthened version of the argument principle (Proposition 8.7), and we cannot get away with the weaker theorem 8.6 fully proved in the course: the function F cannot be extended to a holomorphic function on an open set containing $\bar{H}$; otherwise, its derivative

$$F'(z) = \frac{1}{\sqrt{(z - a_1)(z - a_2)(z - a_3)}}$$

would also extend to a holomorphic function on this open set, while it cannot be extended to a neighborhood of any of the points a_j even as a continuous function. See Remark 8.8.

Now we introduce some notation. Denote the rectangle $F(H)$ by Π and its closure by $\bar{\Pi}$. Further, denote the length of the horizontal side of Π (parallel to the real axis) by u and the length of the vertical side by v. Let $G = F^{-1}\colon \Pi \to H$ be the inverse of F; it is also a conformal map.

Proposition 10.11 *The holomorphic map $G\colon \Pi \to H$ extends to a function meromorphic on the whole complex plane $\mathbb{C}$. This extended function G enjoys the following properties.*

(1) The function G is double periodic with periods $2u$ and $2iv$, that is, $G(z + 2mu + 2inv) = G(z)$ for any z and $m, n \in \mathbb{Z}$.

(2) If p is any vertex of the rectangle Π, then $G(p - z) = G(p + z)$ for all z.

(3) The function G has poles of order 2 at the points $F(\infty) + mu + nv$, for all $m, n \in \mathbb{Z}$, and no other poles.

(4) If p is a vertex of Π different from $F(\infty)$, then at all points $p + mu + inv$, $m, n \in \mathbb{Z}$, the function G is ramified with ramification index 2; at all the other points where G is holomorphic, it is not ramified.

Proof By Proposition 10.10, the function G extends to a continuous bijection from $\bar{\Pi}$ to $\bar{H} = H \cup \ell$. Now consider any side of the rectangle Π; let it lie on a line λ. Since this side is mapped onto an interval of the real axis, we can apply the reflection principle in the form of Corollary 10.6: extend G to a holomorphic function on $\Pi \cup S_\lambda(\Pi) \cup (\lambda \cap \bar{\Pi})$ (which will be also denoted by G) according to the rule $G(S_\lambda(z)) = \overline{G(z)}$, where S_λ is the reflection in the line λ. Note that the extended map G (regarded as a map to $\bar{\mathbb{C}}$) extends by continuity to the boundary of the rectangle $S_\lambda(\Pi)$, and that every side of this new rectangle is also mapped onto an interval of the real axis.

Further, consider any "unused" side of the rectangle Π or any side of the rectangle $S_\lambda(\Pi)$ except the one lying on λ, and apply the reflection principle to it; then G gets extended to another rectangle congruent to the original rectangle Π. Successively reflecting the rectangle in its sides, we can tile the entire plane (with no overlaps) by rectangles congruent to Π, so, iterating this construction, we can extend G to a function holomorphic on the whole plane except, possibly, the vertices of the rectangles, i.e., points of the form $p + mu + inv$ where p is a vertex of Π and $m, n \in \mathbb{Z}$. Now we will show that G extends to a function meromorphic on the whole plane. Indeed, by construction, G is a continuous map from $\mathbb{C}$ to $\bar{\mathbb{C}}$. If now a is a vertex of the tiling that is not mapped to ∞ by G, then G is bounded in a neighborhood of a, so, by Riemann's removable singularity theorem, G extends to a as a holomorphic function; if, on the other hand, $G(a) = \infty$, then $\lim_{z \to a} |G(z)| = +\infty$, so G has a pole at a.

Lemma 10.12 *Let λ be a line containing any side of any rectangle of the tiling. Then $G(S_\lambda(z)) = \overline{G(z)}$ for every z.*   □

Proof of the lemma If $z \in \lambda$, then, by construction, $F(z)$ lies on the real axis and thus coincides with its conjugate. If $z \notin \lambda$, consider the restriction of G to the half-plane determined by λ that contains z and apply the reflection principle to this restriction and the line λ. Then G gets extended to a holomorphic function $\tilde{G}$ in the opposite half-plane, and $\tilde{G}(S_\lambda(z)) = \overline{G(z)}$. Since $\tilde{G}$ coincides with G in the original half-plane and the set $\mathbb{C} \setminus \{a + mu + inv\}$ is connected, the principle of analytic continuation shows that $\tilde{G} = G$ everywhere, which proves the required identity.   □

Now we can verify properties (1)–(4).

Let λ and μ be the lines containing the two vertical sides of the rectangle Π. Clearly, $S_\mu(S_\lambda(z)) = z + 2u$ for every z (the composition of reflections in parallel lines is a translation). Therefore,

$$G(z + 2u) = G(S_\mu(S_\lambda(z))) = \overline{G(S_\lambda(z))} = \overline{(\overline{G(z)})} = G(z).$$

In a similar way we can verify that $2iv$ is also a period of G. This proves property (1).

Further, if p is any vertex of Π (and, moreover, any vertex of any rectangle of the tiling) and $z \in \mathbb{C}$, then the points $p + z$ and $p - z$ are reflections of each other in p; denoting by λ and μ the lines containing the sides of Π that contain p, we have $S_\mu(S_\lambda(p + z)) = p - z$ (the composition of reflections in perpendicular lines is a point reflection). Therefore,

$$G(p - z) = G(S_\mu(S_\lambda(p + z))) = \overline{G(S_\lambda(p + z))} = \overline{(\overline{G(p + z)})} = G(p + z).$$

This proves property (2).

The function G is periodic with periods $2u$ and $2iv$, so, to find all its poles, it suffices to find all its poles in some rectangle with sides of length $2u$ and $2v$ parallel to the coordinate axes. Consider, say, the large rectangle shown in Fig. 10.4. By Lemma 10.12 (or, if you prefer, by construction), it contains only one pole, at the point $F(\infty)$; therefore, all the other poles are exactly at the points of the form $F(\infty) + 2mu + 2inv$ where $m, n \in \mathbb{Z}$. To prove property (3), it remains to verify that these poles are of order 2. Again, since G is periodic, it suffices to establish this for the pole at $F(\infty)$. Note that G has a pole of order 2 at $F(\infty)$ if and only if the function $z \mapsto 1/G(z)$ has a zero of order 2 at $F(\infty)$, or, which is equivalent (since this function has a zero at $F(\infty)$ in any case), the function $z \mapsto 1/G(z)$ has ramification index 2 at $F(\infty)$. To see this, look at Fig. 10.4 once again. Each grey rectangle is conformally mapped by G onto the upper half-plane, each white rectangle is conformally mapped onto the lower half-plane, and the images of points that are reflections of each other in $F(\infty)$ coincide. Since taking the inverse of a complex number is a one-to-one operation, the function $z \mapsto 1/G(z)$ maps each grey rectangle conformally onto the lower half-plane, and each white rectangle, onto the upper half-plane. Therefore, the function $z \mapsto 1/G(z)$ in a neighborhood of $z = F(\infty)$ satisfies the assumptions of Proposition 9.7 for $k = 2$, and thus has ramification index 2 at this point; consequently, the order of the pole of the function G at the point $F(\infty)$ is equal to 2, and this completes the proof of property (3).

To prove property (4), first observe that if a point z does not lie on a side or coincide with a vertex of any rectangle of the tiling, then it lies inside one of such rectangles; but each of them is conformally mapped by G onto the upper or lower half-plane, so the function G is not ramified at this point. If z lies on a side of some rectangle but does not coincide with a vertex, then the rectangle to one side of z is conformally mapped by G onto the upper half-plane and the rectangle to the other side of z is conformally mapped onto the lower half-plane; hence, in some neighborhood of z the function G is one-to-one onto its image, so, again, it is not ramified at z. Finally, if z is a vertex of some rectangle, then the same argument involving Proposition 9.7 shows that the ramification index of f at z is equal to 2. This completes the proof of property (4) and the whole proposition. $\qquad\square$

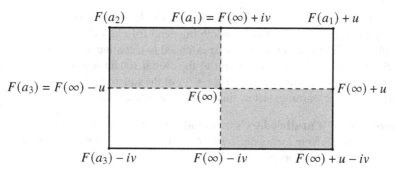

Fig. 10.4

Meromorphic functions on $\mathbb{C}$ that have two periods (linearly independent over $\mathbb{R}$) are called *elliptic functions*. We will meet them (in particular, the function G) again.

10.4 Carathéodory's Theorem

The main result of this section says that in many cases a conformal map between two bounded domains in $\mathbb{C}$ extends to a continuous one-to-one map between their closures. In particular, we obtain a continuous one-to-one map between the boundaries of these domains, so this result is sometimes called the "boundary correspondence" under conformal maps.

For quite arbitrary domains, the boundary correspondence cannot hold for obvious reasons. For example, if $U = \{z: |z| < 1, \operatorname{Im} z > 0\}$ is a semi-disk and V is the unit disk with the interval $[0; 1)$ of the real axis removed, then the function $f: z \mapsto z^2$ maps U conformally onto V, but does not provide a correspondence between the boundaries. Indeed, a continuous extension of f to $\bar{U}$, obviously, exists and is unique (and is given by the same formula $z \mapsto z^2$), but at the boundary it is no longer a bijection, since it "glues together" x and $-x$ for $x \in (0; 1) \subset \mathbb{R}$. The inverse map $f^{-1}: V \to U$ is also conformal, but it cannot even be extended to the boundary.

Hence, we must impose some restrictions on the domains to be mapped. We will prove the boundary correspondence under the assumption that both domains are bounded by piecewise smooth curves. This condition can be significantly weakened, and it is also possible to develop a formalism allowing one to extend Carathéodory's theorem to open sets of more complicated structure, but we steer clear of all these sophisticated issues.

Further, even if the boundary is piecewise smooth, the condition that the map is conformal cannot be dropped. Let, for example,

$$K = \{z \in \mathbb{C}: 0 < \operatorname{Re} z < 1, 0 < \operatorname{Im} z < 1\}$$

be a square and

$$\Delta = \{z \in \mathbb{C}: 0 < \operatorname{Re} z < 1, 0 < \operatorname{Im} z < \operatorname{Re} z\}$$

be a triangle. The map $f: K \to \Delta$ given by the formula $x+iy \mapsto x+ixy$ is, of course, not holomorphic, but still it is a smooth bijection (of class C^∞) whose inverse is also of class C^∞. However, it cannot be extended to a continuous bijection between the closures: its continuous extension to the closed square $\bar{K}$ (given by the same formula) maps the whole side $\{it: 0 \le t \le 1\}$ of the square to the point 0.

After all these warnings, let us finally state a positive result.

Theorem 10.13 (Carathéodory's theorem) *Let $U, V \subset \mathbb{C}$ be bounded domains with piecewise smooth boundaries, and let $f: U \to V$ be a conformal isomorphism. Then f extends to a continuous bijection $\tilde{f}: \bar{U} \to \bar{V}$.*

The proof of Theorem 10.13 is rather lengthy. We begin with three lemmas, two easy and one more involved.

Lemma 10.14 *Let $U \subset \mathbb{C}$, and let $f\colon U \to \mathbb{C}$ be a continuous map. If f is uniformly continuous on U, then it extends uniquely to a continuous map $\tilde{f}\colon \bar{U} \to \mathbb{C}$.*

Sketch of the proof The uniqueness follows from the fact that if $a \in \bar{U} \setminus U$ and $a = \lim z_n$ with $z_n \in U$, then $\tilde{f}(a)$ is necessarily equal to $\lim \tilde{f}(z_n) = \lim f(z_n)$. If f is uniformly continuous, then it can be extended to $\bar{U}$ as follows. If a and $\{z_n\}$ are the same as above, then the sequence $\{z_n\}$, being convergent, is Cauchy; then it follows from the uniform continuity that the sequence $\{f(z_n)\}$ is also Cauchy, and thus converges. Now set $\tilde{f}(a) = \lim f(z_n)$.

All necessary verifications are left to the reader. $\square$

Lemma 10.15 *Let $[a; b] \subset \mathbb{R}$ be an interval and $h\colon [a; b] \to [0; +\infty)$ be a continuous function. Then*

$$\int_a^b (h(x))^2\, dx \geq \frac{1}{b-a}\left(\int_a^b h(x)\, dx\right)^2.$$

Proof This inequality can be obtained by squaring the continuous version of the inequality between the quadratic and arithmetic means:

$$\sqrt{\frac{\int_a^b (h(x))^2\, dx}{b-a}} \geq \frac{\int_a^b h(x)\, dx}{b-a}.$$

One can also apply the Cauchy–Schwarz inequality to the function h and the function identically equal to 1. $\square$

The next lemma is intuitively obvious, but its rigorous proof requires some effort.

Lemma 10.16 *Let $U \subset \mathbb{C}$ be an open set with a piecewise smooth boundary, and let $a \in \bar{U} \setminus U$ be a point of the boundary. Then there exists $\varepsilon > 0$ such that for every r, $0 < r < \varepsilon$, the following conditions hold.*
 (1) *The intersection of the circle $\{z\colon |z - a| = r\}$ with U is an arc.*
 (2) *The intersection $\{z\colon |z - a| < r\} \cap U$ is connected.*

Proof First we prove part (1). The intersection of the boundary of U and a disk of sufficiently small radius centered at a is the union of two arcs of C^1-smooth curves issuing from a. Here C^1 means that the tangent to the curve depends continuously on the point. If a is a smooth point of the boundary, then the tangents at a to both arcs coincide.

It suffices to show that for all sufficiently small r the circle $\gamma_{a,r} = \{z\colon |z - a| = r\}$ intersects each of these arcs at one point and this intersection is transversal (i.e., the tangents to the two intersecting curves at the intersection points are different), see Fig. 10.5.

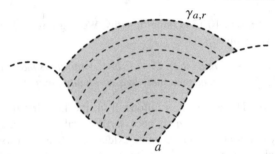

Fig. 10.5 The grey set is the intersection of the domain U and the disk of radius r centered at a. The dotted lines represent the intersections of U with circles of smaller radius.

Let us prove this. We may assume without loss of generality that a is the origin and the arc beginning at a is the graph of a continuously differentiable function (of a real variable) $y = h(x)$, $0 \le x \le x_0$, for which $h(0) = h'(0) = 0$. Take a small positive angle θ (say, $\theta = \pi/10$ is quite enough). Then, since $h'(0) = 0$ and the function $x \mapsto h'(x)$ is continuous, there exists $\varepsilon > 0$ such that for $0 < x < \varepsilon$ the following conditions hold:

(a) the graph of the function $y = h(x)$ lies inside the angle formed by two rays issuing from the origin and making the angles $\pm\theta$ with the abscissa axis;

(b) the angle made by the tangent to the graph of h at any point with abscissa $x \in (0; \varepsilon)$ and the abscissa axis lies in the interval $[-\theta; \theta]$.

We will show that for every $r \in (0; \varepsilon)$ the circle of radius r centered at the origin intersects the graph of h at exactly one point and this intersection is transversal. The existence of at least one intersection point is clear from condition (a). It is also clear from the same condition that the abscissa of every intersection point lies in the interval $[r \cos \theta; r]$; since the angle made by the tangent to the circle at a point with such an abscissa and the abscissa axis lies in the interval $[\pi/2 - \theta; \pi/2 + \theta]$, condition (b) implies that the tangent to the circle at such an intersection point cannot coincide with the tangent to the graph, so every intersection of the circle and the graph is automatically transversal. Finally, we prove that an intersection point is unique. If there are more than one such points, denote two of them by P_1 and P_2; by Rolle's theorem, the graph of h contains a point at which the tangent is parallel to the line $P_1 P_2$. Since the points P_1 and P_2 lie at the arc shown in Fig. 10.6, the angle made by the line $P_1 P_2$ and the abscissa axis lies in the interval $[\pi/2 - \theta; \pi/2 + \theta]$; as we have already observed, the tangent to our graph cannot make such an angle with the abscissa axis, so part (1) is proved.

Now we prove part (2). For this, assume that $0 < r < \varepsilon$, where ε is the same as in the proof of part (1). We will show that the set $W = \{z: |z-a| < r\} \cap U$ is connected. Arguing by contradiction, let $S = W_1 \cup W_2$ where W_1 and W_2 are open, nonempty, and disjoint. By part (1), for every ρ with $0 < \rho < \varepsilon$, the set $\gamma_{a,\rho} = \{z: |z-a| = \rho\} \cap U$ is an open arc of a circle; since such an arc is connected, $\gamma_{a,\rho}$ is entirely contained in W_1 or W_2. Set $I_k = \{\rho \in (0; \varepsilon): \gamma_{a,\rho} \subset W_k\}$ for $k = 1, 2$. By the above, we have $(0; \varepsilon) = I_1 \cup I_2$, and also $I_1 \cap I_2 = \varnothing$. Note that the sets I_1 and I_2 are open in $\mathbb{R}$. Indeed, if, say, $\rho \in I_1$, then $z \in W_1$ for every $z \in \gamma_{a,\rho}$. Since the set $W_1 \subset \mathbb{C}$

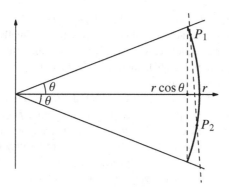

Fig. 10.6

is open, some δ-neigborhood of z (denote it by U_δ) is contained in W_1. Clearly, $U_\delta \cap \gamma_{a,\rho'} \neq \varnothing$ for $|\rho' - \rho| < \delta$, hence for such ρ we have $\gamma_{a,\rho'} \subset W_1$, so $\rho' \in I_1$, which proves that I_1 is open. Since intervals of the real line are connected, it follows that I_1 or I_2 coincides with $(0; \varepsilon)$, i.e, W_1 or W_2 coincides with W. The lemma is proved. $\square$

Proof of Theorem 10.13 Note that it suffices to prove that every conformal isomorphism between bounded domains with piecewise smooth boundaries is uniformly continuous. Indeed, if this is proved and if $f: U \to V$ is such an isomorphism, then, by Lemma 10.14, it extends uniquely to a continuous map $\tilde{f}: \bar{U} \to \bar{V}$, while the inverse isomorphism $g: V \to U$ extends uniquely to a continuous map $\tilde{g}: \bar{V} \to \bar{U}$; moreover, by the uniqueness of such extensions, the compositions $\tilde{f} \circ \tilde{g}$ and $\tilde{g} \circ \tilde{f}$, which are extensions of the identity maps onto V and U, are identity maps.

To prove that the conformal isomorphism $f: U \to V$ is uniformly continuous, we argue by contradiction. Namely, if it is not, then in U there exist sequences of points $\{z_n'\}$ and $\{z_n''\}$ such that $\lim_{n\to\infty} |z_n' - z_n''| = 0$ but $|f(z_n') - f(z_n'')| \geq c$ for all n and some $c > 0$. Since our domains are bounded, by extracting subsequences we can ensure that $\{z_n'\}$ and $\{z_n''\}$ converge (to the same point, since the distances between z_n' and z_n'' tend to zero). Then, extracting subsequences once again, we can ensure that the sequences $\{f(z_n')\}$ and $\{f(z_n'')\}$ also converge (to points at distance at least c from each other). Since both f and its inverse $g: V \to U$ are continuous bijections, the common limit of the sequences $\{z_n'\}$ and $\{z_n''\}$ lies on the boundary of U, while the limits of the sequences $\{f(z_n')\}$ and $\{f(z_n'')\}$ lie on the boundary of V.

Set $\lim_{n\to\infty} z_n' = \lim_{n\to\infty} z_n'' = a$, $\lim_{n\to\infty} f(z_n') = b'$, $\lim_{n\to\infty} f(z_n'') = b''$, and for every $z \in \mathbb{C}$ denote the ε-neighborhood of z by $U_\varepsilon(z)$. Applying Lemma 10.16 to the point a on the boundary of U, we can find ε, ε', and ε'' such that the neighborhoods $U_\varepsilon(a)$, $U_{\varepsilon'}(b')$, and $U_{\varepsilon''}(b'')$ have the properties stated in this lemma. Taking, if necessary, smaller ε' and ε'', we can ensure that

$$\inf_{\substack{w' \in U_{\varepsilon'}(b') \\ w'' \in U_{\varepsilon''}(b'')}} |f(w') - f(w'')| = \delta > 0 \tag{10.4}$$

and $g(V \cup U_{\varepsilon'}(b')) \subset U \cap U_{\varepsilon}(a)$, $g(V \cup U_{\varepsilon''}(b'')) \subset U \cap U_{\varepsilon}(a)$ (recall that $g: V \to U$ is the conformal automorphism inverse to f). Finally, removing finitely many elements from the sequences $\{z_n'\}$ and $\{z_n''\}$, we can ensure that all z_n' and z_n'' lie in $U_{\varepsilon}(a)$, all $f(z_n')$ lie in $U_{\varepsilon'}(b')$, and all $f(z_n'')$ lie in $U_{\varepsilon''}(b'')$; below we assume that these conditions are satisfied.

Since the open set $U_{\varepsilon'}(b') \cap V$ is connected, there exists a continuous path $\gamma': [0; +\infty) \to U_{\varepsilon'}(b') \cap V$ passing through all the points $f(z_n')$. Indeed, let γ' be a continuous path that connects $f(z_1')$ and $f(z_2')$ on the segment $t \in [0; 1]$, connects $f(z_2')$ and $f(z_3')$ on the segment $t \in [1; 2]$, etc. In a similar way we construct a path $\gamma'': [0; +\infty) \to U_{\varepsilon''}(b'') \cap V$ passing through all the points $f(z_j'')$. Set $\mu' = g \circ \gamma'$, $\mu'' = g \circ \gamma''$. The path μ' (respectively, μ'') lies in $U_{\varepsilon}(a) \cap U$ and passes through all the points z_k' (respectively, z_k'').

Now observe that since the Jacobian of f at a point z is equal to $|f'(z)|^2$ (see Exercise 2.12), we have

$$S(f(U_{\varepsilon}(a) \cap U)) = \iint_{U_{\varepsilon}(a) \cap U} |f'(x + iy)|^2 \, dx \, dy, \qquad (10.5)$$

where S stands for the area. Recall that, by our choice of ε, for every $r \in (0; \varepsilon)$ the intersection $\{z: |z - a| = r\} \cap U$ is an arc of a circle; denote the arguments of the endpoints of this arc by $\theta_1(r)$ and $\theta_2(r)$. Now we can rewrite the integral in the right-hand side of (10.5) in polar coordinates (r, φ):

$$S(f(U_{\varepsilon}(a) \cap U)) = \int_0^{\varepsilon} dr \int_{\theta_1(r)}^{\theta_2(r)} |f'(z)|^2 r \, d\varphi, \qquad (10.6)$$

where $z = re^{i\varphi}$.

Since the path μ' passes through all the points z_n', the path μ'' passes through all the points z_n'', and the sequences $\{z_n'\}$ and $\{z_n''\}$ converge to a, it follows that for all sufficiently small r (say, for all $r \leq r_0$) the circle of radius r centered at a intersects both the path μ' and the path μ''. In other words, on the arc $\gamma_{a,r} = \{z: |z-a| = r\} \cap U$ there is a point $re^{i\theta'(r)}$ that is sent by f to a point on the path γ', and a point $re^{i\theta''(r)}$ that is sent by f to a point on the path γ''. Now we estimate the "inner" integral in the right-hand side of (10.6) from below using Lemma 10.15:

$$\int_{\theta_1(r)}^{\theta_2(r)} |f'(z)|^2 r \, d\varphi \geq \int_{\theta'(r)}^{\theta''(r)} |f'(z)|^2 r \, d\varphi$$

$$\geq \frac{1}{\theta'(r) - \theta''(r)} \left(\int_{\theta'(r)}^{\theta''(r)} |f'(z)| r \, d\varphi \right)^2 \geq \frac{1}{2\pi r} \left(\int_{\theta'(r)}^{\theta''(r)} |f'(z)| r \, d\varphi \right)^2 \qquad (10.7)$$

(for $r \leq r_0$).

On the arc $\gamma_{a,r}$ we have $r\,d\varphi = |dz|$, hence the integral $\int_{\theta'(r)}^{\theta''(r)} |f'(z)|r\,d\varphi$ is equal to the length of the f-image of the part of this arc between the points $re^{i\theta'(r)}$ and $re^{i\theta''(r)}$. This image is a curve in V connecting the points $f(re^{i\theta'(r)})$ and $f(re^{i\theta''(r)})$. Since the first of these points lies in $U_{\varepsilon'}(b')$ and the second one lies in $U_{\varepsilon''}(b'')$ (recall that these points lie on the arcs γ' and γ'', respectively), the distance between $f(re^{i\theta'(r)})$ and $f(re^{i\theta''(r)})$ is at least δ, where δ is the constant in the right-hand side of (10.4). A fortiori, the length of every curve connecting the points $f(re^{i\theta'(r)})$ and $f(re^{i\theta''(r)})$ is at least δ. Thus, the right-hand side of (10.7) is at least $\delta^2/2\pi r$. Substituting this estimate into the right-hand side of (10.6), we obtain the following:

$$S(f(U_\varepsilon(a) \cap U)) \geq \int_0^{r_0} dr \int_{\theta_1(r)}^{\theta_2(r)} |f'(z)|^2 r\,d\varphi \geq \int_0^{r_0} \frac{\delta^2}{2\pi r}\,dr.$$

The integral in the right-hand side of this inequality is equal to $+\infty$, so the area of the open set $f(U_\varepsilon(a) \cap U)$ is infinite. However, this is incompatible with the assumptions of the theorem, since the set $f(U_\varepsilon(a) \cap U) \subset V$ is bounded, because the set V is bounded by assumption. The obtained contradiction finally proves Carathéodory's theorem.   □

The reader who has made it this far is encouraged to return to the example at the beginning of this section (a conformal isomorphism between a semi-disk and a disk with a cut) and to think through all the reasons why our proof of Theorem 10.13 does not work for this example.

The trick used in the above proof (a lower bound on the area of the image in terms of lower bounds on the lengths of arcs which relies on the inequality from Lemma 10.15) is called the *length-area principle*.

10.5 Quasiconformal Maps

Having put a lot of effort into the proof of Carathéodory's theorem, we would like to figure out to what extent the fairly strong requirement that f be holomorphic is essential. We already know that it cannot be dropped completely; now we will see that it can be considerably weakened.

Let $U \subset \mathbb{C}$ be an open set and $f: U \to \mathbb{C}$ be a function of class C^1. Assume additionally that the Jacobian of f is positive everywhere (in such cases, f is said to be *orientation-preserving*).

For every point $a \in U$, we have the real derivative $f'(a): \mathbb{C} \to \mathbb{C}$, which is a real-linear map. The map $f'(a)$ sends circles to ellipses.

By the *stretch factor* of an ellipse we mean the ratio of its semi-major axis to the semi-minor axis.

Definition 10.17 Let $U \subset \mathbb{C}$ be an open set and $f: U \to \mathbb{C}$ be a map of class C^1 with an everywhere positive Jacobian. The map f is said to be *quasiconformal* if there exists a constant $C > 0$ such that the stretch factors of all ellipses that are the images of circles under the maps $f'(a): \mathbb{C} \to \mathbb{C}$ for all $a \in U$ do not exceed C.

If all stretch factors from this definition are equal to 1, then all maps $f'(a)$ send circles to circles and thus are compositions of rotations and dilations; this means that f is holomorphic. The stretch factor of f measures the deviation of f from being holomorphic (at a given point). The quasiconformity of f means that it is allowed to be nonholomorphic, but its deviation from being holomorphic cannot be indefinitely large.

If a real-linear map $L: \mathbb{C} \to \mathbb{C}$ sends circles to ellipses with stretch factor C, then, obviously, the inverse map L^{-1} sends circles to ellipses with the same stretch factor. Hence, if $f: U \to V$ is a quasiconformal diffeomorphism between two open subsets in $\mathbb{C}$, then the inverse diffeomorphism $f^{-1}: V \to U$ is also quasiconformal.

Definition 10.17 can be stated in formulas. For this, we first calculate the stretch factor.

Proposition 10.18 *Let $U \subset \mathbb{C}$ be an open set and $f: U \to \mathbb{C}$ be a map of class C^1 with an everywhere positive Jacobian. If $a \in \mathbb{C}$, then the stretch factor of the ellipses that are the images of circles under the map $f'(a): \mathbb{C} \to \mathbb{C}$ is equal to*

$$\frac{1 + \left|\frac{(\partial f/\partial \bar{z})(a)}{(\partial f/\partial z)(a)}\right|}{1 - \left|\frac{(\partial f/\partial \bar{z})(a)}{(\partial f/\partial z)(a)}\right|}.$$

Proof This is a fact from linear algebra. Namely, as we know, the map $f'(a)$ is given by the formula

$$h \mapsto Ah + B\bar{h}, \quad A = \frac{\partial f}{\partial z}(a), \; B = \frac{\partial f}{\partial \bar{z}}(a).$$

The image of the circle $\{e^{i\varphi}: \varphi \in \mathbb{R}\}$ is an ellipse centered at the origin, and its semi-major and semi-minor axes are the points of this ellipse (i.e., the points of the form $Ae^{i\varphi} + Be^{-i\varphi}$) with the largest and smallest absolute values. It is more convenient to consider the squared absolute value. Since

$$|Ae^{i\varphi} + Be^{-i\varphi}|^2 = (Ae^{i\varphi} + Be^{-i\varphi})(\bar{A}e^{-i\varphi} + \bar{B}e^{i\varphi})$$
$$= |A|^2 + |B|^2 + A\bar{B}e^{2i\varphi} + \overline{A\bar{B}e^{2i\varphi}} = |A|^2 + |B|^2 + 2\operatorname{Re}(A\bar{B}e^{2i\varphi}),$$

the maximum and minimum of the left-hand side (for fixed A and B and varying $\varphi \in \mathbb{R}$) are equal to $|A|^2 + |B|^2 \pm 2|A\bar{B}| = |A|^2 + |B|^2 \pm 2|A| \cdot |B|$. Since the Jacobian of f at a is equal to $|A|^2 - |B|^2$ (see Exercise 2.12), we have $|A| > |B|$; taking the square root, we see that the semi-major and semi-minor axes are equal to $|A| \pm |B|$, and thus the stretch factor is equal to

$$\frac{A+B}{A-B} = \frac{1+\frac{B}{A}}{1-\frac{B}{A}},$$

as required. $\qquad\qquad\qquad\qquad\qquad\qquad\qquad\qquad\qquad\qquad\qquad\qquad\square$

Now we immediately obtain an analytic version of the definition of quasiconformity.

Proposition 10.19 *Let $U \subset \mathbb{C}$ be an open set and $f : U \to \mathbb{C}$ be a map of class C^1 with an everywhere positive Jacobian. The map f is quasiconformal if and only if there exists a constant $\mu < 1$ such that $|(\partial f/\partial \bar{z})/(\partial f/\partial z)| \le \mu$ on U.*

Proof Denote the quotient $|(\partial f/\partial \bar{z})/(\partial f/\partial z)|$ by $\beta < 1$; our claim immediately follows from Proposition 10.18 in view of the relation

$$\frac{1+\beta}{1-\beta} \le C \Leftrightarrow \beta \le \frac{C-1}{C+1}.$$

The promised generalization of Carathéodory's theorem reads as follows.

Theorem 10.20 *Let $U, V \subset \mathbb{C}$ be bounded domains with piecewise linear boundaries and $f : U \to V$ be a quasiconformal diffeomorphism of class C^1. Then f extends to a continuous bijection $\tilde{f} : \bar{U} \to \bar{V}$.*

Proof We proceed similarly to the proof of Theorem 10.13 (Carathéodory's theorem for conformal maps), but with slightly more careful estimates.

Up to formula (10.5), both proofs coincide word for word. Instead of (10.5), using the formula for the Jacobian from Exercise 2.12b, we can write the following (keeping the notation from the proof of the "conformal" version of the theorem):

$$S(f(U_\varepsilon(a) \cap U)) = \iint_{U_\varepsilon(a) \cap U} \left(\left|\frac{\partial f}{\partial z}\right|^2 - \left|\frac{\partial f}{\partial \bar{z}}\right|^2 \right) dx\, dy. \qquad (10.8)$$

By assumption, $|\partial f/\partial \bar{z}| \le \mu |\partial f/\partial z|$ for all $z \in U$ for some constant $\mu \in [0; 1)$. Hence,

$$\left|\frac{\partial f}{\partial z}\right|^2 - \left|\frac{\partial f}{\partial \bar{z}}\right|^2 \ge (1 - \mu^2) \left|\frac{\partial f}{\partial z}\right|^2$$

and (10.8) implies the following inequality:

$$S(f(U_\varepsilon(a) \cap U)) \ge (1 - \mu^2) \iint_{U_\varepsilon(a) \cap U} \left|\frac{\partial f}{\partial z}\right|^2 dx\, dy.$$

Rewriting, as before, the integral in the right-hand side of this inequality in polar coordinates and using Lemma 10.15 to estimate from below the integrals of $|\partial f/\partial z|^2$ over arcs, we obtain the following:

$$S(f(U_\varepsilon(a) \cap U)) \ge (1 - \mu^2) \int_0^\varepsilon \frac{dr}{2\pi r} \left(\int_{\theta_1(r)}^{\theta_2(r)} \left|\frac{\partial f}{\partial z}\right| r\, d\varphi \right)^2 \qquad (10.9)$$

($\theta_1(r)$ and $\theta_2(r)$ denote the same arguments as at p. 156). The left-hand side of this inequality, being the area of a bounded set, is finite, so, to obtain a contradiction, it suffices to show that all integrals in the parentheses in the right-hand side of (10.9) are bounded from below by the same positive number.

To do this, observe that if $\gamma: [p; q] \to U$ is an arbitrary piecewise smooth path, then, taking into account that $|\partial f/\partial \bar{z}| < |\partial f/\partial z|$ everywhere on U (since the Jacobian of f is positive everywhere), we obtain the following:

$$\text{length}(f \circ \gamma) = \int_p^q \left| \frac{d(f(\gamma(t)))}{dt} \right| dt = \int_p^q \left| \frac{\partial f}{\partial z}(\gamma(t))\gamma'(t) + \frac{\partial f}{\partial \bar{z}}(\gamma(t))\overline{\gamma'(t)} \right| dt$$

$$\leq \int_p^q \left(\left| \frac{\partial f}{\partial z}(\gamma(t)) \right| + \left| \frac{\partial f}{\partial \bar{z}}(\gamma(t)) \right| \right) |\gamma'(t)| \, dt \leq 2 \int_p^q \left| \frac{\partial f}{\partial z}(\gamma(t)) \right| \cdot |\gamma'(t)| \, dt$$

$$= 2 \int_\gamma \left| \frac{\partial f}{\partial z} \right| |dz|,$$

that is,

$$\int_\gamma \left| \frac{\partial f}{\partial z} \right| |dz| \geq \frac{1}{2} \text{length}(f \circ \gamma). \tag{10.10}$$

On an arc of radius r centered at a and parametrized as $\varphi \mapsto a + re^{i\varphi}$, we have $|dz| = r \, d\varphi$. Now if $\theta'(r)$ and $\theta''(r)$ are defined as at p. 156, then by (10.10) we have

$$\int_{\theta_1(r)}^{\theta_2(r)} \left| \frac{\partial f}{\partial z} \right| r \, d\varphi \geq \int_{\theta'(r)}^{\theta''(r)} \left| \frac{\partial f}{\partial z} \right| r \, d\varphi \geq \frac{1}{2} \text{length}(f \circ \gamma)$$

where γ is the arc of the circle of radius r centered at a between the points $a + re^{i\theta'(r)}$ and $a + re^{i\theta''(r)}$. Since the length of the image of this arc γ is not less than the positive constant δ from (10.4), we see that all integrals in the parentheses in (10.9) are at least $\delta/2 > 0$, which implies that the right-hand side of (10.9) is infinite, a desired contradiction.   □

Exercises

10.1. Show that if a subset $S \subset \bar{\mathbb{C}}$ has no accumulation points, then it is finite.

10.2. Show that a function meromorphic on the whole Riemann sphere is a rational function (i.e., the quotient of two polynomials).

10.3. Show that every conformal automorphism of the Riemann sphere is a linear fractional transformation.

10.4. Let $\gamma\colon [A; B] \to \bar{\mathbb{C}}$ be a closed (continuous) path, and let $a \in \bar{\mathbb{C}}$ be a point that does not lie on γ. Show that there exists a homotopy of closed paths $\Gamma\colon [A; B] \times [0; 1] \to \bar{\mathbb{C}}$ such that $\Gamma(t, 0) = \gamma(t)$, $\Gamma(t, 1)$ is a constant map, and $\Gamma(t, s) \neq a$ for any t and s. (*Hint.* If $a = \infty$, this is quite easy. In the general case, let $f\colon \bar{\mathbb{C}} \to \bar{\mathbb{C}}$ be a linear fractional transformation sending a to ∞; first, construct a homotopy for $f \circ \gamma$.)

10.5. Let $U \subset \bar{\mathbb{C}}$ be a connected open set, $f\colon U \to \mathbb{C}$ be a holomorphic function, C and C' be generalized circles. Assume that $f(U \cap C) \subset C'$. Show that $f(S_C(z)) = S_{C'}(f(z))$ whenever both z and $S_C(z)$ lie in U. Here S_C is the reflection with respect to the generalized circle C.

10.6. Let $U = \{z\colon -1 < \operatorname{Im} z < 1\} \setminus [1; +\infty)$ (a strip cut along a ray of the real axis). Find a conformal map from U onto the upper half-plane. (*Hint.* First, following the examples from Chap. 3, map the strip $\{0 < \operatorname{Im} z < 1\}$ onto the upper half-plane; the reflection principle shows that this map extends to a map from U onto a cut plane.)

10.7. (a) Let $U = \{z\colon \operatorname{Im} z > 0, 1 < |z| < 2\}$ be a semi-annulus, and let $H = \{z\colon \operatorname{Im} z > 0\}$ be the upper half-plane. Show that if $f\colon U \to H$ is a conformal map from U onto H, then f extends to a holomorphic function $F\colon H \to \mathbb{C}$.
(b) Can the map F from part (a) be extended to a function holomorphic on the entire complex plane $\mathbb{C}$?
(The existence of a conformal map from U onto H follows from a theorem that will be proved in Chap. 12.)

10.8. Let $a_1 < a_2 < a_3$ be real numbers. Show, without using complex numbers, that

$$\int_{-\infty}^{a_1} \frac{dt}{\sqrt{|(t - a_1)(t - a_2)(t - a_3)|}} = \int_{a_2}^{a_3} \frac{dt}{\sqrt{|(t - a_1)(t - a_2)(t - a_3)|}}.$$

10.9. Let $a_1 < a_2 < a_3 < a_4$ be real numbers, $z_0 \in \mathbb{R}$, and H be the upper half-plane. Define a function $F\colon H \to \mathbb{C}$ by

$$F(z) = \int_{z_0}^{z} \frac{dz}{\sqrt{(z - a_1)(z - a_2)(z - a_3)(z - a_4)}}.$$

Show that F is a conformal map from the upper half-plane onto its image, and describe this image.

10.10. Construct a conformal map from the upper half-plane onto a right triangle.

10.11. Does there exist a conformal map between the interior of a square and the interior of a rectangle with sides ratio $1 : 2$ such that the boundary correspondence sends vertices to vertices? (*Hint.* The reflection principle.)

10.12. Does there exist a conformal map between the interior of a regular triangle and the interior of a right isosceles triangle such that the boundary correspondence sends vertices to vertices? (*Hint.* Don't jump to an answer!)

10.13. Consider the annuli

$$U_1 = \{z: r_1 < |z| < R_1\} \quad \text{and} \quad U_2 = \{z: r_2 < |z| < R_2\}.$$

Show that U_1 and U_2 are conformally isomorphic if and only if $R_1/r_1 = R_2/r_2$.

10.14. Let

$$K = \{z \in \mathbb{C}: 0 < \operatorname{Re} z < 1, 0 < \operatorname{Im} z < 1\},$$
$$\Delta = \{z \in \mathbb{C}: 0 < \operatorname{Re} z < 1, 0 < \operatorname{Im} z < \operatorname{Re} z\}.$$

Show directly (without appealing to Carathéodory's theorem) that the map $K \to \Delta$ given by the formula $x + iy \mapsto x + ixy$ is not quasiconformal.

10.15. Let $U \subset \mathbb{C}$ be an open set and $f: U \to \mathbb{C}$ be a map of class C^1 with positive Jacobian. The real derivative of f sends circles to ellipses. Express the angle between the real axis and the semi-major axis of the ellipse corresponding to a point a in terms of $\partial f/\partial \bar{z}$ and $\partial f/\partial z$.

10.16. Let $U \subset \mathbb{C}$ be an open sector with vertex at the origin; choose, once for all, a continuous branch of the logarithm in this sector; denote it by φ.
    (a) Is the map $U \to U$ given by the formula $re^{i\varphi} \mapsto r^2 e^{i\varphi}$ quasiconformal?
    (b) The same question for the map $re^{i\varphi} \mapsto r^3 e^{i\varphi/r}$.
    (c) The same question for the map $re^{i\varphi} \mapsto re^{2i\varphi}$.
    (*Hint.* In all three cases, the answer can be obtained without calculations.)

10.17. Show that there is no quasiconformal diffeomorphism from $\mathbb{C}$ onto the unit disk $D = \{z: |z| < 1\}$. (*Hint.* The length-area principle. To begin with, use it to prove that there is no *conformal* isomorphism.)

Chapter 11
Infinite Sums and Products

11.1 The Cotangent as an Infinite Sum

Imagine that we want to construct a meromorphic function on $\mathbb{C}$ that has a pole with residue 1 at each integer point of the real line and no other poles. The reader familiar with trigonometry can at once give one of the possible answers, but let us pretend that we know nothing about trigonometry and try to solve this problem "from scratch." The results thus obtained are interesting in themselves, and they will also serve as a warm-up for the more sophisticated constructions from the subsequent sections.

Denote the function we are seeking by f. At each point $n \in \mathbb{Z}$, the principal part of f must have the form $1/(z - n)$. The most naïve attempt is simply to add all the principal parts together:

$$f(z) = \sum_{n \in \mathbb{Z}} \frac{1}{z - n}. \tag{11.1}$$

Alas, this series is not absolutely convergent, since for fixed z and all sufficiently large n we have

$$\left| \frac{1}{z - n} \right| \geq \frac{1}{2n}$$

(and if we do not require the series to be absolutely convergent, then it is not clear in what order the terms indexed by all integers are to be summed).

Before explaining how this naïve idea can be saved, we first give a rigorous definition of convergence of a series of meromorphic functions.

Definition 11.1 Let $U \subset \mathbb{C}$ be an open set and $\sum f_j$ be a series of functions holomorphic on U except, possibly, some isolated singularities. It is said to *converge compactly on U* if for every compact subset $K \subset U$ the following two conditions hold:

(1) at most finitely many functions f_j have singularities in K;

(2) the series consisting of the functions f_j that have no singularities in K converges on K absolutely and uniformly.

S. Lvovski, *Principles of Complex Analysis*, Moscow Lectures 6,
https://doi.org/10.1007/978-3-030-59365-0_11

It is clear from this definition that the sum of a compactly convergent series of meromorphic functions is itself a meromorphic function whose set of poles is contained in the set of poles of all functions f_j (if some principal parts cancel out, then the set of poles of f can be strictly less then the union of the sets of poles of all f_j).

Now we return to the divergent series (11.1). Let us try to improve its convergence as follows: add a holomorphic function to each term (this does not affect the poles) in such a way that the terms of the resulting series decrease faster. In the present case, such an improvement can be achieved quite easily: to each term $1/(z - n)$ (for nonzero n) we add $1/n$. In the next proposition we adopt the following (commonly used) convention: the symbol $\sum'$ means that the summation extends to all values of the parameter except zero. For instance, $\sum'_{n\in\mathbb{Z}}$ means a sum over all nonzero integers.

Proposition 11.2 *The series*

$$f(z) = \frac{1}{z} + \sum_{n\in\mathbb{Z}}{}'\left(\frac{1}{z - n} + \frac{1}{n}\right) \tag{11.2}$$

converges absolutely and uniformly on compact subsets of $\mathbb{C}$ to a meromorphic function.

Proof If $K \subset \mathbb{C}$ is a compact set, then $|z| \leq C$ for all $z \in K$, where C is a constant. For $z \in K$ and sufficiently large $|n|$, we have

$$\left|\frac{1}{z - n} + \frac{1}{n}\right| = \frac{|z|}{|n^2 - nz|} \leq \frac{C}{n^2|1 - z/n|} = O\left(\frac{1}{n^2}\right),$$

so the series converges absolutely and uniformly on K.   □

It is clear from what we have proved that the function f has a simple pole with residue 1 at each integer point and no other poles. Now we investigate its further properties.

Proposition 11.3 *The function f given by* (11.2) *is odd and periodic with period* 1.

Proof The first claim is obvious because the terms of an absolutely convergent series can be rearranged. To prove the second claim, we find the derivative of f. Since a compactly convergent series of holomorphic functions can be differentiated term by term, we obtain

$$f'(z) = -\sum_{n\in\mathbb{Z}}\frac{1}{(z - n)^2}.$$

The series in the right-hand side converges absolutely and uniformly on compact subsets, since for bounded $|z|$ we have $|1/(z - n)^2| = O(1/n^2)$. Replacing z by $z + 1$ in the series for f' results in a rearrangement of terms, so we certainly have $f'(z + 1) = f'(z)$ for all z. Thus, the function $z \mapsto f(z + 1) - f(z)$ has zero derivative on the connected set $\mathbb{C} \setminus \mathbb{Z}$ and, therefore, is constant (see Corollary 4.12). Hence, $f(z + 1) = f(z) + c$ where c is a constant. It remains to show that $c = 0$. To this end, we find $f(1/2)$. Grouping together the terms with n and $-n$ in (11.2), we obtain the expansion

$$f(z) = \frac{1}{z} + \sum_{n=1}^{\infty} \frac{2z}{z^2 - n^2}. \tag{11.3}$$

Substituting $z = 1/2$ yields

$$f(1/2) = 2 + \sum_{n=1}^{\infty} \frac{1}{\frac{1}{4} - n^2} = 2 - 4 \sum_{n=1}^{\infty} \frac{1}{4n^2 - 1}$$

$$= 2 - 2 \sum_{n=1}^{\infty} \left(\frac{1}{2n-1} - \frac{1}{2n+1} \right) = 2 - 2 \cdot 1 = 0.$$

Therefore, $f(1/2) = 0$ and, since f is odd, $f(-1/2) = 0$ too, which implies that $c = f(-1/2 + 1) - f(-1/2) = 0$ and f is periodic. □

The function f was defined as the sum of a series, but a meromorphic function with period 1 that has simple poles with residue 1 at all integer points can also be named off the top of one's head: this is the function $\pi \cot(\pi z)$. Now we will show that these functions coincide.

Proposition 11.4 *The function $f(z)$ given by (11.2) coincides with $\pi \cot(\pi z)$.*

Proof It will be more convenient to compare not the functions $f(z)$ and $\pi \cot(\pi z)$ themselves, but their derivatives. As we have seen in the proof of Proposition 11.3, $f'(z) = - \sum_{n \in \mathbb{Z}} \frac{1}{(z-n)^2}$, where the series converges absolutely and uniformly on compact subsets of $\mathbb{C}$. It is also obvious that $(\pi \cot(\pi z))' = -\pi^2 / \sin^2(\pi z)$.

Consider the function $g(z) = f'(z) - (\pi \cot(\pi z))'$. It is easy to see that both terms in the right-hand side have poles only at integer points, and the principal part at $n \in \mathbb{Z}$ is equal to $-1/(z-n)^2$. Therefore, g is an entire function. Since both $f'(z)$ and $(\pi \cot(\pi z))'$ have period 1, the function g also has period 1. Now we show that g is a constant; for this, by Liouville's theorem, it suffices to show that g is bounded; by periodicity, it suffices to prove that it is bounded in the strip $\{z : 0 \le \operatorname{Re} z \le 1\}$.

Take a number $\varepsilon > 0$. The entire function g is bounded on the compact set $\{z : 0 \le \operatorname{Re} z \le 1, |\operatorname{Im} z| \le \varepsilon\}$. It remains to check that g is bounded also on the set $\{z : 0 \le \operatorname{Re} z \le 1, |\operatorname{Im} z| \ge \varepsilon\}$. We will show that on the latter set even each of the functions $f'(z)$ and $\pi \cot(\pi z)$ is bounded. Since, clearly, $f'(\bar{z}) = \overline{f'(z)}$ and $\pi^2 / \sin^2(\pi \bar{z}) = \overline{\pi^2 / \sin^2(\pi z)}$, it suffices to prove that they are bounded on the semi-strip

$$\Pi = \{z : 0 \le \operatorname{Re} z \le 1, \operatorname{Im} z \ge \varepsilon\}.$$

First, we show that the function f is bounded on Π. Indeed, for $|n| \ge 2$ and $\operatorname{Re} z \in [0; 1]$ we have

$$\left| \frac{1}{(z-n)^2} \right| \Big/ \left(\frac{1}{n^2} \right) = \frac{1}{|1 - z/n|^2} \le \frac{1}{(1 - \operatorname{Re} z/n)^2} \le \frac{1}{(1 - 1/2)^2} = 4. \tag{11.4}$$

Since the sum of the remaining three terms $\frac{1}{(z-1)^2} + \frac{1}{z} + \frac{1}{(z+1)^2}$ on Π is, obviously, bounded and the series $\sum\limits_{\substack{n \in \mathbb{Z} \\ |n| \geq 2}} (1/n^2)$ is, obviously, convergent, it follows from (11.4) that f' is bounded on Π.

To show that the function $\pi^2/\sin^2(\pi z)$ is bounded on Π, it suffices to prove that $\lim\limits_{y \to +\infty} |\sin \pi(x + iy)| = +\infty$ for $x \in [0; 1]$, where the convergence is uniform with respect to $x \in [0; 1]$; this, in turn, follows from the fact that

$$| \sin(\pi(x + iy))| = \frac{|e^{ix}e^{-y} - e^{-ix}e^y|}{2} \geq \frac{e^y - e^{-y}}{2}.$$

So, the difference $f'(z) - (\pi \cot(\pi z))'$ is a bounded entire function and hence a constant. Therefore, the difference $h(z) = f(z) - \pi \cot(\pi z)$, whose derivative is a constant, is a function of the form $z \mapsto az + b$. Since the function h, being the difference of two functions with period 1, also has period 1, we see that $a = 0$ and hence h is a constant. Since the functions $f(z)$ and $\pi \cot(\pi z)$ are odd, their difference $h(z)$ is also odd. However, a constant odd function is identically zero, so $f(z) - \pi \cot(\pi z) = 0$ for all z, as required.   □

Substituting z/π instead of z and dividing by π, we obtain the following corollary.

Corollary 11.5 *The following expansions hold:*

$$\cot z = \frac{1}{z} + {\sum_{n \in \mathbb{Z}}}' \left(\frac{1}{z - \pi n} - \frac{1}{\pi n} \right), \qquad (11.5)$$

$$\cot z = \frac{1}{z} + \sum_{n=1}^{\infty} \frac{2z}{z^2 - \pi^2 n^2}, \qquad (11.6)$$

where both series converge absolutely and uniformly on compact subsets of $\mathbb{C}$.

In (11.6), each term of the sum is holomorphic on the disk $\{z : |z| < \pi\}$ and can easily be expanded into a power series in this disk. Let us try to write these expansions.

Obviously, we have

$$\frac{2z}{z^2 - \pi^2 n^2} = -\frac{2z}{\pi^2 n^2} \cdot \frac{1}{1 - z^2/(\pi^2 n^2)} = -\frac{2z}{\pi^2 n^2} \left(1 + \frac{z^2}{\pi^2 n^2} + \frac{z^4}{\pi^4 n^4} + \ldots \right)$$

$$= -\frac{2z}{\pi^2 n^2} - \frac{2z^3}{\pi^4 n^4} - \frac{2z^5}{\pi^6 n^6} - \ldots \quad (11.7)$$

(for $|z| < \pi$). Substituting these expansions into (11.6) and combining like terms (the latter is justified since both the series in (11.6) and expansions (11.7) absolutely converge), we obtain the following expression for the Laurent series of the cotangent at the origin:

$$\cot z = \frac{1}{z} - 2\left(\frac{\sum_{n=1}^{\infty}\frac{1}{n^2}}{\pi^2}z + \frac{\sum_{n=1}^{\infty}\frac{1}{n^4}}{\pi^4}z^3 + \frac{\sum_{n=1}^{\infty}\frac{1}{n^6}}{\pi^6}z^5 + \dots\right). \tag{11.8}$$

Clearly, if we now derive an expansion for $\cot z$ directly from the definition $\cot z = i\frac{e^{iz}+e^{-iz}}{e^{iz}-e^{-iz}}$, it will not involve π, so we may hope to obtain some interesting identities by equating the coefficients of the same powers of z. This will indeed be the case, but a difficulty lurks here: the Laurent series coefficients of $\cot z$ cannot be expressed in terms of simple arithmetic functions such as powers, factorials, or binomial coefficients. Nevertheless, this and related sequences of coefficients appear in mathematics quite often, so there is a special definition and notation for one of the sequences in this family.

Consider the function $f(z) = z/(e^z - 1)$, which, obviously, has a removable singularity at the origin. Let n be a positive integer.

Definition 11.6 The value $f^{(n)}(0)$ is called the nth *Bernoulli number* and denoted by B_n.

In other words, the Bernoulli numbers are defined by the formula

$$\frac{z}{e^z - 1} = 1 + \sum_{n=1}^{\infty}\frac{B_n}{n!}z^n.$$

Bernoulli numbers with small indices can easily be calculated by hand. For example, it is not difficult to see that $B_1 = -1/2$; further, it is less straightforward but still easy to verify that the function $z \mapsto \frac{z}{e^z-1} - (-z/2)$ is even, which implies that the Bernoulli numbers with odd indices, except B_1, vanish. Several first Bernoulli numbers are tabulated below:

n	1	2	4	6	8	10	12
B_n	$-\frac{1}{2}$	$\frac{1}{6}$	$-\frac{1}{30}$	$\frac{1}{42}$	$-\frac{1}{30}$	$\frac{5}{66}$	$-\frac{691}{2730}$

Proposition 11.7 *The Laurent series of the function $z \mapsto \cot z$ at the origin has the form*

$$\cot z = \frac{1}{z} + \sum_{m=1}^{\infty}(-1)^m\frac{2^{2m}B_{2m}}{(2m)!}z^{2m-1}.$$

Proof This can be proved by a direct calculation taking into account that $B_1 = -1/2$:

$$\cot z = i\frac{e^{iz}+e^{-iz}}{e^{iz}-e^{-iz}} = i + \frac{2i}{e^{2iz}-1} = i + \frac{2i}{2iz}\left(1 + \sum_{n=1}^{\infty}\frac{B_n}{n!}(2iz)^n\right)$$

$$= \frac{1}{z} - \frac{2^2 B_2}{2!}z + \frac{2^4 B_4}{4!}z^3 - \frac{2^6 B_6}{6!}z^5 + \dots.$$

Comparing (11.8) with the expansion from Proposition 11.7, we obtain the following result.

Proposition 11.8 *For every positive integer m,*

$$1 + \frac{1}{2^{2m}} + \frac{1}{3^{2m}} + \ldots + \frac{1}{n^{2m}} + \ldots = (-1)^{m-1} \frac{2^{2m-1} B_{2m} \pi^{2m}}{(2m)!}.$$

In particular, $\sum_{n=1}^{\infty} \frac{1}{n^2} = \frac{\pi^2}{6}$, $\sum_{n=1}^{\infty} \frac{1}{n^4} = \frac{\pi^4}{90}$, etc.

It is worth mentioning that much less is known about the sums $\sum_{n=1}^{\infty} (1/n^k)$ for odd k ("the values of the Riemann zeta function at positive odd numbers"): for example, it was not proved until the late 1970s that the number $\sum_{n=1}^{\infty} (1/n^3)$ is irrational.

11.2 Elliptic Functions

In the previous section, we have constructed a periodic meromorphic function "from scratch." Now we turn to complex functions having two periods. We have already seen one example of such a function: the conformal map from a rectangle onto the upper half-plane considered in Sec. 10.3.

Definition 11.9 A *lattice* in $\mathbb{C}$ is an additive subgroup $\Gamma \subset \mathbb{C}$ generated by two numbers $\omega_1, \omega_2 \in \mathbb{C}$ that are linearly independent over $\mathbb{R}$.

In other words, $\Gamma = \{m_1\omega_1 + m_2\omega_2 : m_1, m_2 \in \mathbb{Z}\}$ where ω_1 and ω_2 are nonzero complex numbers whose quotient does not belong to $\mathbb{R}$.

Definition 11.10 An *elliptic function* with respect to a lattice $\Gamma \subset \mathbb{C}$ is a meromorphic function f on $\mathbb{C}$ such that $f(z + u) = f(z)$ for all $z \in \mathbb{C}$ and $u \in \Gamma$.

If Γ is generated by ω_1 and ω_2, then an elliptic function with respect to Γ is a meromorphic function with periods ω_1 and ω_2. If the lattice with respect to which f is periodic is clear from the context, then we simply say that f is an elliptic function.

Definition 11.11 Let $\Gamma \subset \mathbb{C}$ be a lattice with generators ω_1 and ω_2. A *fundamental parallelogram* of Γ is an arbitrary set of the form $\{a + s\omega_1 + t\omega_2\}$ where $a \in \mathbb{C}$ is fixed and s and t are arbitrary real numbers from the interval $[0; 1)$.

We emphasize that, according to this definition, a fundamental parallelogram of a lattice is by no means unique: it depends on the choice of both an initial point a and a basis of the lattice (numbers ω_1 and ω_2). If Π is a fundamental parallelogram of a lattice Γ, then $\mathbb{C}$ can be represented as the union of the sets $\Pi + u$ (translations of Π) over $u \in \Gamma$, where the sets $\Pi + u$ are disjoint for different u.

A fundamental parallelogram contains exactly one representative of each coset of Γ in $\mathbb{C}$ (regarded as an additive group). Hence, if f is an elliptic function with respect to Γ, then both ω_1 and ω_2 are its periods. All values assumed by this function are assumed already within a fundamental parallelogram. In the same way, its poles are also "periodic": if f has a pole at a point a, then it has a pole at each point $a + u$, $u \in \Gamma$, with the principal part differing from the principal part at a only by a shift of the argument.

First we consider some general properties of elliptic functions.

Proposition 11.12 *If an elliptic function has no poles, then it is a constant.*

Proof An elliptic function assumes all its values within a fundamental parallelogram. Since this parallelogram is bounded and an elliptic function without poles is continuous, its values are bounded on the fundamental parallelogram and hence on the whole complex plane $\mathbb{C}$. By Liouville's theorem, every bounded entire function is a constant. □

Proposition 11.13 *Let f be an elliptic function with respect to a lattice Γ, and let Π be a fundamental parallelogram of Γ. Then the orders of the zeros and poles of f lying within Π sum to zero (the orders of poles are counted as negative).*

Proof Since this assertion does not depend on the choice of a fundamental parallelogram, we may assume, translating Π if necessary, that its boundary contains neither poles nor zeros of f. Denoting by γ the positively oriented boundary of Π, by Corollary 8.9 we have

$$\sum_{z \in \Pi} \operatorname{ord}_z(f) = \frac{1}{2\pi i} \int_\gamma \frac{f'(z)}{f(z)} \, dz. \tag{11.9}$$

Obviously, the function f'/f is also elliptic with respect to Γ, and this implies that the integral of $(f'(z)/f(z)) \, dz$ over the boundary of Π vanishes: the values of the function f'/f at opposite sides coincide by periodicity, and these sides are traversed in opposite directions. Therefore, in (11.9) the integrals over the opposite sides of the parallelogram cancel and the right-hand side vanishes, so the left-hand side vanishes too, as required. □

Let us now construct examples of elliptic functions. We will proceed as in the previous section. While we were dealing with functions with one period, we have even managed to construct a function with exactly one, and simple, pole (of course, up to periodicity). To know whether this can be done in the case of two periods, solve Exercise 11.4, and now we turn straight to functions with a double pole.

Lemma 11.14 *If $\Gamma \subset \mathbb{C}$ is a lattice and $\alpha \in \mathbb{R}$, then the series $\sum'_{u \in \Gamma} \frac{1}{|u|^\alpha}$ converges for $\alpha > 2$ and diverges for $\alpha \leq 2$.*

Recall that $\sum'$ stands for summation over all elements except zero.

Proof Let $\alpha > 2$. Choose a basis $\langle \omega_1, \omega_2 \rangle$ of the lattice, for every $u \in \Gamma$ set

$$\Pi_u = \{u + s\omega_1 + t\omega_2 \colon s, t \in [0; 1)\},$$

and define a function h on $\mathbb{C}$ as follows: $h(z) = 1/|z|^\alpha$ if $z \in \Pi_u$ with $u \neq 0$, and $h(z) = 0$ if $z \in \Pi_0$ (we could choose any other constant). The series $\sum'_{u \in \Gamma}(1/|u|^\alpha)$ converges if and only if the improper integral $\iint_{|z| \geq 1} h(z)\,dx\,dy$ does. Since, obviously, there exists a constant $C > 0$ such that $h(z) \leq C/|z|^\alpha$, and the integral $\iint_{|z| \geq 1} dx\,dy/|z|^\alpha$ for $\alpha > 2$ converges (which can be verified by passing to polar coordinates), the series $\sum'(1/|u|^\alpha)$ also converges. For $\alpha \leq 2$, the argument is similar. $\square$

Proposition-Definition 11.15 The series

$$\frac{1}{z^2} + \sum_{u \in \Gamma}'\left(\frac{1}{(z-u)^2} - \frac{1}{u^2}\right) \tag{11.10}$$

converges compactly in $\mathbb{C}$ to a meromorphic function in the sense of Definition 11.1. It is denoted by $\wp(z)$ and called the *Weierstrass $\wp$-function*.

Proof If $|z| \leq C$ and, say, $|u| \geq 2C$, then $|z - 2u| \leq 3|u|$ and $|z - u| \geq |u/2|$, whence

$$\left|\frac{1}{(z-u)^2} - \frac{1}{u^2}\right| = \frac{|z| \cdot |z - 2u|}{|u|^2 \cdot |z-u|^2} \leq \frac{12C}{|u|^3},$$

so it follows from Lemma 11.14 and the Weierstrass M-test (Proposition 1.2) that for $|z| \leq C$ the series (11.10) converges absolutely and uniformly with respect to z. Since compact sets are bounded, we are done. $\square$

It is clear from the definition that the function $\wp$ has a pole of order 2 with principal part $1/(z-u)^2$ at each point $u \in \Gamma$ and no other poles.

Proposition 11.16 *The Weierstrass $\wp$-function (11.10) is even and elliptic with respect to the lattice Γ.*

Proof The fact that $\wp$ is even is obvious from the definition and the absolute convergence of the series (11.10).

Differentiating (11.10) term by term, we obtain

$$\wp'(z) = -2\sum_{u \in \Gamma}\frac{1}{(z-u)^3}.$$

Since the terms of an absolutely convergent series can be rearranged, we have $\wp'(z+u) = \wp'(z)$ for every $u \in \Gamma$, so $\wp'$ is an elliptic function. If $\langle \omega_1, \omega_2 \rangle$ is a basis of Γ, then it follows that $\wp(z+\omega_1) = \wp(z)+C$ for all z, for some constant C. Since $\wp$ is even, we have

$$\wp\left(-\frac{\omega_1}{2}\right) = \wp\left(\frac{\omega_1}{2}\right) = \wp\left(-\frac{\omega_1}{2}+\omega_1\right) = \wp\left(-\frac{\omega_1}{2}\right)+C,$$

whence $C = 0$ and ω_1 is a period of $\wp$. In the same way we can show that ω_2 is also a period, so $\wp$ is an elliptic function with respect to Γ. $\square$

It is easy to see that the cotangent function, which in the previous section has also been represented as an infinite sum of fractions, satisfies the differential equation $(\cot z)' = -(\cot z)^2 - 1$. For the function $\wp$, there exists a differential equation of the same type, but more interesting. Writing it down requires some preparation.

Again, let $\langle \omega_1, \omega_2 \rangle$ be a basis of Γ; set $e_1 = \wp(\omega_1/2)$, $e_2 = \wp((\omega_1 + \omega_2)/2)$, $e_3 = \wp(\omega_2/2)$. (The numbers $\omega_1/2$, $(\omega_1 + \omega_2)/2$, and $\omega_2/2$ represent all elements of order 2 in the quotient group $\mathbb{C}/\Gamma$.)

Proposition 11.17 *The functions $\wp$ and $\wp'$ satisfy the identity*

$$(\wp'(z))^2 = 4(\wp(z) - e_1)(\wp(z) - e_2)(\wp(z) - e_3). \qquad (11.11)$$

Proof To prove (11.11), it suffices to show that both sides are elliptic functions with respect to Γ having the same collection of zeros and poles (counting multiplicities) and the same principal part at the origin.

In accordance with this plan, observe that the periodicity of both sides of (11.11) with respect to Γ is obvious. Further, it suffices to compare the zeros and poles within some fundamental parallelogram; we choose it to be the parallelogram with vertices 0, ω_1, $\omega_1 + \omega_2$, and ω_2. Within this parallelogram, the function $\wp$ (and hence each of the functions $\wp - e_j$) has a unique pole at the origin, with principal part $1/z^2$, so the function $\wp'$ also has a unique pole at the origin, with principal part $-2/z^3$. Therefore, within the fundamental parallelogram both sides of (11.11) have a unique pole at the origin, and both have principal part $4/z^6$.

It remains to compare the zeros. Since each of the points e_j, $1 \le j \le 3$, is an element of order 2 in the group $\mathbb{C}/\Gamma$, and the function $\wp$ is even, each of the functions $\wp - e_1$, $\wp - e_2$, and $\wp - e_3$ has zeros of order at least 2 at the points $\omega_1/2$, $(\omega_1 + \omega_2)/2$, and $\omega_2/2$, respectively. Indeed, since, for example, $\omega_1/2 \equiv -\omega_2/2 \pmod{\Gamma}$ and the function $\wp$ is even, we have

$$\wp\left(\frac{\omega_1}{2} + z\right) = \wp\left(-\frac{\omega_1}{2} - z\right) = \wp\left(-\frac{\omega_1}{2} - z + \omega_1\right) = \wp\left(\frac{\omega_1}{2} - z\right),$$

so the power series expansion of $\wp$ at $\omega_1/2$ involves only even powers, while, by the very definition of the number e_1, the function $\wp - e_1$ has a zero at $\omega_1/2$. For the functions $\wp - e_j$, the sum of the orders of all poles within the fundamental parallelogram is equal to 2, as for the function $\wp$ itself, hence Proposition 11.13 shows that the function $\wp - e_j$ has no other zeros within the fundamental parallelogram besides those mentioned above, and the orders of these zeros are exactly 2. Since the function $\wp - e_j$ has a zero of order 2 at the corresponding point of order 2, the function $\wp' = (\wp - e_j)'$ has a simple zero at each of the points $\omega_1/2$, $(\omega_1 + \omega_2)/2$, and $\omega_2/2$, while the function $(\wp')^2$, accordingly, has a zero of order 2 at each of these points. On the other hand, the function $(\wp')^2$ has exactly one pole (at the origin) within the fundamental parallelogram, and its multiplicity is 6. Applying Proposition 11.13 once again, we see that the left-hand side of (11.11) has no other zeros. Thus, both sides of (11.11) have not only the same poles, but also the same zeros, and at the beginning of the proof we have verified that their principal parts at the origin coincide. The proof is completed. $\qquad \square$

Corollary 11.18 (of the proof) *If $\wp$ is the Weierstrass function with respect to a lattice with basis $\langle \omega_1, \omega_2 \rangle$, then $\wp$ has ramification points of order 2 at the points $\omega_1/2$, $\omega_2/2$, and $(\omega_1 + \omega_2)/2$ and no other ramification points.*

The right-hand side of (11.11) can be rewritten in a more explicit form. For this, as in the previous section, we must write the Laurent series of the function $\wp$ at the origin as explicitly as possible.

Definition 11.19 Let $\Gamma \subset \mathbb{C}$ be a lattice and $k \geq 2$ be an integer. The series $\sum'_{u \in \Gamma}(1/u^{2k})$ is called an *Eisenstein series*. Its sum is denoted by G_k.

Note that an Eisenstein series converges absolutely by Lemma 11.14. If $k \geq 3$ is an odd integer, then the series $\sum'(1/u^k)$ still converges absolutely, but its sum vanishes because $1/u^k$ cancels with $1/(-u)^k$.

Proposition 11.20 *The Laurent series of the Weierstrass $\wp$-function has the form*

$$\wp(z) = \frac{1}{z^2} + 3G_2 z^2 + 5G_3 z^4 + 7G_4 z^6 + \ldots = \frac{1}{z^2} + \sum_{k=1}^{\infty}(2k+1)G_{k+1}z^{2k}. \quad (11.12)$$

Proof Set $r = \min\limits_{u \in \Gamma \setminus \{0\}} |u|$. In the disk $\{z: |z| < r\}$, each term of the series (11.10) has the following power series expansion:

$$\frac{1}{(z-u)^2} - \frac{1}{u^2} = -\frac{1}{u^2} + \frac{1}{u^2}\left(\frac{1}{1-(z/u)}\right)^2 = \frac{2z}{u^3} + \frac{3z^2}{u^4} + \frac{4z^3}{u^5} + \ldots$$

Take the sum of these expansions combining terms with the same power of z (this can be done in view of the absolute convergence, see Proposition 1.4). The coefficient of z^m has the form $(m+1)\sum'(1/u^{m+2})$. If m is odd, then it vanishes by the remark above, and if $m = 2k$ is even, then it is equal to $(2k+1)G_{k+1}$. To obtain the Laurent series of $\wp$, it remains to make sure that we do not forget the term $\frac{1}{z^2}$. $\quad\square$

Proposition 11.21 *The Weierstrass $\wp$-function satisfies the differential equation*

$$(\wp'(z))^2 = 4(\wp(z))^3 - 60G_2\wp(z) - 140G_3,$$

where G_2 and G_3 are Eisenstein series.

Proof Substitute the expansion (11.12) into (11.11):

$$\wp'(z) = -\frac{2}{z^3} + 6G_2 z + 20G_3 z^3 + \ldots,$$

$$(\wp'(z))^2 = \frac{4}{z^6} - \frac{24G_2}{z^2} - 80G_3 + \ldots, \quad (11.13)$$

$$\wp(z) - e_i = \frac{1}{z^2} - e_i + 3G_2 z^2 + 5G_3 z^4 + \ldots,$$

$$4 \prod_i (\wp(z) - e_i) = \frac{4}{z^6} - \frac{4(e_1 + e_2 + e_3)}{z^4} + \frac{4(e_1 e_2 + e_1 e_3 + e_2 e_3) + 36 G_2}{z^2}$$

$$+ (-4 e_1 e_2 e_3 - 24 G_2 (e_1 + e_2 + e_3) + 60 G_3) + \dots \quad (11.14)$$

Comparing the coefficients of the same power of z in (11.13) and (11.14), we successively obtain:

$$e_1 + e_2 + e_3 = 0, \quad 4(e_1 e_2 + e_1 e_3 + e_2 e_3) = -60 G_2, \quad 4 e_1 e_2 e_3 = 140 G_3. \quad (11.15)$$

Expanding the right-hand side of (11.11) and substituting from (11.15), we obtain the desired relation.   □

Corollary 11.22 (of the proof) *If $\wp$ is the Weierstrass function corresponding to a lattice with basis $\langle \omega_1, \omega_2 \rangle$, then*

$$\wp\left(\frac{\omega_1}{2}\right) + \wp\left(\frac{\omega_2}{2}\right) + \wp\left(\frac{\omega_1 + \omega_2}{2}\right) = 0.$$

Proof This is the relation $e_1 + e_2 + e_3 = 0$ in (11.15).   □

The elliptic function constructed in Sec. 10.3 outwardly resembles a Weierstrass function: it also has exactly one pole of order 2 within a fundamental parallelogram (in this case, it is a rectangle) and no other poles. Now we will see that it not only resembles the function $\wp$, but can easily be expressed in terms of this function.

Recall what precisely we have done in the previous chapter. For three real numbers $a_1 < a_2 < a_3$, we have introduced a function F defined in the upper half-plane by the formula

$$F(z) = \int_{z_0}^{z} \frac{dz}{\sqrt{(z - a_1)(z - a_2)(z - a_3)}}, \quad (11.16)$$

where $z_0 \in \mathbb{R}$ is fixed, and proved that F is a conformal map from the upper half-plane H onto a rectangle Π with sides

$$u = \int_{a_1}^{a_2} \frac{dt}{\sqrt{(t - a_1)(t - a_2)(t - a_3)}}, \quad v = \int_{a_2}^{a_3} \frac{dt}{\sqrt{|(t - a_1)(t - a_2)(t - a_3)|}} \quad (11.17)$$

(the side u is parallel to the real axis). We have established that the inverse map $G : \Pi \to H$ extends to an elliptic function with respect to the lattice Γ with basis $\langle 2u, 2iv \rangle$, which has poles of order 2 at the points $F(\infty) + 2mu + 2niv$, $m, n \in \mathbb{Z}$. Set $V(z) = G(z + F(\infty))$, where

$$F(\infty) = \int_{z_0}^{+\infty} \frac{dz}{\sqrt{(z-a_1)(z-a_2)(z-a_3)}},$$

$$\sqrt{(z-a_1)(z-a_2)(z-a_3)} = \prod_{j=1}^{3} \sqrt{z-a_j},$$

and a branch of each square root $\sqrt{z-a_j}$ is chosen so that $\sqrt{t-a_j} > 0$ for $t \in \mathbb{R}$, $t > a_j$. Denoting by Γ the lattice with basis $\langle 2u, 2iv \rangle$, we see that V is an elliptic function with respect to Γ, which has poles of order 2 at the points of the lattice and no other poles.

Since G is inverse to F, it follows from (11.16) and the differentiation formula for inverse functions that

$$(G'(z))^2 = (G(z) - a_1)(G(z) - a_2)(G(z) - a_3),$$

whence

$$(V'(z))^2 = (V(z) - a_1)(V(z) - a_2)(V(z) - a_3). \tag{11.18}$$

Now let $\wp$ be the Weierstrass function with respect to the lattice Γ with basis $\langle 2u, 2iv \rangle$. Note that the functions V' and $\wp'$ are elliptic with respect to Γ and have the same collection of zeros and poles (counting multiplicities). Indeed, $\wp'$ has poles of order 3 at the points of the lattice and no other poles, and also zeros of order 1 at $u, iv, u + iv$ and the points congruent to them modulo Γ; this follows from Corollary 11.18 upon recalling that ramification points of order k for a function f are the same thing as zeros of order $k - 1$ for its derivative f'. Further, the claim about the ramification points of the function $V(z) = G(z + F(\infty))$ follows from Proposition 10.11. Now the same argument as in the proof of Proposition 11.17 shows that there exists a nonzero constant α such that $V' = \alpha \wp'$; thus, there exist constants $\alpha \neq 0$ and β such that $V = \alpha \wp + \beta$. To find α and β, substitute this expression for V into (11.18). We obtain the following:

$$(\wp')^2 = 4\alpha \left(\wp + \frac{\beta}{\alpha} - \frac{a_1}{\alpha} \right) \left(\wp + \frac{\beta}{\alpha} - \frac{a_2}{\alpha} \right) \left(\wp + \frac{\beta}{\alpha} - \frac{a_3}{\alpha} \right).$$

Comparing the right-hand side of this identity with the right-hand side of (11.11) regarded as a polynomial in $\wp'$, we obtain, matching the roots, that $\frac{\beta-a_1}{\alpha}$, $\frac{\beta-a_2}{\alpha}$, and $\frac{\beta-a_3}{\alpha}$ are e_1, e_2, and e_3 (in some order). Since $e_1 + e_2 + e_3 = 0$ (see (11.15)), we obtain, summing the roots, that $\beta = (a_1 + a_2 + a_3)/3$. In turn, comparing the coefficients of $(\wp')^3$ in the right-hand sides, we find that $\alpha = 1$. We have derived an expression for the function V, and hence for the function $G(z) = V(z - F(\infty))$, in terms of $\wp$. Let us sum up.

Proposition 11.23 *Let* $a_1 < a_2 < a_3$ *be three real numbers. If we fix another number* $z_0 \in \mathbb{R}$ *and consider the function* $F: H \to \mathbb{C}$ *(where H is the upper half-plane) given by (10.1), then the function G inverse to F is given by the formula*

$$G(z) = \wp(z - C) + \frac{a_1 + a_2 + a_3}{3},$$

where $\wp$ is the Weierstrass function with respect to the lattice with basis $\langle 2u, 2iv \rangle$, the numbers u and v are given by (11.17), and

$$C = \int_{z_0}^{+\infty} \frac{dz}{\sqrt{(z - a_1)(z - a_2)(z - a_3)}}.$$

11.3 Infinite Products

Infinite products are an analog of series in which addition is replaced by multiplication. The properties of series and infinite products are very much alike, but it is important to notice some differences.

Definition 11.24 Let $\{a_n\}$ be a sequence of *nonzero* complex numbers. The infinite product $\prod_{n=1}^{\infty} a_n$ is said to converge to a number a if $a = \lim_{n\to\infty} \prod_{k=1}^{n} a_k$ and $a \neq 0$.

The condition $a \neq 0$ is imposed to avoid annoying complications in subsequent statements, for example, in the following obvious proposition.

Proposition 11.25 *If an infinite product* $\prod_{n=1}^{\infty} a_n$ *converges, then* $\lim_{n\to\infty} a_n = 1$.

We will need infinite products not so much of numbers as of functions. For the time being, we assume that the factors in an infinite product are nowhere vanishing, and that the definition of uniform convergence of an infinite product includes the condition that the limit function is nowhere vanishing too.

Proposition 11.26 *Let* $\{u_n\}$ *be a sequence of bounded complex-valued functions on a set X. Also assume that the functions $1 + u_n$ do not vanish on X and that the series $\sum_{n=1}^{\infty} |u_n|$ converges uniformly on X. Then the infinite product $\prod_{n=1}^{\infty} (1 + u_n)$ also converges uniformly on X.*

Proof The uniform convergence of the series $\sum |u_n|$ implies that the sequence $\{|u_n|\}$ is uniformly convergent to zero. Disregarding, if necessary, finitely many factors in the product $\prod_{n=1}^{\infty} (1 + u_n)$ (which does not affect the convergence, because the factors do not vanish), we may assume that $\sup_{x \in X} |u_n(x)| \leq 1/2$ for every n (instead of $1/2$, any positive number less than 1 will do). Thus, all the numbers $1 + u_n(x)$ lie in the right half-plane $H = \{z : \mathrm{Re}(z) > 0\}$, where we have a well-defined continuous function arg sending a complex number to its argument from the interval $(-\pi/2; \pi/2)$. To prove that the product $\prod(1 + u_n)$ converges uniformly, it suffices to verify the uniform convergence of the product $\prod |1 + u_n|$ and the series $\sum \arg(1 + u_n)$.

The uniform convergence of the product $\prod |1 + u_n|$ is equivalent to the uniform convergence of the series $\sum \log |1 + u_n|$, and the latter does converge, even absolutely. Indeed, there obviously exists a constant $C > 0$ such that $|\log(1 + a)| \le C|a|$ for $-1/2 \le a \le 1/2$; since $1 - |u_k(x)| \le |1 + u_k(x)| \le 1 + |u_k(x)|$ and $|u_k(x)| \le 1/2$ for all $x \in X$, we have $|\log |1 + u_k(x)|| \le C|u_k(x)|$, so the series $\sum \log |1 + u_k|$ converges uniformly by the comparison test.

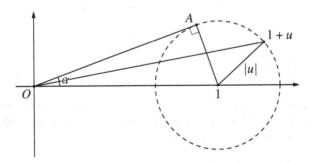

Fig. 11.1 $|\arg(1 + u)| \le \alpha = \arcsin |u|$

To establish the uniform convergence of the series $\sum \arg(1 + u_n)$, observe that (for $|u| < 1$) $|\arg(1 + u)| \le \arcsin |u|$ (see Fig. 11.1); since $\arcsin t / t \to 0$ as $t \to 0$, the uniform convergence of the series $\sum \arg(1 + u_n)$ follows from the uniform convergence of the series $\sum |u_n|$. $\square$

Corollary 11.27 (of the proof) *Under the assumptions of Proposition 11.26, let* $\sup |u_j(x)| < 1$ *for all j. If we define a function* log *on the right half-plane as* $\log(re^{i\varphi}) = \log r + i\varphi$, $-\pi/2 < \varphi < \pi/2$, *then the series* $\sum \log(1 + u_k)$ *and* $\sum |\log(1 + u_k)|$ *converge uniformly on X. If, besides, the values of all partial products* $\prod_{j=1}^{n} (1 + u_j)$ *lie in the right half-plane, then*

$$\sum_{k=1}^{\infty} \log(1 + u_k) = \log \prod_{k=1}^{\infty} (1 + u_k).$$

Proof Indeed, we have established that the series consisting of the real and imaginary parts of $\log(1 + u_k)$ converge absolutely. The second claim follows from the fact that if the numbers $a_1, \ldots, a_n$ and their product $a = a_1 \cdot \ldots \cdot a_n$ lie in the right half-plane, then $\log a = \log a_1 + \ldots + \log a_n$. $\square$

Up to now, we have dealt with infinite products of nonvanishing functions. Later, we will need products of holomorphic functions that do have zeros. To define infinite products for them, we must impose additional conditions similar to those made in the definition of an infinite sum of functions with isolated singularities (Definition 11.1).

Definition 11.28 Let $U \subset \mathbb{C}$ be a connected open set and $\prod f_j$ be a series of functions defined on U none of which is identically zero. We say that this series *converges* *compactly on* U if for every compact subset $K \subset U$ the following two conditions hold:

(1) at most finitely many functions f_j have a zero in K;

(2) the infinite product of the functions f_j with no zeros in K converges uniformly on K.

Clearly, if $\prod f_j$ satisfies this definition, then the function $f(z) = \lim\limits_{n\to\infty} \prod\limits_{j=1}^{n} f_j(z)$ is holomorphic on U and the set of zeros of f is the union of the sets of zeros of f_j with the corresponding multiplicities.

As an example, let us try to represent the sine function as an infinite product (by analogy with how we have represented the cotangent function as an infinite product at the beginning of this chapter). It is easy to see that all zeros of the function sin are the simple zeros at the points πn, $n \in \mathbb{Z}$, so the first thought that crosses one's mind is to try to represent it as a product of linear factors with zeros at these points. The product $\prod(z - \pi n)$, of course, diverges. The more civilized-looking product $z \cdot \prod' \left(1 - \frac{z}{\pi n}\right)$ (where $\prod'$, like $\sum'$, means a product over all indices except zero) is still not absolutely convergent: $\log(1 - (z/n))$ is asymptotically equivalent to $|z/n|$, and the harmonic series diverges. The next attempt is to group the factors with n and $-n$ together. Surprisingly, it leads us not only to a convergent product, but straight to the goal.

Proposition 11.29 *The infinite product*

$$z \cdot \prod_{n=1}^{\infty} \left(1 - \frac{z^2}{\pi^2 n^2}\right)$$

converges to $\sin z$ *compactly in* $\mathbb{C}$.

Proof The convergence is obvious by Proposition 11.26, so $f(z) = z \prod\limits_{n=1}^{\infty} \left(1 - \frac{z^2}{\pi^2 n^2}\right)$ is an entire function with simple zeros at the points πn. To prove that f coincides with sin, it suffices, by the principle of analytic continuation, to show that these two functions coincide in some punctured neighborhood of the origin. It will be more convenient to prove that the entire functions $\sin z/z$ and $g(z) = f(z)/z$ coincide. For this, in turn, we compare the logarithmic derivatives of these functions. Recall that the *logarithmic derivative* of a nonvanishing holomorphic function φ is $(\log \varphi)' = \varphi'/\varphi$; clearly, the logarithmic derivative does not depend on the choice of a branch of the logarithm, and the logarithmic derivative of a product is equal to the sum of the logarithmic derivatives of the factors.

In particular, the logarithmic derivative of the function $\sin z/z$ is, obviously, equal to

$$(\log \sin z)' - (\log z)' = \cot z - \frac{1}{z}.$$

As to the logarithmic derivative of the function

$$g(z) = \prod_{n=1}^{\infty}\left(1 - \frac{z^2}{\pi^2 n^2}\right),$$

in a sufficiently small neighborhood of the origin we have, by Corollary 11.27,

$$\log g(z) = \log\left(1 - \frac{z^2}{\pi^2}\right) + \log\left(1 - \frac{z^2}{2^2\pi^2}\right) + \log\left(1 - \frac{z^2}{3^2\pi^2}\right) + \dots, \qquad (11.19)$$

where the series is absolutely convergent. Expanding all terms in the right-hand side of (11.19) using the formula

$$\log(1 - t) = -t - \frac{t^2}{2} - \frac{t^3}{3} - \dots - \frac{t^n}{n} - \dots \qquad (11.20)$$

(the series is absolutely convergent for $|t| < 1$), we obtain

$$\log g(z) = \left(-\frac{z^2}{\pi^2} - \frac{1}{2}\cdot\frac{z^4}{\pi^4} - \frac{1}{3}\cdot\frac{z^6}{\pi^6} - \frac{1}{4}\cdot\frac{z^8}{\pi^8} - \dots\right)$$
$$+ \left(-\frac{z^2}{2^2\pi^2} - \frac{1}{2}\cdot\frac{z^4}{2^4\pi^4} - \frac{1}{3}\cdot\frac{2^6 z^6}{2^6\pi^6} - \frac{1}{4}\cdot\frac{z^8}{2^8\pi^8} - \dots\right)$$
$$+ \left(-\frac{z^2}{3^2\pi^2} - \frac{1}{2}\cdot\frac{z^4}{3^4\pi^4} - \frac{1}{3}\cdot\frac{z^6}{3^6\pi^6} - \frac{1}{4}\cdot\frac{z^8}{3^8\pi^8} - \dots\right) + \dots.$$

Rearranging the terms (which is justified by Proposition 1.4), we see that the coefficient of z^{2m}, $m \in \mathbb{N}$, in the expansion of $g(z)$ is equal to

$$\frac{1}{m\pi^{2m}}\sum_{k=1}^{\infty}\frac{1}{k^{2m}} = (-1)^m\frac{2^{2m-1}B_{2m}}{m\cdot(2m)!},$$

where B_{2m} is a Bernoulli number (we have used Proposition 11.8). Differentiating yields

$$(\log g(z))' = \sum_{m=1}^{\infty}(-1)^m 2m\cdot\frac{2^{2m-1}B_{2m}}{m\cdot(2m)!}\cdot z^{2m-1} = \sum_{m=1}^{\infty}(-1)^m\frac{2^{2m}B_{2m}}{(2m)!}\cdot z^{2m-1},$$

and this coincides with the expansion for $\cot z - (1/z)$ obtained from (11.7). Thus, the logarithmic derivatives of the functions $g(z)$ and $\sin z/z$ coincide; therefore, their logarithms differ by a constant, so there exists a constant C such that $g(z) = C\sin z/z$. Comparing the values at the origin, we see that $C = 1$. Thus, $f(z) = \sin z$, and we are done. $\square$

11.4 The Mittag-Leffler and Weierstrass Theorems

In the previous three sections, we used infinite sums and products to construct "by hand" meromorphic (respectively, holomorphic) functions with some prescribed poles (respectively, zeros). Now we will see that for any prescribed collection of isolated poles or zeros there exists a meromorphic (or, respectively, holomorphic) function with exactly these poles or zeros.

We begin with the case of prescribed poles. As you will see, one can specify not merely poles with multiplicities, but even principal parts at these poles. Moreover, essential singularities may also be allowed.

Definition 11.30 A *principal part* at a point $a \in \mathbb{C}$ is a series of the form $\sum_{n=1}^{\infty} c_n(z - a)^{-n}$ that converges in a punctured neighborhood of a (cf. Definition 7.7).

It is clear from general properties of power series that if $\sum_{n=1}^{\infty} c_n(z - a)^{-n}$ is a principal part, then the radius of convergence of the series $\sum_{n=1}^{\infty} c_n w^n$ is infinite, so we can also define a principal part at a point a as a function of the form $h(1/(z-a))$ where h is an entire function (which vanishes at the origin, since the series for h has no constant term).

Theorem 11.31 (Mittag-Leffler theorem) *Let $U \subset \mathbb{C}$ be an open set and $S \subset U$ be a subset in U that has no accumulation points. Assume that for each point $a \in S$ we are given a principal part φ_a at a. Then there exists a function holomorphic on $U \setminus S$ such that the principal part of its Laurent series at each point $a \in S$ coincides with φ_a.*

To prove this theorem, one is tempted to define a function f we are seeking as the sum of the given principal parts, but we already know from examples that such a series does not necessarily converge. In Secs. 10.1 and 10.2, we managed to achieve the convergence of the series of principal parts by subtracting appropriate constants, but this does not suffice for the general case: to enhance the convergence, we will subtract functions holomorphic on U (and not necessarily constant). As we will see, the role of these "correction" functions can be played either by polynomials or by rational functions with poles outside U.

We begin with a lemma precisely about the choice of these "correction" functions.

Lemma 11.32 *Let $a, b \in \mathbb{C}$ be two different points, and let φ be a principal part at a of the form $\varphi(z) = h(1/(z - a))$ where h is an entire function. Then for every $r > |a - b|$ and every $\varepsilon > 0$ there exists a rational function R that may have a pole only at b such that $|\varphi(z) - R(z)| \le \varepsilon$ for every z outside the disk of radius r centered at b.*

Proof Set $w = 1/(z - b)$; then it is easy to see that

$$\frac{1}{z-a} = \frac{w}{1+(b-a)w},$$

so $\varphi(z) = h\left(\frac{w}{1+(b-a)w}\right)$. The function $w \mapsto h\left(\frac{w}{1+(b-a)w}\right)$ is holomorphic in the disk $\{w: |w| < 1/|b-a|\}$, so on every closed disk centered at the origin whose radius is strictly less than $1/|b-a|$ it can be approximated with arbitrary precision by a partial sum of its power series. Set $\rho = 1/r$; then $\rho < 1/|b-a|$. If P is a partial sum of the power series for $h\left(\frac{w}{1+(b-a)w}\right)$ at the origin such that

$$\left|h\left(\frac{w}{1+(b-a)w}\right) - P(w)\right| \le \varepsilon \quad \text{for } |w| \le \rho,$$

then, substituting $w = 1/(z-b)$, we obtain

$$\left|\varphi(z) - P\left(\frac{1}{z-b}\right)\right| \le \varepsilon \quad \text{for } |z-b| \ge \frac{1}{\rho} = r.$$

It remains to set $R(z) = P(1/(z-b))$. $\square$

Proof of the Mittag-Leffler theorem First, we show that there exists a sequence of compact subsets $K_j \subset U$ (for all integers $j \ge 1$) and functions R_j holomorphic on U (for all integers $j \ge 2$) with the following properties:

(1) $K_j \subset \text{Int } K_{j+1}$ for all j, and $\bigcup_{j=1}^{\infty} K_j = U$;
(2) if for every $j > 1$ we set $\psi_j = \sum_{a \in S \cap (K_{j+1} \setminus K_j)} \varphi_a$ (this sum is finite, because the

set S, having no accumulation points in U, can have only finitely many points in the compact set $K_j \subset U$), then $\sup_{z \in K_{j-1}} |\psi_j(z) - R_j(z)| \le 1/2^j$.

Indeed, if $U = \mathbb{C}$, then take K_n to be the closed disk of radius n centered at the origin, and if $U \ne \mathbb{C}$, then set

$$K_n = \left\{z \in U: \inf_{w \in \mathbb{C} \setminus U} |z - w| \ge \frac{1}{n} \text{ and } |z| \le n\right\}. \tag{11.21}$$

It is easy to see (check this!) that the sets K_n have the required properties. To choose a function f_j, assume that the set $S \cap (K_{j+1} \setminus K_j)$ consists of m points. For every $a \in S \cap (K_{j+1} \setminus K_j)$, at least one of the following two assertions is true: either $\inf_{w \in \mathbb{C} \setminus U} |a - w| < 1/j$, or $|a| > j$. In the first case, there exists a point $b \in \mathbb{C} \setminus U$ such that $|a - b| < 1/j$; on the other hand, for all $z \in K_{j-1}$ we have $|z - b| \ge 1/(j-1) > 1/j$; applying Lemma 11.32 with $\varepsilon = 1/(m \cdot 2^j)$ and $r = 1/(j-1)$ to the principal part φ_a and the points a and b, we find a function R_a holomorphic on $\mathbb{C} \setminus \{b\}$ (and a fortiori on U) such that $|\varphi_a(z) - R_a(z)| \le 1/(m \cdot 2^j)$ whenever $|z - b| \ge 1/(j-1)$, and hence for all $z \in K_{j-1}$. In the second case, observe that the function φ_a is holomorphic in a neighborhood of the closed disk of radius j centered at the origin; expanding it into a Taylor series at the origin, we see that there exists a polynomial R_a such that $|\varphi_a(z) - R_a(z)| \le 1/(m \cdot 2^j)$ whenever $|z| \le j$, and hence for all $z \in K_{j-1}$. Setting $R_j = \sum\limits_{a \in S \cap (K_{j+1} \setminus K_j)} R_a$, we see that R_j is holomorphic on U and $\sup\limits_{z \in K_{j-1}} |\psi_j(z) - R_j(z)| \le 1/2^j$.

Also, set $\psi_1 = \sum\limits_{a \in K_1} \varphi_a$ and consider the series

$$\psi_1 + \sum_{m=2}^{\infty} (\psi_m - R_m). \tag{11.22}$$

We will show that it converges compactly on U. Indeed, since $\bigcup K_j = U$ and $K_j \subset \mathrm{Int}(K_{j+1})$, it follows that every compact subset $K \subset \mathbb{C}$ is contained in a finite union of several nested open sets $\mathrm{Int}(K_j)$, and hence in some K_n. All terms of the series (11.22) from $(\psi_{n+1} - R_{n+1})$ onwards have no singular points in K_n (and hence in K), and the maxima of their absolute values on K_n (and hence on K) are bounded from above by the convergent series $1/2^n + 1/2^{n+1} + \ldots$. Thus, the series (11.22) converges compactly in $\mathbb{C}$ in the sense of Definition 11.1; obviously, its limit is a function with isolated singularities in S that has principal part φ_a at each point $a \in S$. □

For further reference, we extract a result obtained in the proof of the theorem into a separate statement.

Corollary 11.33 (compact exhaustion lemma) *Let $U \subset \mathbb{C}$ be an open set. Then there exists a sequence of compact subsets $K_j \subset U$, $j \in \mathbb{N}$, such that $K_j \subset \mathrm{Int}(K_{j+1})$ for all j and $\bigcup_j K_j = U$.*
Every compact subset $K \subset U$ is contained in some K_j.

Just as infinite sums allow one to construct a function with prescribed principal parts, infinite products allow one to construct a function with prescribed zeros. The proof of the following theorem uses a method of improving convergence which we have not yet encountered.

Theorem 11.34 (Weierstrass theorem) *Let $U \subset \mathbb{C}$ be an open set, $S \subset \mathbb{C}$ be a subset without accumulation points in U, and assume that for each $a \in S$ we are given a positive integer n_a. Then there exists a function f holomorphic on U that has a zero of order n_a at each point $a \in S$ and no other zeros.*

Proof We may assume without loss of generality that $0 \notin S$: once we have constructed a function that has zeros of prescribed orders at given nonzero points, then, multiplying it by z^k for an appropriate k, we will obtain a zero of prescribed order at 0.

Keeping in mind that $0 \notin S$, consider the family of nested compact subsets $K_n \subset U$ defined, as in the proof of the Mittag-Leffler theorem, by (11.21). For each point $a \in S$ except those in K_1, we define a function φ_a holomorphic on U as follows. Let $a \in K_j$ but $a \notin K_{j-1}$, where $j > 1$; such a j exists because $a \notin K_1$. Since $a \notin K_{j-1}$, we see that either $|a| > j - 1$, or there is a point $b \notin U$ such that $|a - b| > 1/(j-1)$. In the first case set $\varphi_a = z/a$, in the second case set $\varphi_a = (a - b)/(z - b)$.

Lemma 11.35 *Let $a \in S$, $a \in K_j \setminus K_{j-1}$, and let φ_a be the function defined above. Then:*

(1) $1 - \varphi_a$ is a holomorphic function on U which has a single, and simple, zero at a;

(2) there exists a constant $c < 1$ such that

$$\sup_{z \in K_{j-1}} |\varphi_j(z)| \le c.$$

Proof The first claim is obvious. As to the second claim, if $|a| > j - 1$, then $\varphi_a(z) = z/a$ and we can set $c = (j-1)/|a|$; and if $|a| \le j-1$ but $|a-b| < 1/(j-1)$, then we can set $c = \frac{|a-b|}{1/(j-1)}$.   □

Now we arrange the elements of S as a sequence $\{a_j\}$ by taking first all points lying in K_1 (there are finitely many of them, because K_1 is compact and S has no accumulation points in U), then all points lying in $K_2 \setminus K_1$, etc.; a point $a \in S$ with $n_a > 1$ should be repeated n_a times. If the infinite product $\prod_m (1-\varphi_{a_m})$ converged compactly on U, then, multiplying it by the finite product $\prod_{a \in K_1 \cap S} (z - a)^{n_a}$, we would obtain a desired function. However, such a convergence is not guaranteed, so we improve the convergence as follows. For $|t| < 1$, we have the well-defined function $t \mapsto \log(1 - t)$ given by the series (11.20). Multiply each $(1 - \varphi_{a_m})$ by e^{-p_m} where p_m is an appropriate partial sum of the series obtained by substituting φ_{a_m} instead of t into (11.20).

Specifically, if $a_m \in K_j \setminus K_{j-1}$ and $\sup_{z \in K_{j-1}} |\varphi_{a_m}(z)| \le c < 1$, then we choose a positive integer n such that

$$\sum_{k=n}^{\infty} \frac{c^k}{k} \le \frac{1}{2^m}$$

(it does exist because the series $\sum_{k=1}^{\infty} c^k/k$ converges for $|c| < 1$), and set

$$p_m(z) = -\sum_{k=1}^{n-1} \frac{(\varphi_{a_m}(z))^k}{k};$$

then we have

$$\sup_{z \in K_{j-1}} |\log(1 - \varphi_{a_m}(z)) - p_m(z)|$$

$$= \sup_{z \in K_{j-1}} \left| \frac{(\varphi_{a_m}(z))^n}{n} + \frac{(\varphi_{a_m}(z))^{n+1}}{n+1} + \dots \right| \le \sum_{k=n}^{\infty} \frac{c^k}{k} \le \frac{1}{2^m};$$

therefore, the series

$$\sum_{m=2}^{\infty} (\log(1 - \varphi_{a_m}(z)) - p_m(z))$$

converges compactly on U (since every compact subset is contained in some K_j). Exponentiating its partial sums, we see that the infinite product $\prod\limits_{m=1}^{\infty} (1 - \varphi_{a_m}(z)) e^{-P_m(z)}$ also converges compactly on U; multiplying this product by $\prod\limits_{a \in S \cap K_1} (z - a)^{n_a}$, we obtain a desired function.   □

We present also a more special result, which, however, contains an explicit construction of a function with prescribed zeros.

First, we introduce some notation.

Notation 11.36 Set

$$E_k(t) = (1 - t) e^{t + \frac{t^2}{2} + \dots + \frac{t^k}{k}}$$

if k is a positive integer, and

$$E_0(t) = 1 - t.$$

The function $E_k(t)$ is called an *elementary factor*.

If $a \neq 0$, then $E_k(z/a)$ is an entire function with a single, and simple, zero at a.

Proposition 11.37 *Let $\{a_n\}$ be a sequence of nonzero complex numbers such that $0 < |a_1| \leq |a_2| \leq \dots$ (in other words, the elements of the sequence are ordered by increasing absolute value) and $\lim\limits_{n \to \infty} a_n = \infty$. If k is a positive integer such that the series $\sum\limits_{n=1}^{\infty} \frac{1}{|a_n|^{k+1}}$ converges, then the infinite product*

$$h(z) = \prod_{n=1}^{\infty} E_k(z/a_n)$$

converges compactly in $\mathbb{C}$ to an entire function whose set of zeros coincides with the set of numbers a_k and the multiplicity of every zero a_k is equal to the number of its occurrences in the sequence.

Proof It suffices to check that the product $\prod\limits_{n} E_k(z/a_n)$ satisfies the conditions of Definition 11.28. To do this, it suffices to check that for every $R > 0$ the product $\prod\limits_{n \geq n_0} E_k(z/a_n)$ converges uniformly on the disk $\{z \colon |z| \leq R\}$ for some n_0. For this, as in the proof of the Weierstrass theorem, it suffices to establish the uniform convergence of the series

$$\sum_{n \geq n_0} \log |E_k(z/a_n)| = \sum_{n \geq n_0} \operatorname{Re} \log E_k(z/a_n), \qquad (11.23)$$

where $\log t = t - \frac{t^2}{2} + \frac{t^3}{3} - \dots$ for $|t| < 1$. If n_0 is such that $a_{n_0} > R$, then for $n \geq n_0$ we have

$$|\log E_k(z/a_n)| = \left| \frac{z}{a_n} + \frac{1}{2}\frac{z^2}{a_n^2} + \ldots \frac{1}{k}\frac{z^k}{a_n^k} - \sum_{j=1}^{\infty} \frac{1}{j}\frac{z^j}{a_n^j} \right| = \left| \sum_{j=k+1}^{\infty} \frac{1}{j}\left(\frac{z}{a_n}\right)^j \right|$$

$$\leq \frac{1}{k+1} \sum_{j \geq k+1} \left| \frac{z}{a_n} \right|^j = \frac{1}{k+1} \left| \frac{z}{a_n} \right|^{k+1} \frac{1}{1 - \left|\frac{z}{a_n}\right|} \leq \frac{R^{k+1}}{k+1} \cdot \frac{1}{1 - \frac{R}{|a_{n_0}|}} \cdot \frac{1}{|a_n|^{k+1}}.$$

Since, by assumption, the series $\sum \frac{1}{|a_n|^{k+1}}$ converges, it follows that the series (11.23) converges uniformly on the disk of radius R centered at the origin, as required. □

Now we state two other simple corollaries of the Weierstrass theorem.

Corollary 11.38 *Let $U \subset \mathbb{C}$ be an open set, $S \subset \mathbb{C}$ be a subset without accumulation points in U, and assume that for each $a \in S$ we are given an integer n_a. Then there exists a function f meromorphic on U such that $\mathrm{ord}_a(f) = n_a$ for every $a \in S$ and $\mathrm{ord}_a(f) = 0$ for every $a \notin S$.*

To prove this, use the Weierstrass theorem to construct a holomorphic function with zeros of orders n_a for all $n_a > 0$ and a holomorphic function with zeros of orders $-n_a$ for all $n_a < 0$, and then take their quotient.

Corollary 11.39 *Every meromorphic function on an open subset in $\mathbb{C}$ is a quotient of two functions holomorphic on this set.*

To prove this, it suffices to construct a holomorphic function f whose zeros compensate the poles of the given meromorphic function g and observe that the product fg is holomorphic.

11.5 Blaschke Products

The Weierstrass theorem shows that zeros (and their multiplicities) for a holomorphic function can be specified arbitrarily. However, the situation becomes more interesting if additional constraints are imposed on the function. I give an example showing what results can be obtained in this case.

Proposition 11.40 *Let $D = \{z : |z| < 1\}$. Then the following statements are equivalent:*

(1) there exists a bounded holomorphic function $f : U \to \mathbb{C}$ (that is not identically zero) with the set of zeros $\{a_n\}_{n \in \mathbb{N}}$;

(2) the series $\sum_{n=1}^{\infty} (1 - |a_n|)$, in which every term is repeated as many times as the multiplicity of the corresponding zero, converges.

Proof of the implication (1) ⇒ (2) Let $f : U \to \mathbb{C}$ be a bounded holomorphic function with the sequence of zeros $\{a_n\}$, where every zero is repeated as many times as its multiplicity. If f has a zero of order $k > 0$ at the origin, then, replacing $f(z)$ with $f(z)/z^k$, we still have a bounded holomorphic function but without zero at the origin, and the series of the form $\sum(1 - |a_j|)$ for these two functions converge or diverge simultaneously. From this point on, we assume that $f(0) \neq 0$.

Since the convergence of a real positive term series is preserved under permutations of its terms, we may assume that the zeros a_j are arranged in order of increasing absolute value. Further, since $\lim\limits_{x\to 0} |\log(1-x)|/|x| = 1$, the convergence of the series $\sum(1 - |a_n|)$ is equivalent to the convergence of the negative term series $\sum \log |a_n|$, and the latter is equivalent to the fact that the sequence of the exponentials of its partial sums converges to a finite nonzero limit; in other words, to the fact that the infinite product $\prod\limits_{n=1}^{\infty} |a_n|$ converges to a finite nonzero limit. Since the sequence of partial products $|a_1| \cdot |a_2| \cdot \ldots \cdot |a_n|$ is monotone decreasing, it suffices to show that it is bounded from below by a positive constant.

Let $\sup\limits_{z\in U} |f(z)| = C < +\infty$. We will show that

$$\prod_{j=1}^{n} |a_j| \geq \frac{|f(0)|}{C} \quad \text{for all } n. \tag{11.24}$$

Indeed, for each positive integer n consider the function

$$g(z) = \prod_{j=1}^{n} \frac{z - a_j}{1 - \bar{a}_j z}$$

(these functions are called *finite Blaschke products*). By Proposition 3.5, the function g is holomorphic on the disk D, extends to a continuous function on its closure $\bar{D}$, and $|g(z)| = 1$ for $|z| = 1$. I claim that

$$\lim_{r\to 1} \inf_{\theta\in[0;2\pi]} |g(re^{i\theta})| = 1. \tag{11.25}$$

Indeed, the continuous function $z \mapsto |g(z)|$ is uniformly continuous on the compact set $\bar{D}$. Taking into account that $|g(e^{i\theta})| = 1$ for every $\theta \in [0; 2\pi]$, we see that for every $\varepsilon > 0$ there is $\delta > 0$ such that

$$1 - \varepsilon \leq |g(re^{i\theta})| \leq 1 + \varepsilon \quad \text{for } \theta \in [0; 2\pi]$$

whenever $r \in [1 - \delta; 1]$ (by the maximum modulus principle, we can replace $1 + \varepsilon$ by 1, but this is irrelevant). Taking the lower bound over $\theta \in [0; 2\pi]$, we see that for every $\varepsilon > 0$ there is $\delta > 0$ such that $|\inf\limits_{\theta\in[0;2\pi]} |g(re^{i\theta})| - 1| \leq \varepsilon$ whenever $r \in [1 - \delta; 1]$. This is exactly (11.25).

Now set $h(z) = f(z)/g(z)$. By construction, this is a holomorphic function on D. Let $\max(|a_1|, \ldots, |a_n|) < r < 1$. It follows from the maximum modulus principle (Proposition 9.12) and the fact that $|f(z)| \leq C$ for all z that

$$\frac{|f(0)|}{|a_1| \cdot |a_2| \cdot \ldots \cdot |a_n|} = |h(0)| \leq \sup_{|z|=r} |h(z)| \leq \frac{C}{\inf_{|z|=r} |g(z)|};$$

taking the limit as $r \to 1$, we obtain, by (11.25),

$$\frac{|f(0)|}{|a_1| \cdot |a_2| \cdot \ldots \cdot |a_n|} \leq C,$$

which is equivalent to (11.24).   □

To prove the converse statement, that the convergence of the series $\sum(1 - |a_j|)$ implies the existence of a bounded holomorphic function with zeros a_j, we must use infinite products in a more serious way.

Proof of the implication (2) $\Rightarrow$ (1) in Proposition 11.40 We must prove that if $\{a_n\}$ is a sequence of points in the unit disk $U = \{z : |z| < 1\}$ such that $\sum_n (1 - |a_n|) < +\infty$, then there exists a bounded function f holomorphic on U whose set of zeros coincides with the set $\{a_n\}$ and the multiplicity of each zero is equal to the number of its occurrences in the sequence $\{a_n\}$.

To begin with, observe that in the sequence $\{a_n\}$ every number is repeated at most finitely many times, otherwise the series $\sum(1 - |a_n|)$ would diverge. (For the same reason, the sequence $\{a_n\}$ has no accumulation points in U.) In particular, this sequence contains only finitely many zeros, and we may assume without loss of generality that it contains no zeros at all; indeed, if we have already constructed a bounded holomorphic function that has zeros at given points except 0, then we can always multiply it by z^k to obtain a bounded holomorphic function with an additional zero of order k at the origin.

Now consider the infinite product

$$\prod_{n=1}^{\infty} \frac{a_n - z}{1 - \bar{a}_n z} e^{-i \arg a_n} = \prod_{n=1}^{\infty} \frac{a_n - z}{1 - \bar{a}_n z} \frac{|a_n|}{a_n}, \tag{11.26}$$

which is called a *Blaschke product*. The nth factor of this product is an automorphism of the unit disk that sends a_n to 0 and 0 to $|a_n|$, so the absolute value of this factor is bounded by 1, on U it has a unique (simple) zero at a_n, and the absolute values of all partial products in (11.26) are bounded by 1. Hence, if we show that the product (11.26) converges compactly on U, then the limiting function is a desired bounded holomorphic function with zeros at the points a_n. To establish this convergence, note that it suffices to establish the uniform convergence on the disk $D_r = \{z : |z| \leq r\}$ for every $r \in (0; 1)$. It is easy to see that

$$1 - \frac{a_n - z}{1 - \bar{a}_n z} \frac{|a_n|}{a_n} = (1 - |a_n|) \cdot \frac{a_n + |a_n|z}{a_n - |a_n|^2 z} = (1 - |a_n|) \frac{1 + (|a_n|/a_n)z}{1 - \bar{a}_n |a_n| z}.$$

If $|z| \leq r$, then the absolute value of the numerator of the fraction in the right-hand side does not exceed $1 + r$, while the absolute value of the denominator is not less than $1 - r$, so for $z \in D_r$ we have

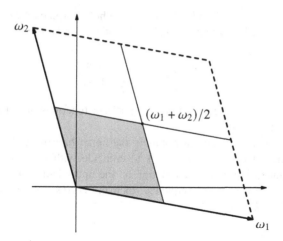

Fig. 11.2

$$\left| 1 - \frac{a_n - z}{1 - \bar{a}_n z} \frac{|a_n|}{a_n} \right| \leq \frac{1 + r}{1 - r}(1 - |a_n|).$$

Since the series $\sum(1 - |a_n|)$ converges by assumption, the infinite product (11.26) converges uniformly on every disk D_r, $0 < r < 1$, by Proposition 11.26. This completes the proof.   □

Exercises

11.1. Write the function $f(z) = 1/\sin z$ as a sum of rational functions with poles at the points πn, $n \in \mathbb{Z}$.

11.2. Show that the signs of the Bernoulli numbers with even indices alternate.

11.3. Find a general formula for the coefficients of the power series expansion of the function $f(z) = \tan z$ at the origin (it will involve Bernoulli numbers).

11.4. Does there exist an elliptic function that has exactly one, and simple, pole in a fundamental parallelogram? (*Hint.* Look at the residue.)

11.5. Let $\Gamma \subset \mathbb{C}$ be a lattice with basis $\langle \omega_1, \omega_2 \rangle$ and $\wp$ be its Weierstrass function. Show that the values of $\wp$ at the points $\omega_1/2$, $(\omega_1 + \omega_2)/2$, and $\omega_2/2$ are different. (*Hint.* Use Proposition 11.13.)

11.6. Show that the Weierstrass function is a conformal map from the "quarter" of the fundamental parallelogram (shown in gray in Fig. 11.2) onto its image. (*Hint.* Use the fact that $\wp$ is even to deduce that its values at points symmetric about the center of the fundamental parallelogram coincide.)

11.7. Express the Eisenstein series G_4 in terms of G_2 and G_3.

11.8. Show that every Eisenstein series G_k with $k > 3$ can be expressed as a polynomial (not depending on the lattice) in G_2 and G_3.

11.9. Let $D = \{z : |z| < 1\}$ and f be a function holomorphic on $\bar{D}$ and continuous on its interior. Show that if $|f(z)| = 1$ on the boundary of D, then either f is a constant, or

$$f(z) = e^{i\theta} \prod_{j=1}^{n} \frac{z - a_j}{1 - \bar{a}_j z}$$

where $\theta \in \mathbb{R}$ and $|a_j| < 1$ for all j (these functions are also called finite Blaschke products).

11.10. Let $\{a_n\}$ be a sequence of points in the upper half-plane. Find a necessary and sufficient condition for $\{a_n\}$ to coincide with the collection of zeros of a bounded holomorphic function on the upper half-plane (if a number is repeated k times in the sequence, then the corresponding zero must have multiplicity k).

Chapter 12
Conformal Maps. Part 2

12.1 The Riemann Mapping Theorem: the Statement and a Sketch of the Proof

The first two sections of this chapter are devoted to the proof of the following important fact.

Theorem 12.1 (Riemann mapping theorem) *If $U \subset \mathbb{C}$ is a simply connected open set that is not the whole complex plane $\mathbb{C}$, then U is isomorphic to the unit disk $D = \{z : |z| < 1\}$.*

This statement calls for some comments. First, $\mathbb{C}$ and U are not conformally isomorphic: by Liouville's theorem, every holomorphic map from $\mathbb{C}$ to U, being a bounded entire function, is a constant.

Further, let us discuss what happens with non-simply connected open sets. Open sets $U_1 \subset \mathbb{C}$ and $U_2 \subset \mathbb{C}$ are said to be *homeomorphic* if there exists a bijective continuous map between them whose inverse is also continuous (this is a special case of the notion of a homeomorphism between topological spaces), so if two open sets are not homeomorphic, then they are, a fortiori, not conformally isomorphic. Hence, the question of whether two domains are conformally isomorphic makes sense only for homeomorphic domains. The Riemann mapping theorem shows that for simply connected sets, the existence of sets that are homeomorphic but not isomorphic is an isolated exception; however, for non-simply connected sets, this is far from being the case. We have already encountered examples of homeomorphic but nonisomorphic non-simply connected open sets in exercises to previous chapters (see Exercises 9.7, 9.10, 9.11, 10.13); in fact, in the non-simply connected case, the existence of such sets is more of a norm than an exception.

Now we turn to the proof of the Riemann mapping theorem. Its first steps are quite simple.

Proposition 12.2 *Every simply connected open subset $U \subset \mathbb{C}$ that is not the whole complex plane $\mathbb{C}$ is isomorphic to a bounded open set.*

© Springer Nature Switzerland AG 2020
S. Lvovski, *Principles of Complex Analysis*, Moscow Lectures 6,
https://doi.org/10.1007/978-3-030-59365-0_12

Proof Since $U \neq \mathbb{C}$, there exists a point $a \in \mathbb{C} \setminus U$. Since U is simply connected, by Proposition 6.17, there exists a holomorphic function $f \colon U \to \mathbb{C}$ that is a single-valued branch of the function $\sqrt{z - a}$. The function f is a conformal map from U onto its image $V = f(U)$: indeed, since $f(z)^2 + a = z$ for every $z \in U$, it follows from $f(z_1) = f(z_2)$ that $z_1 = z_2$. Further, setting $-W = \{-w \colon w \in W\}$, we have $W \cap -W = \emptyset$. Indeed, arguing by contradiction, if $w \in W \cap (-W)$, then $w \neq 0$ (because $W \not\ni 0$) and we have $w = f(z_1)$, $-w = f(z_2)$, where $z_1, z_2 \in U$, $z_1 \neq z_2$; on the other hand, $z_1 = w^2 + a = (-w)^2 + a = z_2$, a contradiction.

The open set $-W$, which is disjoint with W, contains a closed disk of the form $\Delta = \{z \colon |z - b| \leq \varepsilon\}$. Since $\Delta \cap W = \emptyset$, the function $w \mapsto 1/(w - b)$ is a conformal map from W onto a domain contained in the disk of radius $1/\varepsilon$ centered at the origin; thus, W is isomorphic to a bounded set, while, on the other hand, W is isomorphic to U. $\qquad\square$

Every bounded open set can be mapped onto a subset of the unit disk $D = \{z \colon |z| < 1\}$ by a function of the form $z \mapsto az + b$. Hence, we may (and will) assume, without loss of generality, that $U \subset D$.

Throughout the proof, we fix a point $a \in U$.

Notation 12.3 By S we denote the set of holomorphic injective maps $f \colon U \to D$ such that $|f'(a)| \geq 1$.

Key lemma 12.4 *There exists a map $f \in S$ for which the number $|f'(a)|$ is the greatest possible.*

Note that the set S is nonempty: since $U \subset D$, it necessarily contains the map $z \mapsto z$.

The proof of the key lemma occupies the whole next section, and for now we explain how it implies the Riemann mapping theorem.

Proposition 12.5 *A map $f \in S$ for which the number $|f'(a)|$ is the greatest possible is a conformal isomorphism between U and D.*

Proof A map $f \in S$ (i.e., a holomorphic injective map from U to D) for which $|f'(a)|$ is the greatest possible will be said to be *optimal*. By the very definition, such a map necessarily satisfies $|f'(a)| \geq 1$. It suffices to show that an optimal map f satisfies $f(U) = D$.

For this, first observe that if f is optimal, then $f(a) = 0$. Indeed, arguing by contradiction, if $f(a) = b \neq 0$, then f can be "improved" as follows. Let $\varphi \colon U \to U$ be the automorphism given by $\varphi(z) = \frac{z - b}{1 - \bar{b}z}$. I claim that $|(\varphi \circ f)'(a)| > |f'(a)|$. Indeed, by the chain rule,

$$|(\varphi \circ f)'(a)| = |f'(a)| \cdot |\varphi'(f(a))| = |f'(a)| \cdot |\varphi'(b)|,$$

and

$$\varphi'(z) = \frac{1 - \bar{b}z + \bar{b}(z - b)}{(1 - \bar{b}z)^2} = \frac{1 - |b|^2}{(1 - \bar{b}z)^2}, \qquad |\varphi'(b)| = \frac{1}{1 - |b|^2} > 1.$$

Thus, $|(\varphi \circ f)'(a)| > |f'(a)|$, and this cannot be the case since f is optimal, a contradiction.

So, $f(a) = 0$. To establish that $f(U) = D$, assume, arguing by contradiction, that there exists $b \in D \setminus f(U)$; since $f(a) = 0$, it follows that $b \neq 0$. We will show that in this case f can also be "improved." This time, the argument proceeds in three steps. First, consider the same automorphism $\varphi \colon U \to U$ given by $\varphi(z) = \frac{z-b}{1-\bar{b}z}$, and set $\varphi_1 = \varphi \circ f$; since f does not assume the value b, it follows that $\varphi \circ f$ does not assume the value $\varphi(b) = 0$. Further, set $V = (\varphi \circ f)(U)$; the map $\varphi \circ f$ is a conformal isomorphism between U and V, the set U is simply connected, hence the set V is also simply connected. Besides, V does not contain the origin, so on V we have a well-defined branch of the function $\sqrt{z}$; denote it by g. As in the proof of Proposition 12.2 above, we see that $g \colon V \to g(V)$ is a conformal map from V onto its image contained in U, and $g \circ \varphi \circ f$ is a conformal map from U onto its image contained in U.

Note that $\varphi(0) = -b$, so $(g \circ \varphi)(0) = g(-b) = \sqrt{-b}$ (we do not specify which of the two possible values of the square root is meant, since this is irrelevant for what follows). Now compose $g \circ \varphi$ with another disk automorphism φ_1 sending $\sqrt{-b}$ to 0:

$$\varphi_1(z) = \frac{z - \sqrt{-b}}{1 - \sqrt{-b} \cdot z}.$$

Namely, set $f_1 = \varphi_1 \circ g \circ \varphi \circ f$. This is also an injective holomorphic map from U to D, and I claim that $|f_1'(a)| > |f'(a)|$. Indeed, by the chain rule,

$$|f_1'(a)| = |f'(a)| \cdot |\varphi'(f(a))| \cdot |g'(\varphi(f(a)))| \cdot |\varphi_1'(g(\varphi(f(a))))|$$
$$= |f'(a)| \cdot |\varphi'(0)| \cdot |g'(-b)| \cdot |\varphi_1'(\sqrt{-b})|. \quad (12.1)$$

A direct calculation similar to that given above shows that $|\varphi'(0)| = 1 - |b|^2$ and

$$|\varphi_1'(\sqrt{-b})| = \frac{1}{1 - |\sqrt{-b}|^2} = \frac{1}{1 - (\sqrt{|b|})^2} = \frac{1}{1 - |b|}.$$

Finally, the differentiation rule for inverse functions shows that every branch of $\sqrt{z}$ satisfies $(\sqrt{z})' = 1/(2\sqrt{z})$, whence $|g'(-b)| = |1/(2\sqrt{-b})| = 1/(2\sqrt{|b|})$. Then it follows from (12.1) that

$$\frac{|f_1'(a)|}{|f'(a)|} = \frac{1 - |b|^2}{2\sqrt{|b|}(1 - |b|)} = \frac{1 + |b|}{2\sqrt{|b|}} > 1$$

by the arithmetic-geometric mean inequality. Therefore, $|f_1'(a)| > |f'(a)|$, contradicting the optimality of f. The contradiction shows that $f(U) = D$, so f defines a conformal isomorphism between U and D. $\qquad\square$

Indulging in a little demagoguery, we may say that the above proof of the existence of a conformal map from U onto D is constructive. Indeed, it provides the following "algorithm": begin with an embedding of U into D; if it is not an isomorphism, then improve the map by first sending $f(a)$ to 0 and then applying two linear fractional transformations and a square root; if the image of the improved map does not coincide with the whole disk D, then improve it again, etc.

So, it remains only to prove the key lemma 12.4.

12.2 The Riemann Mapping Theorem: Justifications

To prove the key lemma, we must develop an entire theory.

Notation 12.6 Let U be an (arbitrary) open subset in $\mathbb{C}$. By $H(U)$ we denote the set of all holomorphic functions on U.

Definition 12.7 A subset $\mathcal{F} \subset H(U)$ is said to be *closed* if for every sequence of functions $f_n \in \mathcal{F}$ compactly convergent on U to some function f, we have $f \in \mathcal{F}$.

A subset $\mathcal{F} \subset H(U)$ is said to be *compact* if every sequence of functions $f_n \in \mathcal{F}$ contains a subsequence compactly convergent on U to a function $f \in \mathcal{F}$.

If $\mathcal{F} \subset H(U)$ and $\Phi \colon \mathcal{F} \to \mathbb{C}$ is a map, then Φ is said to be *continuous* if for every sequence of functions $f_n \in \mathcal{F}$ compactly convergent on U to a function $f \in \mathcal{F}$, we have

$$\lim_{n \to \infty} \Phi(f_n) = \Phi(f).$$

For example, if $a \in U$, then the maps $f \mapsto f(a)$ and $f \mapsto f'(a)$ are continuous on $H(U)$: the first claim is obvious, and the second one follows from Proposition 5.17.

Remark 12.8 (for the reader familiar with topology). It is easy to see that the set $H(U)$ together with the collection of closed sets defined above is a topological space. One can check that this topology on $H(U)$ can be induced by a metric and that the standard definitions of continuity and compactness are equivalent to those given above. None of this will be used below, all our arguments are direct.

The proofs of the next two propositions reproduce standard arguments probably familiar to most readers.

Proposition 12.9 *If $\mathcal{F} \subset H(U)$ is a compact set and $\mathcal{F}_1 \subset \mathcal{F}$ is a closed subset, then the subset $\mathcal{F}_1 \subset H(U)$ is compact too.*

Proof Let $\{f_n\}$ be a sequence of functions from $\mathcal{F}_1 \subset \mathcal{F}$. Since $\mathcal{F}$ is compact, it has a (compactly) convergent subsequence, and since $\mathcal{F}_1 \subset H(U)$ is closed, the limit of this subsequence lies in $\mathcal{F}_1$.   □

Proposition 12.10 *Let $\mathcal{F} \subset H(U)$ be a compact subset and $\Phi \colon \mathcal{F} \to \mathbb{R}$ be a continuous map. Then Φ attains its maximum on $\mathcal{F}$ at some function $f \in \mathcal{F}$.*

Proof First, we show that Φ is bounded. Indeed, if this is not the case, then there exists a subsequence $\{f_{n_k} = g_k\}_{k \in \mathbb{N}}$ such that $\Phi(g_k) \geq k$. Since $\mathcal{F}$ is compact, the sequence $\{g_k\}$ has a subsequence $\{g_{k_r}\}$ that converges to a function $g \in \mathcal{F}$; by the continuity of Φ, we have $\lim_{r \to +\infty} \Phi(g_{k_r}) = \Phi(g)$, and this is impossible because $\Phi(g)$ is a real number while $\lim \Phi(g_{k_r}) = +\infty$.

Since Φ is bounded on $\mathcal{F}$, set $M = \sup_{f \in \mathcal{F}} \Phi(f) < +\infty$. There exists a sequence $f_n \in \mathcal{F}$ such that $\lim \Phi(f_n) = M$. Extracting a convergent subsequence $f_{n_k} = g_k$, where $\lim g_k = g \in \mathcal{F}$, we see, by continuity, that $\Phi(g) = \lim \Phi(g_k) = M$, so the maximum of Φ is attained at the function $g \in \mathcal{F}$. $\square$

We need the following result on the compactness of subsets in $H(U)$.

Theorem 12.11 (Montel's theorem) *Let $U \subset \mathbb{C}$ be an open set and $M > 0$. Then the set $\{f \in H(U) : \sup_{z \in U} |f(z)| \leq M\}$ is compact in $H(U)$.*

Every proof of Montel's theorem involves extracting uniformly convergent subsequences from a given sequence of functions. It is based on the following general analytic result.

Theorem 12.12 (Arzelà–Ascoli theorem) *Let $K \subset \mathbb{C}$ be a compact set and S be a family of complex-valued continuous functions on K satisfying the following conditions:*

(1) there exists $M > 0$ such that $|f(x)| \leq M$ for all $f \in S$ and $x \in K$;

(2) for every $\varepsilon > 0$ there is $\delta > 0$ such that the inequality $|z_1 - z_2| \leq \delta$, where $z_1, z_2 \in K$, implies the inequality $|f(z_1) - f(z_2)| \leq \varepsilon$ for all $f \in S$.

Then every sequence of functions from S has a subsequence that converges uniformly on K.

A family of functions satisfying condition (1) is said to be *uniformly bounded*, and that satisfying condition (2) is said to be *equicontinuous*.

The Arzelà–Ascoli theorem remains valid upon replacing K with an arbitrary compact metric space.

I omit the proof of this theorem; it can be found in a sufficiently detailed textbook of analysis, see, e.g., [6, Chap. XVI, Sec. 4]). Also, you can prove the Arzelà–Ascoli theorem yourself. A possible plan is as follows: first, use compactness to deduce that there exists a countable subset $S \subset K$ such that $\bar{S} = K$, then use uniform boundedness to deduce that every sequence of functions $\{f_n\}$ from S contains a subsequence $\{g_k\}$ that converges at every point of S, and, finally, use equicontinuity (and compactness) to conclude that if a sequence of functions from S converges at every point of S, then it converges uniformly on the whole set K.

The main part of the proof of Montel's theorem is summarized in the following lemma.

Lemma 12.13 *Let $U \subset \mathbb{C}$ be an open set, and let $\Delta = \{z : |z-a| \leq r\}$ be a closed disk contained in U. If $\{f_n\}$ is a uniformly bounded sequence of functions holomorphic on U (this means that there exists $M > 0$ such that $|f_n(z)| \leq M$ for every $z \in U$ and every n), then the sequence $\{f_n\}$ contains a subsequence that converges uniformly on Δ.*

Proof Clearly, there exists a number $r' > r$ such that the disk $\{z: |z - a| \le r'\}$ is also contained in U. If f is a holomorphic function on U and $\sup_{z \in U} |f(z)| \le M$, then, by (5.17), for every $z \in \Delta$ we have

$$f'(z) = \frac{1}{2\pi i} \int\limits_{|z-a|=r'} \frac{f(\zeta)\, d\zeta}{(\zeta - z)^2}. \tag{12.2}$$

Since for $|\zeta - a| = r'$ and $|z - a| \le r$ we have $|\zeta - z| \ge r' - r$ and $|f(z)| \le M$, it follows from (12.2) and Proposition 4.15 that

$$|f'(z)| \le C = \frac{Mr'}{(r' - r)^2} \quad \text{for } z \in \Delta.$$

Therefore, for $z_1, z_2 \in \Delta$ we have

$$|f(z_1) - f(z_2)| = \left| \int\limits_{[z_1, z_2]} f'(\zeta)\, d\zeta \right| \le C|z_1 - z_2|, \tag{12.3}$$

where $[z_1, z_2]$ stands for the line segment between z_1 and z_2.

If we now denote by $\mathcal{F}$ the family of functions on Δ that are restrictions of holomorphic functions $f: U \to \mathbb{C}$ such that $\sup_{z \in U} |f(z)| \le M$, then $\mathcal{F}$ satisfies the conditions of the Arzelà–Ascoli theorem: by construction, all functions from $\mathcal{F}$ are uniformly bounded in absolute value by the number M, and the equicontinuity follows from the fact that if $z_1, z_2 \in \Delta$ and $|z_1 - z_2| \le \varepsilon/C$, then, by (12.3), we have $|f(z_1) - f(z_2)| \le \varepsilon$ for every $f \in \mathcal{F}$. The existence of a subsequence that converges uniformly on Δ is now obvious. □

Now we can prove Montel's theorem itself. All the hard work has been already done, it remains only to perform some simple manipulations.

Proof of Theorem 12.11 Let $\{f_n\}$ be an arbitrary sequence of holomorphic functions on U that are bounded in absolute value by a number M. To prove Montel's theorem, we must show that $\{f_n\}$ contains a subsequence that converges compactly on U.

First, we will prove a weaker statement: if $K \subset U$ is an arbitrary compact set, then the sequence $\{f_n\}$ contains a subsequence that converges uniformly on K. Indeed, K is contained in the union of finitely many closed disks $\Delta_1, \ldots, \Delta_n$ lying in U. By Lemma 12.13, from $\{f_n\}$ we can extract a subsequence that converges uniformly on Δ_1; from it, a subsequence that converges uniformly on Δ_2; etc. At the final nth step, we obtain a subsequence of the original sequence that converges uniformly on every disk Δ_j, $1 \le j \le n$. Therefore, this subsequence converges uniformly on their union $\Delta_1 \cup \ldots \cup \Delta_n$. Since $K \subset \Delta_1 \cup \ldots \cup \Delta_n$, it converges on K, too.

Now let $U = \bigcup_{j=1}^{\infty} K_j$ where K_j are compact sets and $K_j \subset \text{Int}(K_{j+1})$ for all j. Since every compact subset $K \subset U$ is contained in some K_j (see Lemma 11.33), a sequence of functions converges compactly on U if and only if it converges uniformly on each set K_j.

As proved above, from $\{f_n\}$ we can extract a subsequence $f_{11}, f_{12}, \ldots$ that converges uniformly on K_1; from $\{f_{12}, f_{13}, \ldots\}$ (observe that this sequence does not contain f_{11}!), we can extract a subsequence $f_{21}, f_{22}, \ldots$ that converges uniformly on K_2; from $\{f_{23}, f_{24}, \ldots\}$, we can extract a subsequence that converges uniformly on K_3; etc.: every sequence $\{f_{(m+1),j}\}_{j \in \mathbb{N}}$ converges uniformly on K_{m+1} and is a subsequence in $\{f_{mj}\}_{j \in \mathbb{N}}$. Then the diagonal sequence $f_{11}, f_{22}, \ldots$ converges uniformly on every set K_m, because from the term f_{mm} onwards it is a subsequence of the sequence $\{f_{mj}\}_{j \in \mathbb{N}}$, which converges uniformly on K_m. This completes the proof of Montel's theorem. $\qquad\square$

Now we return to the key lemma. Let D be the unit disk, $U \subset D$ be a bounded simply connected domain, $a \in U$ be a point, S be the set of injective holomorphic maps $f: U \to D$, and $\mathcal{F}$ be the set of holomorphic functions f on U such that $\sup_{z \in U} |f(z)| \leq 1$. By Montel's theorem, the set $\mathcal{F}$ is closed; hence if we show that S is closed in $\mathcal{F}$, then we will conclude that S is also compact (by Proposition 12.9) and the key lemma will follow from Proposition 12.10.

Proposition 12.14 *Let $U \subset \mathbb{C}$ be a connected open set and $\{f_n\}$ be a sequence of injective holomorphic functions on U. If this sequence converges compactly on U to a nonconstant function f, then the function $f: U \to \mathbb{C}$ is also injective.*

Proof Arguing by contradiction, assume that $f(a) = f(b)$ where $a, b \in U$, $a \neq b$. Since U is connected, the points a and b can be joined by a polygonal curve that lies entirely in U (to prove this, first join a and b by an arbitrary continuous curve, and then replace it by a polygonal curve as in the proof of the homotopy enhancement lemma 6.4); moreover, we may assume that this polygonal curve does not cross itself. For every sufficiently small $\varepsilon > 0$, the union of the disks of radius ε centered at all points of the polygonal curve is a domain with a piecewise smooth boundary (this boundary consists of line segments and circular arcs). Denote this domain by V_ε and its boundary by γ_ε (see Fig. 12.1).

Set $f(a) = f(b) = v$. Since the function f is holomorphic and nonconstant on U and the set U is connected, the set $f^{-1}(v)$ has no accumulation points in U. Hence, for all sufficiently small ε, the curve γ_ε does not contain points at which f assumes the value v. Since $f_n \to f$ uniformly on γ_ε, for all sufficiently large n we have $f_n(z) \neq v$ for all $z \in \gamma_\varepsilon$. Since $f_n' \to f'$ uniformly on γ_ε too, we have

$$\lim_{n \to \infty} \frac{f_n'(z)}{f_n(z) - v} = \frac{f'(z)}{f(z) - v} \quad \text{uniformly on } \gamma_\varepsilon.$$

Therefore,

$$\lim_{n \to \infty} \frac{1}{2\pi i} \int_{\gamma_\varepsilon} \frac{f_n'(z)\, dz}{f_n(z) - v} = \frac{1}{2\pi i} \int_{\gamma_\varepsilon} \frac{f'(z)\, dz}{f(z) - v}. \tag{12.4}$$

Since $f(a) = f(b) = v$, the function $f - v$ has at least two zeros in V_ε, so, by Corollary 8.9, the right-hand side of (12.4) is an integer not less than 2. On the other

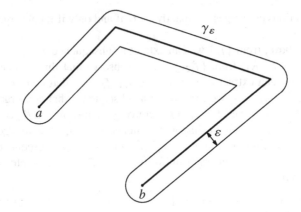

Fig. 12.1

hand, since all functions f_n are injective, the functions $f_n - v$ can have at most one zero, which is necessarily of multiplicity 1, even in the whole domain U, so all terms of the sequence in the left-hand side of (12.4) are equal to 0 or 1, a contradiction.

Completing the proof of the key lemma Let $\{f_n\}$ be a sequence of functions from S that converges compactly on U to a function f. Since $|f_n'(a)| \geq 1$ for all n, we have $|f'(a)| \geq 1$, so f is not a constant and hence, by Proposition 12.14, is injective. Clearly, $|f(z)| \leq 1$ for all $z \in U$; since f is not a constant, $|f(z)|$ can never equal 1 by the open mapping theorem. Thus, $f(U) \subset D$ and $f \in S$, so the set S is closed in $\mathcal{F}$, which was all that remained to be proved. $\square$

We have shown that a function that maps U conformally onto D maximizes the value $|f'(a)|$ under some additional conditions (injectivity). In fact, one can show (see Exercise 12.3) that (if $U \neq \mathbb{C}$ is a simply connected domain and $a \in U$) a map $f : U \to D$ that maximizes $|f'(a)|$ without additional conditions is also a conformal isomorphism.

12.3 The Schwarz–Christoffel Formula

In Chap. 10, we have seen that integrals like (10.1) can be used to define a conformal map from a half-plane onto some rectangles. Now we will show that a conformal map from the upper half-plane onto a quite arbitrary polygon is also given by an integral of this type.

Throughout this section, $H = \{z : \operatorname{Im} z > 0\}$ stands for the upper half-plane.

Theorem 12.15 (Schwarz–Christoffel theorem) *Let $U \subset \mathbb{C}$ be the interior of a polygon $A_1 A_2 \ldots A_n$ (which is assumed to be bounded and non-self-crossing, but not necessarily convex). Denote the interior angle of this polygon at the vertex A_j by $\pi \alpha_j$, $0 < \alpha_j < 2$.*

Let $f \colon H \to U$ be a conformal isomorphism. If the preimages of the vertices $A_1, \ldots, A_n$ under the boundary correspondence are points $a_1, \ldots, a_n$ on the real axis (and not the point ∞), then there exist constants C_1 and C_2 such that f has the form

$$f(z) = C_1 \int_{z_0}^{z} (z - a_1)^{\alpha_1 - 1} \cdot \ldots \cdot (z - a_n)^{\alpha_n - 1} \, dz + C_2. \qquad (12.5)$$

If the preimages of the vertices $A_1, \ldots, A_{n-1}$ under the boundary correspondence are points $a_1, \ldots, a_{n-1}$ on the real axis and the preimage of the vertex A_n is the point ∞, then there exist constants C_1 and C_2 such that f has the form

$$f(z) = C_1 \int_{z_0}^{z} (z - a_1)^{\alpha_1 - 1} \cdot \ldots \cdot (z - a_{n-1})^{\alpha_{n-1} - 1} \, dz + C_2. \qquad (12.6)$$

This statement calls for some comments. First, a conformal map $f \colon H \to U$ does always exist by the Riemann mapping theorem 12.1. Further, the integrals in (12.5) and (12.6) are taken over an arbitrary path in the upper half-plane from z_0 to z, where z_0 is an arbitrary point in the upper half-plane (or even on the real axis, see Sec. 10.3). Changing the initial point z_0 alters the integral by an additive constant, which anyway gets absorbed into the indefinite constant C_2, so it does not affect the form of the formulas. Finally, $(z - a_j)^{\alpha_j - 1}$ is understood as $e^{(\alpha_j - 1) \log(z - a_j)}$ where $\log(z - a_j)$ is a single-valued branch of the logarithm in the upper half-plane. Upon choosing another branch, the integrand and the integral get multiplied by $e^{2\pi i k (\alpha_j - 1)}$, and this factor gets absorbed into the indefinite constant C_1, so, again, changing the branches of the logarithm does not affect the form of the formulas.

The remainder of this section will be devoted to the proof of Theorem 12.15. We begin with the case where the vertex A_n corresponds to the point ∞, the general case being even easier.

First of all, let us agree on notation. We may assume without loss of generality that $a_1 < a_2 < \ldots < a_{n-1}$. Set $\overline{\mathbb{R}} = \mathbb{R} \cup \{\infty\}$ and $a_n = \infty$. The extension of f to a continuous bijection from $H \cup \overline{\mathbb{R}}$ to $\overline{U}$ will be denoted by $\tilde{f} \colon H \cup \overline{\mathbb{R}} \to \overline{U}$. Finally, the lower half-plane will be denoted by $L = \{z \in \mathbb{C} \colon \operatorname{Im} z < 0\}$.

Now observe that we can apply the reflection principle (Proposition 10.5) to the map $f \colon H \to U$ as follows. Choosing an arbitrary interval $(a_j, a_{j+1}) \subset \overline{\mathbb{R}}$ (the case $j = n$ is not excluded: in this case we set $a_{j+1} = a_1$, $A_{j+1} = A_1$), the map $F \colon H \cup (a_j, a_{j+1}) \cup L \to \mathbb{C}$ given by

$$F(z) = \begin{cases} \tilde{f}(z), & z \in H \cup (a_j; a_{j+1}), \\ S_{A_j A_{j+1}}(f(\bar{z})), & z \in L \end{cases}$$

(here $S_{A_j A_{j+1}}$ stands for the reflection in the line $A_j A_{j+1}$) is holomorphic and, moreover, unramified on $H \cup (a_j, a_{j+1}) \cup L$. Thus, the holomorphic function f can be continued analytically along every path whose intersection with the real axis is contained in one of the intervals (a_j, a_{j+1}).

Since the restriction of the function F to the lower half-plane L is, obviously, a conformal isomorphism between L and the reflection of the original polygon in the real axis, we see that the restriction of F to L can also be continued analytically along every path whose intersection with the real axis is contained in one of the intervals (a_j, a_{j+1}). Combining these analytic continuations, we deduce that f can be continued analytically along every path in $\mathbb{C}$ that does not pass through the points $a_1, \ldots, a_{n-1}$, and every germ obtained from such a map is a germ of a function with nonzero derivative (see Corollary 10.6).

What is the effect of analytic continuation along a closed path on the germ of f at a point of the upper half-plane? One can easily see that every closed path in $\mathbb{C} \setminus \{a_1, \ldots, a_{n-1}\}$ is homotopic to a piecewise linear path (see Exercise 6.6), which can, obviously, be chosen so that no component of this path lies on the real axis. Then this closed path crosses the real axis an even number of times. Any two successive crossings of the real axis (say, on intervals (a_k, a_{k+1}) and (a_j, a_{j+1})) correspond to the germ of f turning into the germ of the function

$$z \mapsto S_{A_k, A_{k+1}}(S_{A_j, A_{j+1}}(f(\bar{z}))) = S_{A_k, A_{k+1}}(S_{A_j, A_{j+1}}(f(z))).$$

However, it is well known that the composition of reflections in two lines is either a rotation or a translation. Hence, the germ of f turns into the germ of $Af + B$ where A and B are constants, $A \neq 0$ (moreover, $|A| = 1$, but this is irrelevant). Since every germ of f in the upper half-plane can be continued analytically along every path that does not pass through $a_1, \ldots, a_{n-1}$, and since

$$\frac{(Af + B)''}{(Af + B)'} = \frac{f''}{f'},$$

we see that f''/f' extends to a holomorphic function on $\mathbb{C} \setminus \{a_1, \ldots, a_{n-1}\}$.

Lemma 12.16 *Each of the isolated singularities a_j, $1 \leq j \leq n - 1$, of the function f''/f' is a simple pole with residue $\alpha_j - 1$.*

Proof As above, let $\tilde{f}$ be the continuous extension of f to $H \cup \mathbb{R}$. For all sufficiently small $\varepsilon > 0$, the map $\tilde{f}$ takes the set $U_\varepsilon = \{z \colon |z - a_j| < \varepsilon, \operatorname{Im} z \geq 0\}$ to a neighborhood of the vertex A_j in the closure of the polygon U (Fig. 12.2).

In the angle obtained from the angle $A_{j-1} A_j A_{j+1}$ by moving the vertex A_j to the origin, choose a branch of the logarithm and consider on $U_\varepsilon \setminus \{a_j\}$ the function

$$g \colon z \mapsto (\tilde{f}(z) - A_j)^{1/\alpha_j} := e^{\log(\tilde{f}(z) - A_j)/\alpha_j}. \tag{12.7}$$

This function is holomorphic on $\operatorname{Int}(U_\varepsilon)$, extends by continuity to $U_\varepsilon \setminus \{a_j\}$, defines a conformal isomorphism from $\operatorname{Int}(U_\varepsilon)$ onto its image, and also a continuous bijection from the set $(a_j - \varepsilon, a_j) \cup (a_j, a_j + \varepsilon)$ onto a punctured neighborhood of the origin on a line passing through the origin. By the reflection principle, the function g can be continued analytically to a holomorphic function defined in a punctured neighborhood of the point a_j and satisfying the relation $g(\bar{z}) = \overline{g(z)}$. Therefore, the continued function g is bounded in a punctured neighborhood of a_j and thus can be

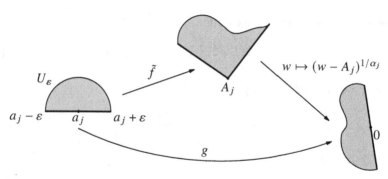

Fig. 12.2

extended to this point as a holomorphic function with $g(a_j) = 0$. By construction, in a punctured neighborhood of a_j the function g is one-to-one onto its image, so it is unramified at a_j, that is, $g(z) = (z - a_j)h(z)$ where h is holomorphic at a_j and $h(a_j) = 0$. Now we have (all calculations refer to restrictions of functions to an open set on which the corresponding logarithms are well defined; the resulting identity can then be extended to the whole domain of definition by the principle of analytic continuation):

$$f(z) = A_j + (g(z))^{\alpha_j};$$
$$f'(z) = \alpha_j(g(z))^{\alpha_j - 1}g'(z);$$
$$\log f'(z) = \log \alpha_j + (\alpha_j - 1)\log g(z) + \log g'(z);$$
$$\frac{f''(z)}{f'(z)} = (\log f'(z))' = (\alpha_j - 1)\frac{g'(z)}{g(z)} + \frac{g''(z)}{g'(z)}. \qquad (12.8)$$

Since g has a simple zero at a_j, the function g'/g has at this point a simple pole with residue 1 and the function g' does not vanish. Hence, the second term in (12.8) is holomorphic at a_j, while the first term has a simple pole with residue $\alpha_j - 1$. Therefore, the function f''/f' also has at a_j a similar pole, which proves the lemma.□

Having analyzed the isolated singularities of the function f''/f', we now investigate its behavior at infinity.

Lemma 12.17 $\lim\limits_{|z| \to \infty} (f''(z)/f'(z)) = 0.$

Proof We proceed as in the proof of the previous lemma. Namely, since $\tilde{f}(\infty) = A_n$ and the interior angle at the vertex A_n is equal to $\pi\alpha_n$, set $g(z) = (f(z) - A_n)^{1/\alpha_n}$. As above, g is a conformal map from a "semi-neighborhood of infinity" of the form $\{z : |z| > R, \text{Im } z > 0\}$, where R is a sufficiently large number, onto a "semi-neighborhood of the origin" (the intersection of a neighborhood of the origin with the upper half-plane). As above, analytic continuation using the reflection principle shows that g is a holomorphic and injective function on some punctured neighborhood of infinity, with $\lim\limits_{|z| \to \infty} g(z) = 0$. Hence, the Laurent series of g in a neighborhood of infinity has the form

$$g(z) = \frac{c_{-1}}{z} + \frac{c_{-2}}{z^2} + \dots, \qquad c_1 \neq 0. \tag{12.9}$$

As above, we obtain the identity

$$\frac{f''(z)}{f'(z)} = (\alpha_j - 1)\frac{g'(z)}{g(z)} + \frac{g''(z)}{g'(z)}; \tag{12.10}$$

since (12.9) implies that

$$g'(z) = -\frac{c_{-1}}{z^2} + \dots, \qquad g''(z) = \frac{2c_{-1}}{z^3} + \dots,$$

both terms in the right-hand side of (12.10) tend to zero as $|z| \to \infty$, which proves the lemma. □

Lemmas 12.16 and 12.17 imply that the function

$$h(z) = \frac{f''(z)}{f'(z)} - \sum_{j=1}^{n-1} \frac{\alpha_j - 1}{z - a_j}$$

is an entire function that tends to zero as $|z| \to \infty$; then it follows from Liouville's theorem that h is constant, and hence identically zero, so

$$(\log f'(z))' = \frac{f''(z)}{f'(z)} = \sum_{j=1}^{n-1} \frac{\alpha_j - 1}{z - a_j}$$

(again, we first obtain the identity on an open set where a branch of the logarithm is well defined, and then extend it to the whole domain of definition by the principle of analytic continuation). Now we successively obtain

$$\log f'(z) = \sum_{j=1}^{n-1}(\alpha_j - 1)\log(z - a_j) + C_0;$$

$$f'(z) = e^{C_0}(z - a_1)^{\alpha_1 - 1} \dots (z - a_{n-1})^{\alpha_{n-1} - 1};$$

$$f(z) = C_1 \int_{z_0}^{z} (z - a_1)^{\alpha_1 - 1} \cdot \dots \cdot (z - a_{n-1})^{\alpha_{n-1} - 1} \, dz + C_2,$$

where $C_1 = e^{C_0}$; this is exactly the formula we wanted to establish.

It remains to consider the case where $\tilde{f}(\infty)$ lies on a side of the polygon and not at a vertex. In this case, the proof up to Lemma 12.16 inclusive proceeds exactly in the same way, and we deduce that f''/f' extends to a function holomorphic on the whole complex plane $\mathbb{C}$ except for simple poles at $a_1, \dots, a_n$ with residues $\alpha_1 - 1, \dots, \alpha_n - 1$.

In order to show that $\lim_{|z|\to\infty} (f''(z)/f(z)) = 0$, set $A_{n+1} = \tilde{f}(\infty)$ and repeat the proof of Lemma 12.17 assuming that the polygon has a fictitious vertex at A_{n+1} with interior angle π. The proof proceeds in the same way, the only simplification being that now we need not raise to the power $1/\alpha_{n+1} = 1$ and can apply the reflection principle to the function f itself. The rest of the proof goes through as before.

Thus, Theorem 12.15 is proved.

12.4 The Hyperbolic Metric

In this section, we discover the true meaning of the Schwarz lemma.

Let us agree on the following notation: the open disk of radius R centered at the origin will be denoted by D_R. Instead of D_1, we write just D.

Definition 12.18 Let $U \subset \mathbb{C}$ be an open set. For every point $a \in U$ denote by $\Psi_U(a)$ the supremum of $|\psi'(0)|$ over all holomorphic maps $\psi: D_1 = D \to U$ such that $\psi(0) = a$. Then the number $\rho_U(a) = 1/\Psi_U(a)$ is called the *hyperbolic density* on the set U at the point a. (We do not exclude the case $\Psi_U(a) = +\infty$; in this case, $\rho_U(a) = 0$.)

Clearly, $\rho_{\mathbb{C}}(a) = 0$ for every $a \in \mathbb{C}$: indeed, for every $R > 0$, the map $D \to \mathbb{C}$ given by $z \mapsto a + Rz$ has derivative R at the origin!

However, the identically zero hyperbolic density is an exception rather than the rule. Later, we will prove that for most domains in $\mathbb{C}$ this density is everywhere positive; meanwhile, here is the first reassuring example.

Proposition 12.19 *If $U \subset \mathbb{C}$ is a simply connected domain that is not the whole plane $\mathbb{C}$, then for $a \in U$ we have $\rho_U(a) = |F'(a)|$ where $F: U \to D$ is a conformal map that sends a to the origin.*

Proof Such a map F does exist by the Riemann mapping theorem. If $\psi: D \to U$ is a holomorphic map such that $\psi(0) = a$, then $(F \circ \psi)(D) \subset D$ and $(F \circ \psi)(0) = 0$; the Schwarz lemma shows that $|(F \circ \psi)'(0)| \leq 1$, with equality attained if $F \circ \psi$ is a conformal automorphism of the disk D. Since $(F \circ \psi)'(0) = F'(a)\psi'(0)$, it follows that $1/|\psi'(0)| \leq |F'(a)|$, and equality can be attained, because if we set $\psi = F^{-1}$, then $F \circ \psi: D \to D$ is not merely a conformal automorphism, but even the identity map. $\qquad\square$

It makes sense to write explicit formulas for the hyperbolic density in especially popular simply connected domains.

Proposition 12.20 *If $D = \{z: |z| < 1\}$ is the unit disk, then $\rho_D(a) = 1/(1 - |a|^2)$ for every $a \in D$.*

Proof It suffices to find the derivative at a of the automorphism

$$\varphi(z) = (z - a)/(1 - \bar{a}z),$$

which sends a to the origin.   □

Calculating the hyperbolic density on the upper half-plane is left as an exercise to the reader; we simply record the answer.

Proposition 12.21 *If $H = \{z : \operatorname{Im} z > 0\}$ is the upper half-plane, then*

$$\rho_H(a) = 1/(2 \operatorname{Im} a)$$

for every $a \in H$.

Definition 12.22 Let $U \subset \mathbb{C}$ be a domain and $\gamma : [p; q] \to U$ be a piecewise smooth path. Then the *hyperbolic length* of γ is the number

$$\text{h-length}_U(\gamma) = \int_\gamma \rho_U(z)\,|dz|. \tag{12.11}$$

(It is not excluded that the hyperbolic length of a path may turn out to be zero; one can show that the integral in the right-hand side of (12.11) always exists.)

The main property of the hyperbolic metric is that holomorphic maps do not increase hyperbolic lengths.

Proposition 12.23 *Let $U, V \subset \mathbb{C}$ be connected open sets, and let $f : U \to V$ be a holomorphic map.*
(1) *For every $a \in U$, we have $\rho_V(f(a)) \cdot |f'(a)| \le \rho_U(a)$.*
(2) *For every piecewise smooth path $\gamma : [p; q] \to U$, we have*

$$\text{h-length}_V(f \circ \gamma) \le \text{h-length}_U(\gamma).$$

Proof Part (2) obviously follows from part (1) and the change of variable formula for definite integrals. To prove part (1), first note that if $f'(a) = 0$, then there is nothing to prove. If $f'(a) \ne 0$ and $\psi : D \to U$ is a holomorphic map such that $\psi(0) = a$, then $f \circ \psi : D \to V$ is a holomorphic map such that $(f \circ \psi)(0) = f(a)$. Thus, we have

$$\rho_V(f(a)) \le \frac{1}{|(f \circ \psi)'(0)|} = \frac{1}{|\psi'(0)|} \cdot \frac{1}{|f'(a)|};$$

taking the supremum of the denominator over all $\psi : D \to U$ such that $\psi(0) = a$, we see that $\rho_V(f(a)) \le \rho_U(a)/|f'(a)|$; it remains to get rid of the denominator.   □

Corollary 12.24 *If $V \subset U$ are connected open subsets in $\mathbb{C}$, then $\rho_V(a) \ge \rho_U(a)$ for every point $a \in V$.*

Proof Apply Proposition 12.23 (1) to the embedding $V \hookrightarrow U$.   □

Corollary 12.25 *If $f : U \to V$ is a conformal isomorphism, then:*
(1) $\rho_V(f(a)) = \rho_U(a)/|f'(a)|$ *for every $a \in U$;*

(2) *if* $\gamma\colon [p; q] \to U$ *is a piecewise smooth path, then*

$$\text{h-length}_V (f \circ \gamma) = \text{h-length}_U (\gamma).$$

(In other words, conformal isomorphisms preserve hyperbolic lengths.)

Proof Apply Proposition 12.23 to the maps f and f^{-1}. $\square$

Definitions 12.18 and 12.22 may be beautiful, but, in fairness, the notion of hyperbolic length makes little sense while we work only with simply connected domains: estimates on the absolute value of the derivative of a map from D to D are known already from the Schwarz lemma (if we bring conformal automorphisms of the disk into the picture, see Exercise 9.13), and if the disk D is replaced with an arbitrary simply connected domain, then it suffices to first map it conformally onto D, which is actually the meaning of Proposition 12.19. To justify introducing the notion of hyperbolic metric, we must learn to work with it on non-simply connected sets. Here is the first step in this direction.

Proposition 12.26 *Let* $D^* = \{z\colon 0 < |z| < 1\}$ *be the punctured unit disk. Then for every* $a \in D^*$,

$$\rho_{D^*}(a) = 1/(2|a| \log(1/|a|)).$$

Proof Let $\psi\colon D \to D^*$ be a holomorphic map such that $\psi(0) = a$. Since f does not assume the value 0, every germ of the function $z \mapsto \log \psi(z)$ can be continued analytically along every path in D and, since the disk D is simply connected, the result of this analytic continuation depends only on the initial and final points of the path. Hence, on D we have a single-valued branch of the function $\log \circ \psi$; therefore, denoting by $H = \{z\colon \operatorname{Im} z > 0\}$ the upper half-plane, there exists a holomorphic map $h\colon D \to H$ such that $\psi(z) = e^{ih(z)}$ and $h(0) = b$ where $e^{ib} = a$. We have $\psi'(0) = ie^{ib} \cdot h'(0) = iah'(0)$, whence

$$\frac{1}{|\psi'(0)|} = \frac{1}{|a|} \cdot \frac{1}{|h'(0)|}$$

and

$$\rho_{D^*}(a) = \sup_{\substack{\psi\colon D \to D^* \\ \psi(0)=a}} \frac{1}{|\psi'(0)|} = \frac{1}{|a|} \cdot \sup_{\substack{h\colon D \to H \\ h(0)=b}} \frac{1}{|h'(0)|} = \frac{\rho_H(b)}{|a|}.$$

It remains to recall that $\rho_H(b) = 1/(2\operatorname{Im} b)$ (Proposition 12.21) and that if $e^{ib} = a$, then $\operatorname{Im} b = -\log|a| = \log(1/|a|)$. $\square$

Ideally, we would like to similarly find the hyperbolic density for an arbitrary non-simply connected domain $U \subset \mathbb{C}$. For this, it would suffice to find an analog of the exponential, namely, an unramified and surjective holomorphic map $F\colon H \to U$ with the property that every germ of the inverse map F^{-1} can be continued analytically along every path in U. In particular, the existence of such a map would immediately imply that the hyperbolic density on U is everywhere positive (and equal to the quotient of the hyperbolic density on H and the absolute value of the derivative of F).

One can prove that such a map F does indeed exist provided that the complement $\mathbb{C} \setminus U$ contains at least two points: this is the content of the so-called Poincaré–Koebe uniformization theorem.[1] A proof of this theorem is beyond the scope of this book. The properties of the hyperbolic metric that are needed in what follows will be proved by a more old-fashioned method due to Edmund Landau. Namely, we will establish nice properties of the hyperbolic metric in the "extreme" case where the complement of U consists of two points.

Theorem 12.27 (Landau's theorem) *Let $p_1, p_2 \in \mathbb{C}$ be two different points and $U = \mathbb{C} \setminus \{p_1, p_2\}$. Then $\rho_U(a) > 0$ for all $a \in U$. Moreover, there exist positive constants C and M such that for $a \in U$ and $|a| \geq M$,*

$$|\rho_U(a)| \geq C/(|a| \log |a|).$$

Proof Since the sets $\mathbb{C} \setminus \{p_1, p_2\}$ and $\mathbb{C} \setminus \{0, 1\}$ are related by a linear transformation, and since, by Corollary 12.25, under a linear automorphism the hyperbolic density gets multiplied by a positive constant, it suffices to prove the theorem for $U = \mathbb{C} \setminus \{0, 1\}$, and from now on we assume that this is the case.

We begin with a quite elementary lemma whose proof is left to the reader.

Lemma 12.28 *If $\sqrt{w} - \sqrt{w-1} = t$ with $t \neq 0$, then $w = ((t^2+1)/2t)^2$; in particular, if*

$$\sqrt{w} - \sqrt{w-1} = \sqrt{n} \pm \sqrt{n-1},$$

then $w = n$.

Here $\sqrt{w}$ (respectively, $\sqrt{w-1}$) stands for a complex number whose square is equal to w (respectively, $w-1$).

Now consider a point $a \in U = \mathbb{C} \setminus \{0, 1\}$, and let $f: D \to U$ be a holomorphic map such that $f(0) = a$ and $f'(0) \neq 0$ (such a map does always exist: if U contains a disk of radius ε centered at a, then set $f(z) = a + \varepsilon z$). To obtain a lower bound on the hyperbolic density at the point a, we must estimate $|f'(0)|$ from above. This can be done as follows.

Since the function f does not assume the value 0 in D, and the disk D is simply connected, we have a well-defined holomorphic function $z \mapsto \log f(z)$ on U. A branch of the logarithm is chosen so as to satisfy the relation

$$\operatorname{Im} \log f(0) = \operatorname{Im} \log a \in [0; 2\pi]. \tag{12.12}$$

In D, the function $\log f(z)$ assumes neither the value 0, nor the value $2\pi i$ (if $\log f(z)$ is equal to one of these numbers, then $f(z) = 1$, which is impossible because $f(U) \subset \mathbb{C} \setminus \{0, 1\}$). Thus, the functions $z \mapsto \sqrt{\log f(z)/2\pi i}$ and $z \mapsto \sqrt{(\log f(z)/2\pi i) - 1}$ are well defined on D. For every $z \in D$, the values of these functions at the point z are different (otherwise, $f(z) = f(z) - 1$, which is absurd),

---

[1] A map $F: H \to U$ whose existence is stated in the theorem is nothing else than the universal cover of the domain U; the informative part of the Poincaré–Koebe theorem is that, under the above condition, this universal cover is isomorphic to the upper half-plane.

hence their difference does not vanish on D; thus, on D we have a well-defined function

$$z \mapsto F(z) = \log\left(\sqrt{\frac{\log f(z)}{2\pi i}} - \sqrt{\frac{\log f(z)}{2\pi i} - 1}\right).$$

We will show that the function $F: D \to \mathbb{C}$ does not assume values of the form $\pm \log(\sqrt{n} - \sqrt{n-1}) + 2\pi i k$ for $n \in \mathbb{N}$, $k \in \mathbb{Z}$ (here the logarithm is understood in the elementary sense: $\log x$ is the real number whose exponential is equal to x). Indeed, if

$$F(z) = \pm \log(\sqrt{n} - \sqrt{n-1}) + 2\pi i k,$$

then

$$\sqrt{\frac{\log f(z)}{2\pi i}} - \sqrt{\frac{\log f(z)}{2\pi i} - 1} = \sqrt{n} \mp \sqrt{n-1},$$

whence, by Lemma 12.28, we have $\log f(z) = 2\pi i n$ and $f(z) = 1$, which is impossible.

The "forbidden points" $\pm \log(\sqrt{n} - \sqrt{n-1}) + 2\pi i k$ lie on the horizontal lines $\{z: \operatorname{Im} z = 2\pi k\}$, $k \in \mathbb{Z}$. Since

$$\lim_{n \to \infty} \log(\sqrt{n} - \sqrt{n-1}) = -\infty$$

and

$$\lim_{n \to \infty} \left(\log(\sqrt{n+1} - \sqrt{n}) - \log(\sqrt{n} - \sqrt{n-1})\right) = 0,$$

the real parts of forbidden points (both positive and negative) on each of these lines can have arbitrarily large absolute values, while the distances between neighboring forbidden points are bounded from above. Therefore, there exists a number $\rho > 0$ such that every open disk of radius $\geq \rho$ contains at least one forbidden point. Applying Bloch's theorem 9.17 to the function F, we deduce that $F(D)$ contains an open disk of radius $|F'(0)|/40$. Since $F(D)$ cannot contain forbidden points, we see that $|F'(0)|/40 \leq \rho$, whence $|F'(0)| \leq 40\rho$. It follows from Lemma 12.28 that if $u = F(z)$, then $f(z) = G(u)$ where $G(u) = e^{-\frac{\pi i}{2}(e^{2u} + e^{-2u})}$. The chain rule implies that $|f'(0)| = |F'(0)| \cdot |G'(F(0))|$, whence $|f'(0)| \leq 40\rho |G'(F(0))|$. It remains to estimate $|G'(F(0))|$ from above. For this, note that $G'(u) = -\pi i G(u)(e^{2u} - e^{-2u})$; since $G(F(0)) = f(0) = a$, we deduce that

$$|G'(F(0))| = \pi |G(F(0))| \cdot |e^{2F(0)} - e^{-2F(0)}| = \pi |a| \cdot |e^{2F(0)} - e^{-2F(0)}|. \quad (12.13)$$

Since

$$e^{2F(0)} = \left(\sqrt{\frac{\log a}{2\pi i}} - \sqrt{\frac{\log a}{2\pi i} - 1}\right)^2,$$

we have

$$e^{2F(0)} - e^{-2F(0)} = \left(\sqrt{\frac{\log a}{2\pi i}} - \sqrt{\frac{\log a}{2\pi i} - 1}\right)^2 - \left(\sqrt{\frac{\log a}{2\pi i}} + \sqrt{\frac{\log a}{2\pi i} - 1}\right)^2$$

$$= -4\sqrt{\frac{\log a}{2\pi i}}\sqrt{\frac{\log a}{2\pi i} - 1}. \quad (12.14)$$

It follows from (12.12) that $|(\log a)/2\pi i| \leq (|\log|a|| + 2\pi)/2\pi < |\log|a|| + 7$ (we do not seek sharp constants in inequalities), $|(\log a)/2\pi i - 1| < |\log|a|| + 8$, so the absolute value of the right-hand side in (12.14) does not exceed

$$4\sqrt{(|\log|a|| + 7)(|\log|a|| + 8)},$$

and from (12.13) we deduce that

$$|G'(F(0))| \leq 4\pi|a|\sqrt{(|\log|a|| + 7)(|\log|a|| + 8)},$$

whence $|f'(0)| \leq 160\rho\pi|a|\sqrt{(|\log|a|| + 7)(|\log|a|| + 8)}$ and

$$\rho_U(a) \geq \frac{1}{|f'(0)|} \geq \frac{1}{160\rho\pi|a|\sqrt{(|\log|a|| + 7)(|\log|a|| + 8)}}.$$

The right-hand side of this inequality is always positive, which proves the first claim of the theorem; it is also clear that for all a with sufficiently large absolute value, the right-hand side is not less than $C/(|a|\log|a|)$ for some positive constant C. This proves the second claim. $\square$

Corollary 12.29 *Let $V \subset \mathbb{C}$ be a domain such that $\mathbb{C} \setminus V$ contains at least two different points. Then $\rho_V(a) > 0$ for all $a \in V$.*

Proof If $\mathbb{C} \setminus V$ contains two different points p_1 and p_2, then, setting $U = \mathbb{C} \setminus \{p_1, p_2\}$ and observing that $V \subset U$, we have, by Corollary 12.24, that $\rho_V(a) \geq \rho_U(a) > 0$ for every $a \in V$. $\square$

To take full advantage of the hyperbolic metric, we need more than Corollary 12.29: besides positivity, we must also establish that the hyperbolic metric is complete (for domains on which it is not identically zero, of course). We do not address this issue (to take a small step in this direction, solve Exercise 12.9); instead, we deduce two interesting corollaries from what we have learned about the hyperbolic density on the plane with two points removed (i.e., from Landau's theorem).

The first result can be obtained quite easily.

Proposition 12.30 (Picard's little theorem) *A nonconstant entire function assumes all complex values except possibly one.*

Proof If an entire function f does not assume values p_1 and p_2, then, setting $U = \mathbb{C} \setminus \{p_1, p_2\}$, we deduce from Proposition 12.23 that $\rho_U(f(a))|f'(a)| \leq \rho_{\mathbb{C}}(a)$. Since $\rho_{\mathbb{C}}(a) = 0$ (the hyperbolic density on $\mathbb{C}$ is identically zero) and $\rho_U(a) > 0$, it follows from Landau's theorem that $|f'(a)| = 0$. Since the derivative of f vanishes everywhere, this function is a constant. □

Now we prove a significant strengthening of Picard's little theorem; its proof requires applying all properties of the hyperbolic metric that we have learned.

Theorem 12.31 (Picard's great theorem) *In a punctured neighborhood of an essential singularity, a holomorphic function assumes all complex values except possibly one.*

Proof Arguing by contradiction, assume that a holomorphic function f, in a punctured neighborhood of an essential singularity, does not take at least two values.

We may assume without loss of generality that the essential singularity is at the origin and the punctured neighborhood has the form $D^* = \{z \colon 0 < |z| < 1\}$; the general case can be reduced to this one by a linear change of variable. If in D^* the function f does not take the values p_1 and p_2, then we can regard it as a holomorphic map from D^* to $U = \mathbb{C} \setminus \{p_1, p_2\}$. By Landau's theorem, there exist numbers $C > 0$ and $M > 0$ such that $\rho_U(a) \geq C/(|a| \log |a|)$ whenever $|a| \geq M$.

Lemma 12.32 *Let γ be a piecewise smooth path in U passing through a point with absolute value at most M and a point with absolute value $L > M$. Then*

$$\text{h-length}_U(\gamma) \geq C \int_M^L \frac{dx}{x \log x}.$$

Proof It suffices to prove this inequality for the part of γ starting at the point with absolute value M and ending at the point with absolute value L. Let this part be parametrized as $\gamma \colon [p; q] \to U$, $|\gamma(p)| = M$, $|\gamma(q)| = L$. Note that $|d\gamma(t)/dt| \geq d(|\gamma(t)|)/dt$. Indeed, this follows from the fact that

$$(|\gamma|)' = |(\sqrt{(\gamma, \gamma)})'| = \left| \frac{(\gamma, \gamma')}{\sqrt{(\gamma, \gamma)}} \right| = |\gamma'| \cdot |\cos \varphi| \leq |\gamma'|$$

(here $(\cdot, \cdot)$ is the inner product and φ is the angle between the vectors γ and γ'); one may also look at Fig. 12.3. Now we have

$$\text{h-length}_U(\gamma) = \int_p^q \rho_U(\gamma(t))|\gamma'(t)| \, dt \geq C \int_p^q \frac{(|\gamma(t)|)' \, dt}{|\gamma(t)| \log |\gamma(t)|} = C \int_M^L \frac{dx}{x \log x},$$

as required. □

By the Casorati–Weierstrass theorem, there exists a sequence of complex numbers $z_n \in D^*$ such that $\lim\limits_{n \to \infty} z_n = 0$ and $\lim\limits_{n \to \infty} f(z_n) = 0$ (instead of 0, any number a with $|a| < M$ would do; and if we slightly modify the argument, then any finite number at all). Set $|z_n| = r_n$; we may, and will, assume that the sequence $\{r_n\}$ is monotone decreasing.

By Proposition 12.26, the hyperbolic length of the circle $\{z: |z| = r\} \subset D^*$ (with respect to the hyperbolic metric on D^*) is equal to

$$\int_{|z|=r} \frac{|dz|}{2|z| \log(1/|z|)} = \frac{2\pi r}{2r \log(1/r)} = \frac{\pi}{|\log r|};$$

therefore, it tends to zero as $r \to 0$. In particular, if $\gamma_n = \{z \in D^*: |z| = r_n = |z_n|\}$, then

$$\lim_{n \to \infty} \text{h-length}_{D^*}(\gamma_n) = 0.$$

Together with Proposition 12.23 this implies that

$$\lim_{n \to \infty} \text{h-length}_U(f(\gamma_n)) = 0;$$

in particular, the hyperbolic lengths of all curves $f(\gamma_n)$ are bounded.

I claim that there exists $R > 0$ such that (for sufficiently large n) all curves $f(\gamma_n)$ are contained in the disk of radius R centered at the origin. Indeed, for all sufficiently large n we have $|f(z_n)| \leq M$. If $L_n = \sup\limits_{z \in \gamma_n} |f(z)|$ and $L_n > M$, then, by Lemma 12.32,

$$\text{h-length}(f(\gamma_n)) \geq C \int_M^{L_n} \frac{dx}{x \log x}.$$

Since the integral $\int_M^\infty dx/(x \log x)$ diverges, the boundedness of the sequence $\{\text{h-length}(f(\gamma_n))\}$ implies the boundedness of the sequence $\{L_n\}$. Now we may set $R = \sup\limits_n L_n$.

So, for all sufficiently large n we have $|f(z)| \leq R$ whenever $|z| = r_n$. By the maximum modulus principle applied to the annuli $\{z: r_{n+1} \leq |z| \leq r_n\}$, it follows that $|f(z)| \leq R$ whenever $r_{n+1} \leq |z| \leq r_n$, and hence (since $r_n = |z_n| \to 0$) for all z sufficiently close to the origin. Thus, by Riemann's theorem, the singularity of f at the origin is removable, a contradiction!   □

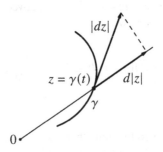

Fig. 12.3 $|dz| \geq d|z|$

Exercises

12.1. Show that the complex plane and the unit disk are homeomorphic. (*Hint.* See Fig. 12.4.)

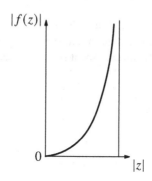

Fig. 12.4

12.2. Show that the open sets $\mathbb{C} \setminus \{0\}$ and $D \setminus \{0\}$ are not conformally isomorphic (D is the unit disk).

12.3. Let $U \subset \mathbb{C}$ be a simply connected open set that is not the whole plane $\mathbb{C}$, and let $a \in U$. By $D = \{z : |z| < 1\}$ we denote the unit disk. Show that there exists a holomorphic map $f : U \to D$ for which $|f'(a)|$ is the greatest among all holomorphic maps from U to D (no injectivity is assumed!), and that such a map is a conformal isomorphism between U and D.

12.4. Does there exist a function f holomorphic on the unit disk $D = \{z : |z| < 1\}$ for which

$$\sup_{z \in D} \operatorname{Re} f(z) = 1, \quad \inf_{z \in D} \operatorname{Re} f(z) = -1,$$

$$\sup_{z \in D} \operatorname{Im} f(z) = 2017, \quad \inf_{z \in D} \operatorname{Im} f(z) = -2017?$$

12.5. Let $A_1 A_2 A_3 A_4 A_5$ and $B_1 B_2 B_3 B_4 B_5$ be mutually equiangular pentagons on the plane (i.e., $\angle A_j = \angle B_j$ for all j). Assume that there is a conformal isomorphism between these pentagons such that the boundary correspondence sends vertices to vertices. Show that the pentagons are similar.

12.6. Show that the hyperbolic density on the open subset $\mathbb{C} \setminus \{0\} \subset \mathbb{C}$ is identically zero.

12.7. Find the hyperbolic density on the annulus $\{z : r < |z| < R\}$.

12.8. Let $U \subset \mathbb{C}$ be an open set, $U \neq \mathbb{C}$. Show that $\rho_U(a) \leq 1/\operatorname{dist}(a, \mathbb{C} \setminus U)$ for every $a \in U$ (here $\operatorname{dist}(a, X) = \inf_{b \in X} |a - b|$ is the distance from the point a to the set X).

12.9. Let $U \subset \mathbb{C}$ be a domain such that $\mathbb{C} \setminus U$ contains at least two points, and let $\{a_n\}$ be a sequence of points of U that converges to a point on the boundary of U. Then $\lim_{n \to \infty} \rho_U(a_n) = +\infty$. Prove this:

(a) for the case where $U = \mathbb{C} \setminus \{0, 1\}$ (if $a_n \to 0$, then this follows from our proof of Landau's theorem; if $a_n \to 1$, use an appropriate automorphism of U);

(b) for an arbitrary domain U (use Corollary 12.24).

12.10. Show that in a punctured neighborhood of an essential singularity, a holomorphic function takes all values, except possibly one, infinitely often.

Chapter 13
A Thing or Two About Riemann Surfaces

The study of functions of (one) complex variable eventually makes it necessary to consider Riemann surfaces. In the main part of the book, we have made every effort to avoid this, though on several occasions have come quite close to this subject; in the concluding chapter, we will finally address it. A detailed discussion of Riemann surfaces is beyond the scope of this book; I will merely give basic definitions and examples, state several classical theorems, and prove them in the simplest nontrivial case.

The prerequisites for this chapter are somewhat more demanding than for all the previous ones. I will assume that the reader has some acquaintance with the definition of a smooth manifold, with differential forms (forms of degree 1 suffice, Stokes' theorem is not needed), and with topology (fundamental group, coverings, triangulations, and the Euler characteristic of surfaces; homologies and cohomologies are not needed).

Coverings and (a little of) fundamental groups will be used without explanation, some things about the Euler characteristic will be recalled, and as to differential forms, even a definition will be given (in the special case that we need), which, hopefully, will allow the reader to understand this chapter.

13.1 Definitions, Simplest Examples, General Facts

The theory of Riemann surfaces stands in the same relation to the theory of complex functions as the analysis on manifolds does to the "ordinary" analysis for functions of several variables. Here is the main definition.

Definition 13.1 A *Riemann surface* is a second-countable Hausdorff topological space X endowed with the following additional structure:

(1) X is represented as a union of open subsets U_α, called *coordinate neighborhoods*;

(2) for every coordinate neighborhood U_α there is a homeomorphism $z_\alpha: U_\alpha \to V_\alpha$, where $V_\alpha \subset \mathbb{C}$ is an open subset, called a *local chart*, or a *local coordinate*;

© Springer Nature Switzerland AG 2020
S. Lvovski, *Principles of Complex Analysis*, Moscow Lectures 6,
https://doi.org/10.1007/978-3-030-59365-0_13

(3) for any two coordinate neighborhoods U_α and U_β with a nonempty intersection, the map

$$\varphi_{\alpha\beta}\colon z_\alpha(U_\alpha \cap U_\beta) \to z_\beta(U_\alpha \cap U_\beta),$$

defined as $x \mapsto z_\beta(z_\alpha^{-1}(x))$, is a conformal isomorphism of open subsets in $\mathbb{C}$.

Of course, any open subset in $\mathbb{C}$ is a Riemann surface. Less trivial examples will be given below.

This definition is completely analogous to that of a smooth manifold (more exactly, a smooth manifold of dimension 1), with $\mathbb{R}$ replaced by $\mathbb{C}$ and smooth functions replaced by holomorphic ones. But while there are literally just a couple of one-dimensional smooth manifolds (only the line, the circle, and their disjoint unions), Riemann surfaces are quite various and make an interesting subject of study.

If, however, we are interested in precise statements rather than analogies, then, identifying $\mathbb{C}$ and $\mathbb{R}^2$, we see that every Riemann surface is a two-dimensional smooth manifold of class C^∞: since holomorphic functions are infinitely complex-differentiable, they are smooth maps of class C^∞.

Holomorphic functions on Riemann surfaces and holomorphic maps between them are defined in exactly the same way as smooth functions on smooth manifolds and smooth maps between them: a function or a map is holomorphic if in local coordinates it is given by a holomorphic function. In more precise terms, this can be stated as follows.

Definition 13.2 If X is a Riemann surface, then a continuous function $f\colon X \to \mathbb{C}$ is said to be *holomorphic* if for every point $a \in X$ and some (equivalently: any) coordinate neighborhood $U_\alpha \ni a$, the composition $f \circ z_\alpha^{-1}\colon V_\alpha \to \mathbb{C}$ is holomorphic. Holomorphic functions on open subsets of a Riemann surface are defined in a similar way.

Actually, in the above definition it would be more correct to write, more awkwardly, $(f|_{U_\alpha}) \circ z_\alpha^{-1}\colon V_\alpha \to \mathbb{C}$ (since the domain of definition of f is wider than the image of the map z_α^{-1}). And, for that matter, condition (3) of Definition 13.1 should involve compositions not of the maps z_α themselves and their inverses, but of their restrictions to smaller open subsets. Such refinements are usually left to the reader. As a rule, they present no difficulties: on each occasion, one must merely restrict the domains of definition of maps so that the inverse maps or compositions under consideration become formally possible.

Remark 13.3 We also comment on the practical use of Definition 13.2 and others like it. In a large number of cases, working with a function f on a Riemann surface X, we deal with its coordinate notation; formally, this means that as long as we stay within a coordinate neighborhood U_α, our calculations involve the composition $f \circ z_\alpha^{-1}\colon V_\alpha \to \mathbb{C}$, where $z_\alpha\colon U_\alpha \to V_\alpha$ is a local chart. In practice, this composition with z_α^{-1} is not specified or mentioned in any way: it's just that, keeping in mind that the local chart identifies $U_\alpha \subset X$ with $V_\alpha \subset \mathbb{C}$, one deals with a function of the variable z_α, and the values of this variable (as long as $z_\alpha \in V_\alpha$) are identified with the points of the Riemann surface lying in U_α.

We leave it to the reader to give a formal definition of a holomorphic map between Riemann surfaces (it is completely analogous to that of a smooth map between smooth manifolds). Still, to be on the safe side, we give the definition of an isomorphism.

Definition 13.4 A map $f\colon X \to Y$ between two Riemann surfaces is called an *isomorphism* if it is holomorphic, bijective, and the inverse map $f^{-1}\colon Y \to X$ is also holomorphic. If there is an isomorphism between two Riemann surfaces, then they are said to be *isomorphic*.

For instance, every coordinate neighborhood is isomorphic to an open subset of the complex plane. See also Exercise 13.1.

Now we analyze several examples.

Example 13.5 The familiar Riemann sphere $\overline{\mathbb{C}} = \mathbb{C} \cup \{\infty\}$ can be regarded as a Riemann surface. To this end, represent $\overline{\mathbb{C}}$ as the union of the following two subsets: $U_0 = \mathbb{C} \subset \overline{\mathbb{C}}, U_1 = (\mathbb{C} \backslash \{0\}) \cup \{\infty\}$. Define local charts $\varphi_0 \colon U_0 \to \mathbb{C}$ and $\varphi_1 \colon U_1 \to \mathbb{C}$ as follows:

$$\varphi_0(z) = z; \qquad \varphi_1(z) = \begin{cases} 1/z & \text{if } z \neq \infty; \\ 0 & \text{if } z = \infty. \end{cases}$$

Finally, introduce a topology on $\overline{\mathbb{C}}$ by declaring a subset $V \subset \overline{\mathbb{C}}$ to be open if the images $\varphi_0(V \cap U_0)$ and $\varphi_1(V \cap U_1)$ are open in $\overline{\mathbb{C}}$. Since the transition map φ_{01} has the form $z \mapsto 1/z$, all conditions of Definition 13.1 are satisfied.

One can easily check that all topological statements about the Riemann sphere from the main part of the book agree with this definition. For example, if $\{z_n\}$ is a sequence of complex numbers, then it converges to the point $\infty \in \overline{\mathbb{C}}$ in the sense of the topology defined in Example 13.5 if and only if $\lim z_n = \infty$, and punctured neighborhoods defined in Sec. 7.4 are indeed punctured neighborhoods in the sense of this topology.

If a holomorphic function f on (an open subset of) a Riemann surface X has an isolated zero at a point a, then its multiplicity is defined in local coordinates as the multiplicity of the zero of the function $f \circ z_\alpha^{-1}$, where z_α is some local chart whose domain of definition contains a. For example, the function $f(z) = 1/z^2$ extended to the point $\infty \in \overline{\mathbb{C}}$ by the formula $f(\infty) = 0$ has a zero of order 2 at this point: since $\varphi_1(z) = 1/z$ (where φ_1 is the local chart on $\overline{\mathbb{C}}$ defined in Example 13.5), we have $\varphi_1^{-1}(w) = 1/w$ and $f \circ \varphi_1^{-1}(w) = w^2$ for $w = 1/z$; this function does indeed have a zero of order 2 at the origin.

The order of a zero thus defined does not depend on the choice of a local chart. Since this order is equal to the ramification index, the easiest way to verify this is to use Proposition 9.7: the characterization of the ramification index provided by this proposition does not involve local charts at all, and hence does not depend on the choice of these local charts.

If $U \subset X$ is an open subset of a Riemann surface $X, a \in U$, and f is a holomorphic function on $U \setminus \{a\}$, then, as in the case of holomorphic functions on $\mathbb{C}$, we can introduce the notion of a removable singularity, pole, or essential singularity of f

at the point a. For example, f is said to have an essential singularity at a if the function $f \circ z_\alpha^{-1}$ has such a singularity. This definition also does not depend on the choice of a local chart, since the type of a singular point can be defined without appealing to charts, from the behavior of $f(z)$ as $z \to a$ (whether f is bounded, tends to infinity, or tends to any given limit). One can easily see that for the Riemann surface $\overline{\mathbb{C}}$ these definitions agree with the definitions of removable singularities, poles, and essential singularities at infinity given in Sec. 7.4.

As follows from the result of Exercise 1.18, the Riemann surface $\overline{\mathbb{C}}$ is compact (for arbitrary topological spaces, the existence of a convergent subsequence for any given sequence is a weaker condition than compactness, but it is known that smooth manifolds, and hence Riemann surfaces, are metrizable, so for $\overline{\mathbb{C}}$ this property suffices). Therefore, $\overline{\mathbb{C}}$ is a one-point compactification of the plane, so $\overline{\mathbb{C}}$ is homeomorphic to the sphere.

Having extended the notion of a pole to Riemann surfaces, we can now define meromorphic functions.

Definition 13.6 Let X be a Riemann surface. A *meromorphic function* on X is a holomorphic function on $X \setminus S$, where $S \subset X$ is a (possibly empty) subset without accumulation points, that has a pole at each point $x \in S$.

Clearly, the quotient of two holomorphic functions is a meromorphic function; every holomorphic function is meromorphic (in this case, $S = \varnothing$); the sum, product, and quotient of two meromorphic functions are also meromorphic, so the meromorphic functions on a given Riemann surface form a field.

Example 13.7 (plane affine curves) Let $F = F(z, w)$ be a polynomial in two variables (with complex coefficients); define

$$V(F) = \{(z, w) \in \mathbb{C}^2 : F(z, w) = 0\}$$

(the set of zeros of F). Now assume that at each point of the set $V(F)$, at least one of the partial derivatives of F does not vanish. In this situation, the set $V(F) \subset \mathbb{C}^2$ is called a *smooth plane affine curve*.

The complex analog of the implicit function theorem says that if F is a polynomial in two variables and $F(a, b) = 0$ but, say, $(\partial F / \partial w)(a, b) \neq 0$, then there exist neighborhoods $U \ni a$, $V \ni b$ and a holomorphic function $\varphi \colon U \to V$ such that for $(z, w) \in U \times V$ the relations $(z, w) \in V(F)$ and $w = \varphi(z)$ are equivalent (and similarly for the case $(\partial F / \partial z)(a, b) \neq 0$). One can prove this statement by reducing it to the real version: identify $\mathbb{C}$ with $\mathbb{R}^2$ and $\mathbb{C}^2$ with $\mathbb{R}^4$, obtain a C^∞-map $\varphi \colon U \to V$, and then verify that it is holomorphic, since its derivative is complex-linear (because the derivative of a polynomial regarded as a map from $\mathbb{R}^4 = \mathbb{C}^2$ to $\mathbb{R}^2 = \mathbb{C}$ is complex-linear). Another way is to obtain the power expansion of φ by the method of undetermined coefficients and then prove that the radius of convergence of this series is positive.

Anyway, due to the implicit function theorem, we can define a Riemann surface structure on $V(F)$. Namely, if $F(a, b) = 0$ and $(\partial F / \partial w)(a, b) \neq 0$, then we can set

$$U_\alpha = (U \times V) \cap V(F), \quad V_\alpha = U$$

(here $U \ni a$ and $V \ni b$ are open subsets in $\mathbb{C}$ whose existence is guaranteed by the implicit function theorem) and take the map $(z, w) \mapsto z$ (the projection to the z axis) for a local chart $z_\alpha : U_\alpha \to V_\alpha$; if, on the other hand, it is the partial derivative with respect to z that does not vanish, then, in a similar way, we take the projection to the w axis for a local chart (if both partial derivatives do not vanish, then we take both corresponding local charts). With this choice of local charts, condition (3) from Definition 13.1 is satisfied. Indeed, if both z_α and z_β are projections to the same axis, then the "transition map" $z_\beta \circ z_\alpha^{-1}$ is just the identity map $z \mapsto z$ or $w \mapsto w$, which is doubtless holomorphic; if, on the other hand, z_α and z_β are projections to different axes, then the transition map is the expression of z in terms of w or w in terms of z, which is a holomorphic function by the implicit function theorem.

Note that the functions $(z, w) \mapsto z$ and $(z, w) \mapsto w$ are holomorphic on $V(F)$.

Example 13.8 (elliptic curves). Let $\Gamma \subset \mathbb{C}$ be a lattice (see Definition 11.9). Set $X = \mathbb{C}/\Gamma$ (the quotient of the additive group $\mathbb{C}$ by the subgroup Γ) and equip X with the quotient topology (by definition, this means that a subset in X is open if and only if its preimage in $\mathbb{C}$ is open). The natural map $p \colon \mathbb{C} \to X$ is a covering; for coordinate neighborhoods, we take sufficiently small open subsets $U_\alpha \subset X$ over which the covering decomposes into a direct product, and for a local chart on U_α, an arbitrary section of the covering over U_α; the transition maps are then just translations by elements of Γ (i.e., $z \mapsto z + u$ for fixed $u \in \Gamma$), so they are, or course, holomorphic. Thus, X is a Riemann surface (homeomorphic to the two-dimensional torus); such Riemann surfaces are called *elliptic curves* (in general, different lattices give rise to nonisomorphic Riemann surfaces!).

One can easily see that meromorphic functions on an elliptic curve $\mathbb{C}/\Gamma$ are the same thing as meromorphic functions on $\mathbb{C}$ that are periodic with respect to the lattice Γ (i.e., elliptic functions with respect to Γ).

The principle of analytic continuation extends to Riemann surfaces.

Proposition 13.9 *Let X and Y be Riemann surfaces with X connected, and let $f, g \colon X \to Y$ be holomorphic maps. If f coincides with g on a subset $S \subset X$ that has an accumulation point, then $f = g$ on X.*

Proof Let us say that a subset $U \subset X$ is nice if it is connected, lies in a coordinate neighborhood in X, and the set $f(U) \subset \mathbb{C}$ lies in a coordinate neighborhood in Y. Every point $x \in X$ is contained in a nice open set. Indeed, if $V \ni f(x)$ and $U \ni x$ are coordinate neighborhoods, then the set $U \cap f^{-1}(V)$ has all the required properties except, possibly, connectedness, so it remains to replace it with any connected neighborhood of p contained in $U \cap f^{-1}(V)$.

If now p is an accumulation point of S and $U \ni p$ is a nice open set, then $f = g$ on U by the "ordinary" principle of analytic continuation (Proposition 5.21). We will show that $f = g$ on X. Let $q \in X$ be an arbitrary point; since X is connected, we can join p and q by a continuous curve. Cover this curve by finitely many nice open sets. We obtain a chain of nice open subsets $U_1, \ldots, U_n \subset X$ such that $U_1 \ni p$, $U_n \ni q$, and each set has a nonempty intersection with the next one. However, if $U', U'' \subset X$ are nice open sets, $f = g$ on U', and $U' \cap U'' \neq \varnothing$, then $f = g$ on $U' \cap U''$ too (since the set $U' \cap U'' \subset U''$ has accumulation points in U''). Thus, induction on j shows that $f = g$ on every set V_j, including V_n. In particular, $f(q) = g(q)$, as required. $\square$

Corollary 13.10 *Let X be a connected Riemann surface and $f, g \colon X \to \mathbb{C}$ be holomorphic functions. If $f = g$ on a subset $S \subset X$ that has an accumulation point, then $f = g$ on X.*

Above we have already extended the definition of meromorphic functions to arbitrary Riemann surfaces; this definition has the following funny reformulation.

Proposition 13.11 *There is a natural one-to-one correspondence between the set of meromorphic functions on a Riemann surface X and the set of holomorphic maps $X \to \overline{\mathbb{C}}$ other than the constant map to the point ∞.*

Proof Let f be a meromorphic function on X and S be the set of poles of f. Extend f to the whole X by setting $f(x) = \infty$ for $x \in S$; we obtain a map $\tilde{f} \colon X \to \overline{\mathbb{C}}$. To verify that $\tilde{f}$ is holomorphic, it suffices to consider only the points from S (the images of other points fall in the coordinate neighborhood $U_0 \subset \overline{\mathbb{C}}$ (see Example 13.5), so no question arises as to them). But if $x \in S$, then the composition of $\tilde{f}$ with the local chart $\varphi_1 \colon U_1 \to \mathbb{C}$ has the form $z \mapsto 1/f(z)$; if f has a pole, then $1/f$ can be extended to a holomorphic function in a neithborhood of the pole, so $\tilde{f}$ is holomorphic at the points of S too.

Conversely, let $F \colon X \to \overline{\mathbb{C}}$ be a holomorphic map; set $S = F^{-1}(\infty)$. If S has an accumulation point in X, then Proposition 13.9 shows that F is the constant map to ∞, contradicting the assumption. Hence, S has no accumulation points. Outside S, the map F is an ordinary holomorphic function; in a neighborhood of every point $s \in S$, the composition of F with the local chart φ_1 on the Riemann sphere is holomorphic, so the function $1/f$ is holomorphic and vanishes at s; hence, the function f itself has a pole at s, as required. $\square$

Here is a simple but instructive corollary of the maximum modulus principle.

Proposition 13.12 *Every holomorphic function on a compact connected Riemann surface is a constant.*

Proof Let f be a holomorphic function on a compact Riemann surface X, and assume that the function $x \mapsto |f(x)|$ attains a maximum value at a point $x \in X$. Then, by the maximum modulus principle, f is a constant in a neighborhood of x, and hence, by Proposition 13.9, on the whole Riemann surface X. $\square$

Finally, we discuss the behavior of Riemann surfaces under coverings.

Proposition 13.13 *Let X be a Riemann surface, Y be a (Hausdorff second-countable) topological space, and $p \colon Y \to X$ be a covering map. Then there is a unique Riemann surface structure on Y with respect to which p is holomorphic.*

Proof of the existence part Cover X by sufficiently small coordinate neighborhoods U_α such that the covering p decomposes into a direct product over each of them; declare the connected components of the preimages of all U_α to be the coordinate neighborhoods on Y; if $z_\alpha \colon U_\alpha \to V_\alpha \subset \mathbb{C}$ is a local chart on $U_\alpha \subset X$, then declare $z_\alpha \circ p$ to be the local chart on each connected component of $p^{-1}(U_\alpha)$. Clearly, the transition maps then coincide with those for the local charts on X, i.e., are holomorphic, so Y becomes a Riemann surface and p becomes holomorphic. □

It remains to prove the uniqueness of a Riemann surface structure on Y for which p is holomorphic. It is a consequence of the following two facts.

Proposition 13.14 *Let X be a Riemann surface, and let $p_1 \colon Y_1 \to X, p_2 \colon Y_2 \to X$ be holomorphic maps between Riemann surfaces that are coverings. If $f \colon Y_1 \to Y_2$ is a continuous map such that $p_2 \circ f = p_1$, then f is holomorphic.*

Proof Let $y_1 \in Y_1$ be an arbitrary point; we will show that the map f is holomorphic in a neighborhood of y_1. Set $y_2 = f(y_1)$ and $x = p_1(y_1) = p_2(y_2)$. Let U be an open subset in X containing x over which both coverings p_1 and p_2 decompose into a direct product. Denote by U_i' ($i = 1, 2$) the connected component of $p_i^{-1}(U)$ that contains y_i. Then $f(U_1') = U_2'$, and we must show that the restriction of f to U_1' is holomorphic; however, this restriction coincides with the composition $s \circ p_1$ where $s \colon U \to Y_2$ is a section of the covering p_2 over U whose image coincides with U_2'. Since p_1 and s are holomorphic, their composition is holomorphic too. □

Corollary 13.15 *If two coverings over a Riemann surface are isomorphic topologically, then they are also isomorphic as Riemann surfaces.*

(Coverings $p_1 \colon Y_1 \to X$ and $p_2 \colon Y_2 \to X$ are said to be isomorphic topologically if there exists a homeomorphism $f \colon Y_1 \to Y_2$ such that $p_2 \circ f = p_1$.)

Proof It suffices to observe that, by Proposition 13.14, the maps f and f^{-1} are holomorphic. □

Corollary 13.15 immediately implies the uniqueness in Proposition 13.13.

Now we apply Proposition 13.15 to prove a quite specific result that we will need in the next section.

Proposition 13.16 *Let $D^* = \{z \in \mathbb{C} \colon 0 < |z| < 1\}$ be a punctured disk and k be a positive integer. Then every (connected) Riemann surface that is a covering of degree k over D^* is isomorphic to D^*, and the isomorphism sends the covering map to the map $p \colon D^* \to D^*$ defined by the formula $z \mapsto z^k$.*

Proof Since the fundamental group of D^* is $\mathbb{Z}$, it has exactly one subgroup of index k, so all coverings of degree k over D^* are isomorphic topologically; by Corollary 13.15, they are also isomorphic analytically, so all of them are isomorphic to the covering defined by the function p from the statement. □

13.2 The Riemann Surface of an Algebraic Function

So far we have seen only two examples of compact Riemann surfaces: the Riemann sphere and elliptic curves. These examples do not exhaust the list of compact Riemann surfaces. Now we consider a classical construction that provides many new examples of compact Riemann surfaces (actually, even all such examples; this fact will be discussed later).

So, let $F = F(z, w)$ be an irreducible polynomial in two variables with complex coefficients. In contrast to Example 13.7, now we do not impose the condition that at every point of the set $V(F) \subset \mathbb{C}^2$ at least one of the partial derivatives of F does not vanish, so the set $V(F) \subset \mathbb{C}^2$ is not necessarily a Riemann surface (see Exercise 13.2). We will also assume that F depends on both variables (to exclude degenerate cases such as $F = \text{const} \cdot w$). Our aim is to associate with F a compact Riemann surface.

Write the polynomial F in the form

$$F(z, w) = P_n(z)w^n + P_{n-1}(z)w^{n-1} + \ldots + P_0(z) \tag{13.1}$$

where P_j are polynomials in z. Denote by $p\colon (z, w) \to z$ and $q\colon (z, w) \to w$ the projections of the curve C to the coordinate axes. For a "generic" point $z \in \mathbb{C}$, the set $p^{-1}(z)$ contains n elements, because F is a polynomial in w of degree n. We want to determine for which points z the number of preimages is less than n (it cannot be greater than n: since the degree of F with respect to w is equal to n, this would mean that at some point $z_0 \in \mathbb{C}$ all polynomials P_i vanish, but then the polynomial F is divisible by $z - z_0$, contradicting the irreducibility).

First, the set of points with less than n preimages includes the points $z \in \mathbb{C}$ at which $P_n(z) = 0$ (since for w we obtain an equation of degree $n - 1$); second, even if $P_n(z) \neq 0$, the equation for w has less than n roots for points $z \in \mathbb{C}$ at which the discriminant of F, regarded as a polynomial in w, vanishes.

Lemma 13.17 *The discriminant of F regarded as a polynomial in w cannot be identically zero.*

Proof Since F is irreducible in the polynomial ring $\mathbb{C}[z, w]$, Gauss's lemma shows that it is irreducible also as an element of the ring $\mathbb{C}(z)[w]$ (the ring of polynomials in the variable w over the field of rational functions $\mathbb{C}(z)$); therefore, the greatest common divisor of F and its derivative (with respect to w) is equal to 1, so its discriminant does not vanish as an element of $\mathbb{C}(z)$, and hence cannot be identically zero as a function of z.   □

So, the discriminant of F may vanish only at finitely many points $z \in \mathbb{C}$. Denote by Σ the (finite) set of points $z \in \mathbb{C}$ at which either $P_n(z) = 0$, or the discriminant of F vanishes; all the other points have exactly n preimages under p. Set $X_0 = p^{-1}(\mathbb{C} \setminus \Sigma)$ and denote by

$$p_0\colon X_0 \to \mathbb{C} \setminus \Sigma \quad \text{and} \quad q_0\colon X_0 \to \mathbb{C}$$

the restrictions of p and q to X_0.

Proposition 13.18 *The map* $p_0 \colon X_0 \to \mathbb{C} \setminus \Sigma$ *is a covering of degree* n; *if* X_0 *is endowed with the Riemann surface structure induced from* $\mathbb{C} \setminus \Sigma$ *by this covering (as in Proposition 13.13), then the maps* p_0 *and* q_0 *are holomorphic.*

Proof First, we show that $(\partial F/\partial w)(z, w) \neq 0$ whenever $(z, w) \in X_0$. Indeed, if this partial derivative vanished, then w would be a multiple root of the polynomial $F(z, \cdot)$ and the discriminant of this polynomial would vanish, i.e., the discriminant of F regarded as a polynomial in w over $\mathbb{C}(z)$ would vanish at z, contradicting the fact that $z \notin \Sigma$.

Since $(\partial F/\partial w)(z, w) \neq 0$, we may use the implicit function theorem to express w locally as a holomorphic function of z. More specifically, there exist neighborhoods $U_{z,w} \ni z$ and $V_{z,w} \ni w$ and a holomorphic function $\varphi_{z,w} \colon U_{z,w} \to V_{z,w}$ such that for $(s, t) \in U_{z,v} \times V_{z,w}$ the equality $F(s, t) = 0$ holds if and only if $t = \varphi_{z,w}(s)$. Now set

$$U_z = \bigcap_{w \in p_0^{-1}(z)} U_{z,w},$$

and for each $w \in p_0^{-1}(z)$ set $U'_{v,w} = \{(z, \varphi_{z,w}(z)) \colon z \in U_z\}$; it is clear that p maps each of the neighborhoods $U'_{z,w}$ homeomorphically onto U_z, with the sets $U'_{z,w}$ pairwise disjoint (if two such sets intersected at a point $(s, t) \in X_0$, then s would have less than n preimages, which cannot happen because $s \notin \Sigma$). Since for every $z \in \mathbb{C} \setminus \Sigma$ there are exactly n (= the number of preimages of z) open sets $U'_{z,w}$, it is proved that p_0 is a covering.

The holomorphicity of the map $p_0 \colon X_0 \to \mathbb{C} \setminus \Sigma$ is obvious from the construction of a Riemann surface structure on a covering, and the holomorphicity of the map $q_0 \colon X_0 \to \mathbb{C}$ is clear from the implicit function theorem: in local coordinates in a neighorhood of (z, w), this map looks as the holomorphic function $\varphi_{v,w}$. □

Before moving on, we prove a technical result.

Proposition 13.19 *The set* $X_0 \subset \mathbb{C}^2$ *is connected.*

Proof Arguing by contradiction, assume that X_0 is the union of two disjoint open sets C_1 and C_2; clearly, they are coverings over $\mathbb{C} \setminus \Sigma$ of degrees k_1 and k_2, respectively, with $k_1 + k_2 = n$. For every point $z \in \mathbb{C} \setminus \Sigma$, let $u_1(z), \dots, u_{k_1}(z)$ be its preimages (under p_0) lying in C_1 and $v_1(z), \dots, v_{k_2}(z)$ be its preimages lying in C_2 (the numbering is arbitrary). Set

$$R_1(z, w) = (w - u_1(z)) \dots (w - u_{k_1}(z)) = w^{k_1} + L_{k_1-1}(z)w^{k_1-1}(z) + \dots + L_0(z),$$
$$R_2(z, w) = (w - v_1(z)) \dots (w - v_{k_2}(z)) = w^{k_2} + Q_{k_2-1}(z)w^{k_2-1}(z) + \dots + Q_0(z).$$

Obviously, the functions L_i and Q_j are holomorphic on $\mathbb{C} \setminus \Sigma$ (because locally they can be written as symmetric functions of the holomorphic functions $\varphi_{z,w}$ from the proof of Proposition 13.18); if we prove that all Q_i and L_j are rational functions of z, then we arrive at a contradiction, since, obviously, $R_1(z, w)R_2(z, w) = F(z, w)/P_n(z)$, and we would obtain that the polynomial $F(z, w)/P_n(z) \in \mathbb{C}(z)[w]$ is reducible over $\mathbb{C}(z)$, which, as we have already observed, contradicts the irreducibility of the polynomial F in $\mathbb{C}[z, w]$ in view of Gauss's lemma.

We precede the proof of the rationality of Q_i and L_j by a simple lemma.

Lemma 13.20 *Every root of the polynomial $w^n + a_{n-1} w^{n-1} + \ldots + a_0$ does not exceed in absolute value the number $M = n \cdot \max(|a_{n-1}|, \ldots, |a_0|, 1)$.*

Proof If $|w| > M$, then $|a_j w^j| < |w^n|/n$ for every $j \in [1; n]$, so $|w^n|$ is greater than the sum of the absolute values of the other terms, and w cannot be a root.   □

Now observe that $u_i(z)$ and $v_j(z)$ are roots of a monic polynomial whose coefficients are rational functions of z with poles in Σ; these rational functions have at most polynomial growth as z approaches points from Σ or infinity; it follows from the above lemma that $u_j(z)$ and $v_j(z)$ also grow at most polynomially as z approaches points from Σ or infinity. Since $Q_i(z)$ and $L_j(z)$ are symmetric functions of $u_{i_k}(z)$ and $v_{j_k}(z)$, they also grow at most polynomially as z approaches points from Σ or infinity. Therefore, they are meromorphic functions on the whole Riemann sphere, and hence rational functions (Proposition 7.29). We have arrived at a contradiction.   □

Now we are ready to construct a compact Riemann surface corresponding to the equation $F(z, w) = 0$ (the classical authors said: "a Riemann surface corresponding to w regarded as an algebraic function of z"). Everything comes down to adding finitely many points to X_0.

Namely, embed $\mathbb{C}$ into $\overline{\mathbb{C}}$, set $S = \Sigma \cup \{\infty\}$, and consider X_0 as a covering over $\overline{\mathbb{C}} \setminus S$. For each point $x \in S$, consider an open set $U \subset \overline{\mathbb{C}}$ containing x, isomorphic to the disk $\{z \in \mathbb{C} : |z| < 1\}$, and containing no other points from S. Set $U^* = U \setminus \{x\}$; obviously, U^* is isomorphic to a punctured disk. We have the decomposition into connected components $p_0^{-1}(U^*) = \bigcup_{j=1}^m V_j^*$, where V_j^* is an open connected set that is a covering over U^* of degree e_j and the sets V_j^* are pairwise disjoint. Proposition 13.16 shows that each of the sets $V_j^* \subset X_0$ is isomorphic to a punctured disk, and the restriction of the map p_0 to V_j^* is isomorphic to raising to the power e_j. For each of the sets V_j^*, add a point ξ_j to X_0; having done this for each of the points $x \in S$, we obtain a set X which is the union of X_0 and finitely many points. This is exactly our compact Riemann surface. We must only equip it with a topology and local charts.

Topology on X. On $X_0 \subset X$, we preserve the original topology; it remains to define bases of neighborhoods of the new points, and this can be done as follows: if a new point ξ_j corresponds to a set V_j^*, then for a basis of neighborhoods of ξ_j we take the collection of sets W such that $W \setminus \xi_j$ is contained and open in V_j^*. Now we can extend the map $p_0 \colon X_0 \to \overline{\mathbb{C}} \setminus S$ to a map $\bar{p} \colon X \to \overline{\mathbb{C}}$ as follows: if a new point ξ_j is as above and V_j^* is a component of the preimage of a punctured neighborhood of x, then we set $\bar{p}(\xi_j) = x$. One can easily check that the map $\bar{p}$ is continuous and the topological space X is compact.

Local charts on X. To the collection of local charts on X_0 already in place, we add the following ones: let ξ_j be a new point corresponding to a set V_j^*; let $V_j = V_j^* \cup \{\xi_j\}$, consider an isomorphism $\varphi \colon V_j^* \to D^*$ onto a punctured disk, and extend it to a map $V_j \to D$ onto the whole disk by setting $\xi_j \mapsto 0$. One can easily see that the resulting map is a homeomorphism; we declare V_j to be the coordinate neighborhood of the point ξ_j, and the constructed map, to be the corresponding local chart.

Obviously, the axioms of a Riemann surface are satisfied; it is also clear that the map $\bar{p} \colon X \to \overline{\mathbb{C}}$ is holomorphic (outside the new points, this is obvious; and in the local charts we have added, this map is raising to a power).

It remains to observe that the Riemann surface X is connected, since a dense subset X_0 of X is connected by Proposition 13.19, and the construction is completed.

In algebraic geometry, the constructed Riemann surface is called the "smooth completion of the affine algebraic curve defined by the equation $F(z, w) = 0$."

We illustrate the working of this construction by an example.

Example 13.21 Let $F(z, w) = w^2 - (z-a_1)(z-a_2)(z-a_3)(z-a_4)$, where $a_1, \ldots, a_4$ are pairwise different complex numbers; less formally, consider the equation

$$w^2 = (z - a_1)(z - a_2)(z - a_3)(z - a_4).$$

Clearly, the discriminant of F regarded as a polynomial in w vanishes at the points $a_1, \ldots, a_4$, so in our case $S = \{a_1, a_2, a_3, a_4, \infty\}$. Thus,

$$X_0 = \{(z, w) \colon w^2 = (z - a_1)(z - a_2)(z - a_3)(z - a_4),\ w \neq 0\}.$$

It remains to determine how many and which points must be added to X_0.

Let us look, for example, at what must be added to X_0 over the point $a_1 \in S$ (i.e., how many points are obtained from the preimage of a punctured neighborhood of a_1). For $\varepsilon > 0$, set $U = \{z \colon |z - a_1| < \varepsilon\}$, $U^* = U \setminus \{a_1\}$; choose ε so small that the points a_2, a_3, and a_4 do no lie in U. Then on U we have a well-defined function

$$\psi(z) = \sqrt{(z - a_2)(z - a_3)(z - a_4)}$$

(one of the branches of the square root). Set $D^*_{\sqrt{\varepsilon}} = \{w \colon 0 < |w| < \sqrt{\varepsilon}\}$. The map $\alpha \colon p^{-1}(U) \to D^*_{\sqrt{\varepsilon}}$ given by the formula $\alpha(z, w) = w/\psi(z)$ defines a homeomorphism between $p^{-1}(U^*)$ and $D^*_{\sqrt{\varepsilon}}$: indeed, the inverse map is given by the formula $t \mapsto (a_1 + t^2, t\psi(a_1 + t^2))$. If we additionally consider the map $\beta \colon D^*_{\sqrt{\varepsilon}} \to U^*$ given by the formula $\beta(t) = a_1 + t^2$, then the diagram

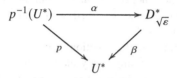

is commutative (recall that here this means that $\beta \circ \alpha = p$). Thus, the map α defines an isomorphism between the coverings $p \colon p^{-1}(U^*) \to U^*$ and $\beta \colon D^*_{\sqrt{\varepsilon}} \to U^*$. In particular, the set $p^{-1}(U)$, being homeomorphic to the punctured disk $D^*_{\sqrt{\varepsilon}}$, is connected, so exactly one new point must be added over a_1, denote it by $\tilde{a}_1$. Correspondingly, for a local chart (or a local coordinate) at this point, we can take the function $t = w/\psi(z)$. Since $t^2 = z - a_1$ by construction, we can use a less rigorous but clearer notation: $t = \sqrt{z - a_1}$ (and writing $z = a_1 + t^2$ does not even spoil the rigor). However, such a choice of a local chart is by no means unique: it is clear from the implicit function theorem (Proposition 9.2) that every holomorphic function of t having a nonzero derivative at the origin will also do. In particular, for a local coordinate at $\tilde{a}_1$ we can also take w (since $w = t\psi(z)$ where $z = a_1 + t^2$).

Clearly, for a_2, a_3, and a_4 we have the same picture: over each of them one point is added, and at each of them we can take $\sqrt{t - a_j}$ or w for a local coordinate.

Now let us look at what happens over the point ∞. A local coordinate on $\overline{\mathbb{C}}$ in a neighborhood of ∞ is $t = 1/z$; hence, over a punctured neighborhood of infinity the equation can be rewritten in the form

$$w^2 = \frac{(1 - a_1 t)(1 - a_2 t)(1 - a_3 t)(1 - a_4 t)}{t^4}.$$

Choose a branch of the root

$$\psi(t) = \sqrt{(1 - a_1 t)(1 - a_2 t)(1 - a_3 t)(1 - a_4 t)}$$

in a neighborhood of the origin; then over a neighborhood U^* of infinity, the map $\pi^{-1}(U^*) \to U^*$ has two sections: $t \mapsto \psi(t)/t^2$ and $t \mapsto -\psi(t)/t^2$. Hence, $\pi^{-1}(U^*)$ consists of two components, over ∞ two points are added (denote them by ∞_1 and ∞_2), and for a local coordinate at each of these points we can take $t = 1/z$. All this is illustrated in (inevitably schematic) Fig. 13.1.

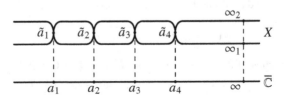

Fig. 13.1 $w^2 = (z - a_1)(z - a_2)(z - a_3)(z - a_4)$

To get comfortable with this situation, we analyze the function w. On $X_0 \subset X$ it is a bona fide holomorphic function, so on X it can be regarded as a function with isolated singularities at six points: $\tilde{a}_1, \ldots, \tilde{a}_4$, ∞_1, and ∞_2. Let us look at the type of singularities at these points.

As we know, at the point $\tilde{a}_1$ the function w is a local coordinate and vanishes (since $w = t\psi(z)$), that is, the singularity at $\tilde{a}_1$ is removable. This zero must be simple, otherwise w would not be one-to-one onto its image in a neighborhood of $\tilde{a}_1$. Another way to find $\mathrm{ord}_{\tilde{a}_1} w$ is to observe that $w = t\psi(z)$ where $\psi(z) \neq 0$ at $\tilde{a}_1$ (because, for example, the square of $\psi(z)$, equal to $(z - a_2)(z - a_3)(z - a_4)$, does not vanish at this point), and t is a local coordinate at $\tilde{a}_1$ which vanishes at this point.

Arguing in the same way, we see that w has zeros of order 1 at the points $\tilde{a}_2$, $\tilde{a}_3$, and $\tilde{a}_4$.

Note also that w has, obviously, no zeros on the set $X_0 = X \setminus \{\tilde{a}_1, \ldots, \tilde{a}_4, \infty_1, \infty_2\}$.

Now we look at what happens "at infinity," that is, at the points ∞_1 and ∞_2. At each of these points, we can take $t = 1/z$ for a local coordinate, with the function (more exactly, functions!) t vanishing at these points. Since at each of these points we have $w = \pm\psi(t)/t^2$ where $\psi(0) \neq 0$, it follows that the function w has a pole of order 2 at each of the two "infinities."

We can see this also in another way. Namely, since on X_0 we have

$$|w| = \sqrt{|(z - a_1)(z - a_2)(z - a_3)(z - a_4)|},$$

it follows that $\lim_{z \to \infty} (|w|/|z|^2) = 1$, so as z approaches any of the two points over ∞, the function $|w|$ grows as $|z|^2 = 1/|t|^2$; therefore, at these points w does indeed have poles of order 2.

Note that since the meromorphic function w on the compact Riemann surface X has four simple zeros and two poles of order 2, the orders of its zeros and poles sum to zero (if the orders of poles are counted as negative). As we will see later, this is no accident, but a manifestation of a general rule.

The construction described in this section is remarkable in that it is actually universal: it allows one to obtain any compact and connected Riemann surface. This statement is called "Riemann's existence theorem" (see p. 237).

13.3 Genus; the Riemann–Hurwitz Formula

As we have already mentioned, every Riemann surface is a smooth two-dimensional manifold of class C^∞. Moreover, since, according to Exercise 2.12a, the Jacobian of a holomorphic function f, regarded as a map from (an open subset in) $\mathbb{R}^2$ to $\mathbb{R}^2$, at a point a is equal to $|f'(a)|^2$, the Jacobians of transition maps on a Riemann surface regarded as a real manifold are positive. Thus, a Riemann surface is not merely a two-dimensional manifold, but this manifold is also necessarily orientable.

It is well known that every compact and connected two-dimensional smooth orientable manifold (a compact orientable surface) is homeomorphic to a sphere with handles; the number of these handles is called the *genus* of the surface. In particular, every compact and connected Riemann surface has a well-defined genus. For instance, the genus of the Riemann sphere is 0, and the genus of every elliptic curve is 1, since elliptic curves are homeomorphic to the torus.

Every compact (and connected) Riemann surface of genus 0 (i.e., homeomorphic to the sphere) is necessarily isomorphic to the Riemann sphere $\overline{\mathbb{C}}$ (this result is not easy to prove, and we will not do it here). On the contrary, for every $g > 0$ there exist many nonisomorphic compact Riemann surfaces of genus g.

Both the Riemann sphere and elliptic curves were defined, more or less from the beginning, as spheres with a known number of handles. A natural question is how one can find the genus of a Riemann surface defined in another way, e.g., by the construction from the previous section? To answer this question, we need to study the topological structure of "ramified coverings," i.e., nonconstant holomorphic maps between compact Riemann surfaces.

To begin with, we extend another couple of properties of holomorphic maps to the case of Riemann surfaces.

Proposition 13.22 *If X and Y are Riemann surfaces, X is connected, and $f : X \to Y$ is a nonconstant holomorphic map, then for every open set $U \subset X$ the set $f(U)$ is open in Y (in other words, f is an open map).*

Proof Let $a \in X$; we must show that $f(a)$ is an interior point in $f(X)$. Let $U \ni a$, $V \ni f(a)$ be open sets isomorphic to open subsets in $\mathbb{C}$ (i.e., subsets of appropriate coordinate neighborhoods) for which $f(U) \subset V$ and U is connected. If the restriction of f to U is constant, then Proposition 13.9 implies that f is constant, contradicting the assumption. On the other hand, if $f|_U$ is not constant, then the open mapping theorem for holomorphic functions implies that $f(a)$ is an interior point in V, and hence in $f(X)$. $\qquad\square$

Proposition 13.23 *Let $f : X \to Y$ be a nonconstant holomorphic map between Riemann surfaces with X connected. Then for every point $a \in X$ one of the following two conditions holds:*

(1) there exist neighborhoods $U \ni a$ and $V \ni f(a)$ such that $f(U) = V$ and f induces an isomorphism between U and V;

(2) there exist a positive integer $k > 1$ and a fundamental system of neighborhoods of a with the following property: if $U \ni a$ is a neighborhood from this system and $f(U) = V$, then every point in V except $f(a)$ has exactly k preimages in U (under f), and the point $f(a)$ has only one preimage in U, namely, the point a.

Here the derivative of f at a point a can be understood either as the derivative of f written in local coordinates (different choices of local coordinates on X and Y may produce different numbers, but whether or not the derivative vanishes does not depend on this choice), or as the derivative of f regarded as a smooth map between smooth manifolds (it is well known that the derivative in this sense is a linear map from the tangent space to X at a to the tangent space to Y at $f(a)$; in Zorich's textbook, such a derivative is called a differential). Recall also that a fundamental system of neighborhoods of a point a in a topological space X is a family of open sets $\{U_\alpha\}$, $U_\alpha \ni a$, such that every open neighborhood $W \ni a$ contains some U_α.

Proof Upon passing to local coordinates, this follows from Propositions 9.2 and 9.7: case (1) occurs if the derivative of f at a does not vanish, and case (2) occurs if it does vanish. $\qquad\square$

Definition 13.24 Let $f\colon X \to Y$ be a nonconstant holomorphic map between connected Riemann surfaces. One says that a point $a \in X$ is a *ramification point* of f, or that f is *ramified* at a, if f and a satisfy condition (2) from Proposition 13.23. The number k from this condition ("the number of preimages close to a of a point close to $f(a)$") is called the *ramification index* of f at a.

Actually, if f is not ramified at a point $a \in X$, then the definition of the ramification index is still applicable: in this case, it is equal to 1.

Informally, the ramification index of a holomorphic map f is the ramification index of the holomorphic function which arises when we write f in local coordinates.

Proposition 13.25 *If $f\colon X \to Y$ is a nonconstant holomorphic map between connected Riemann surfaces, then the set of ramification points of f has no accumulation points in X.*

Proof Arguing by contradiction, assume that $p \in X$ is an accumulation point of ramification points. If $U \ni p$ and $V \ni f(p)$ are neighborhoods isomorphic to open subsets in $\mathbb{C}$ with U connected, then, passing to local coordinates, we may regard f as a holomorphic map from U to V where $U, V \subset \mathbb{C}$. Then, by the principle of analytic continuation, $f' = 0$ everywhere on U, hence f is constant on U. Returning to the original meaning of the symbols U, V, and f, we see that f is constant on a nonempty open set $U \subset X$, and hence, by Proposition 13.9, on the whole Riemann surface X, a contradiction.   □

Now we turn to the study of ramified coverings proper, starting with something simple.

Proposition 13.26 *Let $f\colon X \to Y$ be a nonconstant holomorphic map between compact and connected Riemann surfaces. Then $f(X) = Y$ and the fibers of f are finite.*

Proof Since X is compact, $f(X)$ is closed in Y; by the open mapping theorem (more exactly, by Proposition 13.22), $f(X)$ is open in Y. Since Y is connected, it follows that $f(X) = Y$.

If the set $f^{-1}(y)$, where $y \in Y$, has an accumulation point, then, by the principle of analytic continuation (Proposition 13.9), f is constant, a contradiction. Therefore, the set $f^{-1}(y)$ is discrete in X, and hence it is finite by the compactness of X.   □

Definition 13.27 If $f\colon X \to Y$ is a nonconstant holomorphic map between compact connected Riemann surfaces, then a point $b \in Y$ is called a *branch point* of f if it is the image of a ramification point of f belonging to X.

Remark 13.28 Some authors use the terms "ramification point" and "branch point" interchangeably.

Proposition-Definition 13.29 Let $f\colon X \to Y$ be a nonconstant holomorphic map between compact and connected Riemann surfaces. If $B \subset Y$ is the set of branch points of f, then the map

$$\tilde{f} = f|_{X \setminus f^{-1}(B)} : X \setminus f^{-1}(B) \to Y \setminus B$$

is a finite covering.

The degree of this covering is called the *degree* of f and denoted by $\deg f$.

Nonconstant holomorphic maps between compact and connected Riemann surfaces are called *ramified coverings*.

To prove this, we need a topological lemma.

Lemma 13.30 *Let* $f : M \to N$ *be a continuous map between Hausdorff spaces satisfying the following properties.*

(1) *Every point* $x \in M$ *has a neighborhood* $U \ni x$ *such that the set* $f(U) \subset N$ *is open and the restriction* $f|_U : U \to f(U)$ *is a homeomorphism.* (In such cases, f is said to be a *local homeomorphism.*)

(2) *For every closed set* $F \subset M$, *its image* $f(F) \subset N$ *is closed.* (In such cases, f is said to be a *closed map.*)

(3) *All fibers of* f *are finite.*

Then f *is a covering.*

Proof Let $y \in N$ be an arbitrary point; set $f^{-1}(y) = \{x_1, \dots, x_r\}$ (the preimage is finite by condition (3)). By condition (1), for every integer $j \in [1; r]$ there is a neighborhood $U_j \ni x_j$ such that f induces a homeomorphism between U_j and $f(U_j)$; since M is Hausdorff, we may assume, passing to smaller neighborhoods if necessary, that the neighborhoods U_j are pairwise disjoint.

The subset $M \setminus \bigcup_j U_j$ is closed in M, so condition (2) implies that $f(M \setminus \bigcup_j U_j)$ is closed in N. Thus, the set $V = N \setminus f(M \setminus \bigcup_j U_j)$ is open in N. By construction, $V \ni x$ and $f^{-1}(V) \subset \bigcup_j U_j$. Now, setting $U'_j = f^{-1}(V) \cap U_j$ for each j, we see that f induces a homeomorphism between U'_j and V, and $f^{-1}(V)$ is the disjoint union of the sets U'_j. This means exactly that f is a covering.   □

Proof of Proposition 13.29 Set $M = X \setminus f^{-1}(B)$ and $N = Y \setminus B$; it suffices to verify that the map $\tilde{f}$ satisfies the conditions of Lemma 13.30.

Condition (1) holds by the inverse function theorem and the fact that, by construction, $X \setminus f^{-1}(B)$ contains no ramification points of f.

To verify condition (2), assume that a set $F \subset M = X \setminus f^{-1}(B)$ is closed in M; if $\bar{F}$ is the closure of F in X, then $\bar{F}$ is compact by the compactness of X, so $f(\bar{F})$ is compact, and hence closed, in Y; thus, the set $\tilde{f}(F) = f(\bar{F}) \cap N$ is closed in N.

Finally, condition (3) follows from Proposition 13.26.   □

Note that the degree of a map between Riemann surfaces defined above is equal to its degree in the topological sense, as defined for a map between oriented compact manifolds of the same dimension.

It is clear from the above that the degree of a map is equal to the number of preimages of a "generic point" (which in our situation means "of all points except finitely many"). Here is a more precise statement.

Proposition 13.31 *Let* $f: X \to Y$ *be a nonconstant holomorphic map between compact and connected Riemann surfaces, and let* $y \in Y$ *and* $f^{-1}(y) = \{x_1, \ldots, x_k\}$. *Denoting the ramification index of* f *at* x_j *by* e_j, *we have* $\sum\limits_{j=1}^{k} e_j = \deg f$.

Proof Let $\deg f = n$. It is clear that the points x_j have disjoint neighborhoods $U_j \ni x_j$ such that in appropriate local coordinates the restriction of f to U_j is raising to the power e_j and the restriction of f to $U_j \setminus \{x_j\}$ is an e_j-fold covering onto its image. Since $f^{-1}(y') \subset \bigcup_j U_j$ for all y' sufficiently close to y, the equality $\sum e_j = n$ follows from the fact that a point y' that is sufficiently close to y and does not coincide with it has n preimages. □

Now we can return to the problem of finding the genus of a Riemann surface. Recall that every compact surface can be "triangulated," i.e., cut into finitely many triangles (faces), intervals (edges), and points (vertices). Denote by V, E, and F the number of vertices, edges, and faces, respectively; then the number $V - E + F$, called the *Euler characteristic* of the surface, does not depend on the triangulation.

In particular, the Euler characteristic of a sphere with g handles is $2 - 2g$. Let me briefly remind you of how this number can be obtained.

It is well known that the Euler characteristic of the sphere is 2 (the famous Euler's formula), so it suffices to show that attaching a handle reduces the Euler characteristic by 2. A handle is an open cylinder, and attaching a handle consists in removing two disjoint disks out of the surface and attaching two circles at the bases of the cylinder to the two boundaries of the removed disks. Disks may be replaced with triangles from a triangulation of the surface, and an open cylinder, with an open triangular prism; moreover, this open prism can be triangulated as shown in Fig. 13.2. Upon attaching a handle, the numbers V, E, and F undergo the following changes: the number of vertices (V) does not change at all, the number of edges (E) increases by 6 (we have added three "lateral edges" and three "diagonals" from the triangulation of the handle), and the number of faces (F) increases by 4: two faces corresponding to the removed triangles have disappeared, but six faces from the triangulation of the handle have been added. Thus, the number $V - E + F$ decreases by 2, as was claimed.

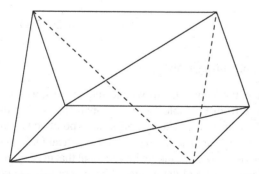

Fig. 13.2 A triangulated handle

So, the Euler characteristic of a Riemann surface of genus g is equal to $2-2g$, hence the genus and Euler characteristic determine each other uniquely. The following important fact is very helpful in finding the Euler characteristic.

Proposition 13.32 (Riemann–Hurwitz formula) *Let $f \colon X \to Y$ be a nonconstant holomorphic map between compact and connected Riemann surfaces. Assume that* $\deg f = n$, *the geni of X and Y are equal to $g(X)$ and $g(Y)$, respectively, and f is ramified at m points of X, with ramification indices $e_1, \ldots, e_m$. Then*

$$2g(X) - 2 = n(2g(Y) - 2) + \sum_{j=1}^{m}(e_j - 1). \qquad (13.2)$$

Proof Triangulate the surface Y in such a way that the images of all ramification points are among the vertices of the triangulation. This can always be done: for example, first consider an arbitrary triangulation of Y and then subdivide each of the triangles so that all points from the set B become vertices.

Denote the number of vertices, edges, and faces of the obtained triangulation by V, E, and F, respectively; then the Euler characteristic of the surface Y is equal to $V - E + F = 2 - 2g(Y)$. Taking the preimages of the faces, edges, and vertices under f, we obtain a triangulation of the surface X in which the number of edges is $E' = nE$, the number of faces is $F' = nF$ (our choice of a triangulation means that f is a covering over each face and over each edge, and faces and edges are simply connected). On the other hand, the number of vertices is $nV - \sum_{j=1}^{m}(e_j - 1)$: if the preimage of some vertex contains ramification points with indices $e_{i_1}, \ldots, e_{i_r}$, then the number of preimages of this vertex is equal to n minus $(e_{i_1} - 1) + \ldots + (e_{i_r} - 1)$ (formally, by Proposition 13.31; informally, because a ramification point of index e is a point where "e sheets meet"), and our choice of a triangulation means that every ramification point on X is contained in the preimage of some vertex.

Now we find the Euler characteristic of the surface X: it is equal to

$$2 - 2g(X) = V' - E' + F' = nV' - nE' + nF' - \sum_{j=1}^{m}(e_j - 1)$$

$$= n(2 - 2g(Y)) - \sum_{j=1}^{m}(e_j - 1);$$

it remains to multiply this equality by (-1). $\square$

To illustrate how the Riemann–Hurwitz formula works, we find the genus of the Riemann surface X from Example 13.21. The map $(z, w) \mapsto z$ induces a ramified covering $X \to \bar{\mathbb{C}}$ of degree 2. Clearly, all branch points on $\bar{\mathbb{C}}$ are $a_1, \ldots, a_4$, all ramification points on X are $\tilde{a}_1, \ldots, \tilde{a}_4$, and all ramification indices are equal to 2. Denoting the genus of X by g and taking into account that the degree of the map under consideration is equal to 2 and the genus of the sphere is 0, we obtain from (13.2) that

$$2 - 2g = 2 \cdot (2 - 2 \cdot 0) - \sum_{j=1}^{4}(2 - 1) = 0.$$

Thus, $g = 1$.

We conclude this section by proving a useful corollary of Proposition 13.31.

Proposition 13.33 *If f is a meromorphic function on a compact and connected Riemann surface X, then $\sum_{a \in X} \mathrm{ord}_a(f) = 0$.*

(The sum is finite, since the order of a function at a point that is neither zero nor pole is zero.)

Proof If f is a constant, then all orders are zero and there is nothing to prove. Otherwise, regarding f as a holomorphic map $f \colon X \to \overline{\mathbb{C}}$, denote its degree by d. Let $a_1, \ldots, a_m$ and $b_1, \ldots, b_n$ be the zeros and poles of f, respectively. Clearly, $\mathrm{ord}_{a_j} f$ coincides with the ramification index of the map $f \colon X \to \overline{\mathbb{C}}$ at a_j (for every j), and $\mathrm{ord}_{b_k} f$ coincides with the ramification index of the map $f \colon X \to \overline{\mathbb{C}}$ at b_k with the sign reversed. Hence, $\sum_{j=1}^{m} \mathrm{ord}_{a_j} f = d$ and $\sum_{k=1}^{n} \mathrm{ord}_{b_k} f = -d$. Since

$$\sum_{a \in X} \mathrm{ord}_a(f) = \sum_{j=1}^{m} \mathrm{ord}_{a_j} f + \sum_{k=1}^{n} \mathrm{ord}_{b_k} f,$$

we are done.   □

13.4 Differential Forms and Residues

Holomorphic differential forms on complex manifolds are defined similarly to differential forms on smooth manifolds. We will use the following "working" definition, which makes no claim to such virtues as being invariant and coordinate-free.

Definition 13.34 Let X be a Riemann surface, and let $X = \bigcup U_\alpha$ be a cover of X by coordinate neighborhoods with local charts $z_\alpha \colon U_\alpha \to \mathbb{C}$. A *holomorphic form* on X is a collection of expressions $f_\alpha \, dz_\alpha$, for each α, where $f_\alpha \colon U_\alpha \to \mathbb{C}$ are holomorphic functions satisfying the following condition: on $U_\alpha \cap U_\beta$,

$$f_\alpha \, dz_\alpha = f_\beta \, dz_\beta, \quad \text{i.e., } f_\alpha = f_\beta \frac{dz_\beta}{dz_\alpha}. \tag{13.3}$$

Meromorphic forms are defined analogously, with the difference that f_α are assumed to be meromorphic rather than holomorphic.

Here the expression dz_β / dz_α should be understood as the derivative of the function $z_\beta \circ (z_\alpha)^{-1}$ (transition map) with respect to z_α, and each function f_α, as a function of the variable z_α (see Remark 13.3).

When speaking of Riemann surfaces, holomorphic and meromorphic differential forms are sometimes called *differentials*.

Example 13.35 If we consider the form ω on the Riemann sphere $\overline{\mathbb{C}}$ given by $z\,dz$ in the coordinate neighborhood $U_0 = \overline{\mathbb{C}} \setminus \{\infty\}$, then its representation in the coordinate neighborhood $U_1 = \overline{\mathbb{C}} \setminus \{0\}$ can be obtained as follows: since for a local coordinate in U_1 we can take $w = 1/z$, substitute $z = 1/w$ and $dz = (1/w)'\,dw = -dw/w^2$ into the expression for ω to obtain

$$z\,dz = \frac{1}{w} \cdot \left(-\frac{dw}{w^2} \right) = -\frac{dw}{w^3}.$$

For a meromorphic differential form, we have well-defined notions of zeros and poles, and also multiplicities of these zeros and poles. Namely, if $\omega = f_\alpha\,dz_\alpha$ in a coordinate neighborhood U_α with local chart z_α and if $p \in U_\alpha$, then we set $\operatorname{ord}_p(\omega) = \operatorname{ord}_p(f_\alpha)$. If p lies also in another coordinate neighborhood U_β, where $\omega = f_\beta\,dz_\beta$, then $\operatorname{ord}_p(f_\alpha) = \operatorname{ord}_p(f_\beta)$, since f_α is obtained from f_β by multiplication by the derivative of the transition map, and such a derivative does not vanish.

For instance, it is clear from Example 13.35 that the form $z\,dz$ on $\mathbb{C}$, regarded as a meromorphic form on $\overline{\mathbb{C}}$, has a pole of order 3 at the point ∞.

There are two important operations related to differentials on a Riemann surface: multiplying a form by a function and taking the differential of a function.

We begin with multiplying a differential form by a function. If ω is a meromorphic form on a Riemann surface X given by $f_\alpha\,dz_\alpha$ in a coordinate neighborhood U_α, and if g is a meromorphic function on X, then $g\omega$ is the meromorphic form given by $g f_\alpha\,dz_\alpha$ in U_α. Since the relations $g f_\alpha\,dz_\alpha = g f_\beta\,dz_\beta \cdot (dz_\beta/dz_\alpha)$ follow from (13.3), $g\omega$ is indeed a differential form.

The second important operation related to differential forms is taking the differential of a function.

Let f be a meromorphic function on a Riemann surface X covered by coordinate neighborhoods U_α with local charts $z_\alpha : U_\alpha \to \mathbb{C}$.

Definition 13.36 The *differential* of f is the meromorphic form df that in a coordinate neighborhood U_α is given by $(df_\alpha/dz_\alpha)\,dz_\alpha$.

In this definition, each f_α is regarded as a function not of a point of the set $U_\alpha \subset X$, but of the complex variable z_α. It would be more correct to write

$$\omega = \frac{d(f \circ (z_\alpha)^{-1})}{dz_\alpha}\,dz_\alpha.$$

The differential is well-defined, since the local representations of df satisfy relation (13.3) by the chain rule:

$$\frac{df}{dz_\alpha} = \frac{df}{dz_\beta}\frac{dz_\beta}{dz_\alpha}.$$

Example 13.37 Let X be the compact Riemann surface from Example 13.21. Denote by $\pi\colon X \to \overline{\mathbb{C}}$ the map corresponding to the meromorphic function z (or to the map $(z, w) \mapsto z$).

Let us find the zeros and poles of the form dz. Over $\mathbb{C}\setminus\{a_1, \ldots, a_4\}$, the map π is an unramified covering, thus in a neighborhood of each point of $\pi^{-1}(\mathbb{C} \setminus \{a_1, \ldots, a_4\})$ we can take the function z for a local coordinate. Hence, on this set dz has no zeros (and certainly no poles, since the function z is holomorphic on this set). In a neighborhood of each of the points a_j, for a local coordinate we can take a function t such that $t^2 = z - a_j$; therefore, $z = t^2 + a_j$, $dz = 2t\,dt$, and the form dz has a zero of order 1 at this point. Finally, consider the points ∞_1 and ∞_2. For a local coordinate in a neighborhood of each of these points, we can take the function $t = 1/z$; since $z = 1/t$, we have $dz = -dt/t^2$, and the form dz has a pole of order 2 at each of the points ∞_1 and ∞_2.

Let us look again at the form dz/w. As we have seen in Example 13.21, the function w has a zero of order 1 at each of the points a_j and a pole of order 2 at each of the points ∞_1 and ∞_2. When multiplying dz by $1/w$, the zeros cancel the zeros and the poles cancel the poles. Thus, dz/w is a holomorphic form that has neither zeros nor poles.

So, on the Riemann surface X we have found a nontrivial holomorphic form. In contrast, on the Riemann sphere this cannot be done (see Exercise 13.17). The general result is as follows:

On a compact and connected Riemann surface, the vector space of holomorphic forms is finite-dimensional over $\mathbb{C}$, and its dimension is equal to the genus of the surface.

The fact that the space of holomorphic forms on a compact Riemann surface is finite-dimensional can be proved by elementary methods (see Exercise 13.18), but the proof of a truly interesting result, that the dimension coincides with the genus, is beyond the scope of this book. Still, for elliptic curves this result is quite elementary, as is clear from the following example.

Example 13.38 Let us look at what holomorphic differential forms can exist on an elliptic curve $X = \mathbb{C}/\Gamma$. In Example 13.8, we have constructed on X local charts with very simple transition maps: $z_\alpha = z_\beta + u$ where $u \in \Gamma$ is a constant. Thus, $dz_\alpha/dz_\beta = 1$, $dz_\alpha = dz_\beta$, and if ω is a holomorphic form on X given by $f_\alpha\,dz_\alpha$ on U_α, then it follows from the condition $f_\alpha\,dz_\alpha = f_\beta\,dz_\beta$ that $f_\alpha = f_\beta$ on $U_\alpha \cap U_\beta$, i.e., all f_α "merge" into a single holomorphic function f on X. Since every holomorphic function on a compact Riemann surface is a constant, we see that every holomorphic form on X has the form $\omega = c\,dz$ where $c \in \mathbb{C}$ is a constant and dz is the common notation for all dz_α that agree with one another on the intersections of their domains.

For both meromorphic forms from Example 13.37, the orders of the zeros and poles sum to zero. It might seem that this is always the case, by analogy with the orders of the zeros and poles of a function, but actually this is a coincidence stemming from the fact that the genus of the surface X is equal to 1. Now we are going to discuss what this sum really equals.

First, we show that on a fixed compact Riemann surface, this sum is the same for all meromorphic forms. The main observation here is that meromorphic forms can be divided by one another:

Proposition 13.39 *Let ω and η be meromorphic forms on a connected Riemann surface X; if η is not identically zero, then there exists a meromorphic function h such that $\omega = h\eta$.*

Proof Let $X = \bigcup U_\alpha$ be a cover of X by coordinate neighborhoods with local charts $z_\alpha \colon U_\alpha \to \mathbb{C}$, and let ω and η on U_α be equal to $f_\alpha\, dz_\alpha$ and $g_\alpha\, dz_\alpha$, respectively. Dividing relations (13.3) for ω and η by one another, we see that on $U_\alpha \cap U_\beta$ the quotients f_α/g_α and f_β/g_β coincide, so the functions f_α/g_α "merge" into a single meromorphic function h; clearly, $\omega = h\eta$. $\square$

Corollary 13.40 *If ω and η are nonzero meromorphic forms on a connected compact Riemann surface X, then $\sum\limits_{p\in X} \operatorname{ord}_p \omega = \sum\limits_{p\in X} \operatorname{ord}_p \eta$.*

Proof By Proposition 13.39, there exists a meromorphic function h such that $\omega = h\eta$. Therefore,

$$\sum_{p\in X} \operatorname{ord}_p \omega = \sum_{p\in X} \operatorname{ord}_p h + \sum_{p\in X} \operatorname{ord}_p \eta.$$

By Proposition 13.33, the first term in the right-hand side vanishes. $\square$

So, the sum of the orders of the zeros and poles of a meromorphic form depends not on this form, but only on the Riemann surface. Let us figure out how exactly it depends on the surface.

Proposition 13.41 *On a compact Riemann surface of genus g, the orders of the zeros and poles of any meromorphic form sum to $2g - 2$ (the orders of poles are counted as negative).*

We will give two proofs of this important fact: the first one involves more algebra, and the second one, more topology.

First proof of Proposition 13.41 In this first proof, we assume that on the Riemann surface in question there exists at least one nonconstant meromorphic function. The existence of such a function follows from Riemann's existence theorem (see p. 237), which we have stated without proof and which says that every compact Riemann surface is isomorphic to a Riemann surface corresponding to some algebraic equation $F(z, w) = 0$; if so, for a meromorphic function we can take, for example, the function z or w. With this in mind, let's get down to business.

Corollary 13.40 shows that it suffices to find the sum of the orders of the zeros and poles for one arbitrary meromorphic form ω; choose ω to be the form df where f is the above-mentioned nonconstant meromorphic function on X. We regard f as a holomorphic map from X to $\overline{\mathbb{C}}$. Denote the degree of f by n. Let $p_1, \ldots, p_r$ be the ramification points of f that are not poles; denote the ramification index of f at p_j by e_j.

If f has ramification index e at a point p with $f(p) = b \neq \infty$, then $\mathrm{ord}_p(df) = e - 1$. Indeed, let t be a local coordinate at p for which $t(p) = 0$. Then in these coordinates we have $f(t) = b + c_e t^e + c_{e+1} t^{e+1} + \dots$ with $c_e \neq 0$, hence the derivative of f has a zero of order $e - 1$ at p. (In particular, if f is unramified at a point $p \in f^{-1}(\mathbb{C})$, then df has neither zero no pole at p.) It remains to consider the points $p \in f^{-1}(\infty)$, i.e., the poles of f. Denote these points by $q_1, \dots, q_m$, and let h_j be the order of the pole at q_j, which, obviously, coincides with the ramification index. Clearly, $\mathrm{ord}_{q_j}(df) = -h_j - 1$. Since $\sum h_j = \deg f = n$, we have

$$\sum_{p \in X} \mathrm{ord}_p(df) = \sum(e_j - 1) - \sum(h_j + 1)$$

$$= \sum(e_j - 1) + \sum(h_j - 1) - 2\sum f_j = \sum(e_j - 1) + \sum(h_j - 1) - 2n.$$

On the other hand, by the Riemann–Hurwitz formula,

$$2n - \left(\sum(e_j - 1) + \sum(h_j - 1) \right) = 2 - 2g.$$

Therefore, $\sum_p \mathrm{ord}_p(df) = 2g - 2$, as required. $\qquad\square$

The second proof does not rely on Riemann's existence theorem, which we do not prove; instead, it uses a fact from elementary topology which we also do not prove, but which is perhaps simpler. To begin with, we need to recall what is meant by the index of a singular point of a vector field.

Fix an orientation on the plane $\mathbb{R}^2$. Let $U \subset \mathbb{R}^2$ be an open subset, $p \in U$, and v be a nonvanishing vector field on $U \setminus \{p\}$. Then the index of the point p with respect to the vector field v is defined as follows. Consider a circle in U centered at p and equip it with a one-to-one smooth parametrization $\gamma \colon [a; b] \to U \setminus \{p\}$ that makes it positively oriented[1] with respect to the chosen orientation on $\mathbb{R}^2$. Then the index of p with respect to v is defined as the index of the closed curve $t \mapsto v(\gamma(t))$ around the origin.

Here is the main example we are interested in.

Lemma 13.42 *Let us identify $\mathbb{R}^2$ with the complex plane $\mathbb{C}$ endowed with the natural orientation[2]. Let $U \subset \mathbb{C}$ be an open subset, $p \in U$, and $\varphi \colon U \setminus \{p\} \to \mathbb{C}$ be a holomorphic function with a removable singularity or a pole at p. Denote by v the vector field on $U \setminus \{p\}$ obtained by assigning the vector $\varphi(z)$ to each point z. Then the index of p with respect to v is equal to $\mathrm{ord}_p \, \varphi$.*

Proof This is a special case of the argument principle (Theorem 8.6). $\qquad\square$

---

[1] This orientation is defined as follows: for every point $z \in \gamma$, the pair consisting of a positively oriented tangent vector to γ at the point z and the vector $\overline{zp}$ (in this order) is positively oriented. In the previous chapters, in this situation we said, without further ado, that γ is oriented counterclockwise.

[2] Informally, the same orientation that we have used throughout; formally, the orientation with respect to which the basis $\langle 1, i \rangle$ of $\mathbb{C}$ over $\mathbb{R}$ is positive.

If $\varphi \colon U \to U'$ is an orientation-preserving diffeomorphism, $\varphi(p) = p'$, and v' is the vector field on U' that is the image of v under φ, then the index of p with respect to v is equal to the index of p' with respect to v'. Then the index of a point p with respect to a vector field v is well defined also in the case where X is an orientable surface, $U \subset X$ is an open subset, $p \in U$, and v is a nonvanishing vector field on $U \setminus p$. Denote this index by $\mathrm{Ind}_v\, p$.

Now we state without proof a topological fact that we are going to use.

Proposition 13.43 *Let X be a compact orientable surface, $S \subset X$ be a finite subset, and v be a smooth nonvanishing vector field on $X \setminus S$. Then the sum $\sum\limits_{p \in S} \mathrm{Ind}_v\, p$ is equal to the Euler characteristic of X.*

Second proof of Proposition 13.41 Let ω be a meromorphic form on X, and let S be the set of its zeros and poles. If $X = \bigcup U_\alpha$ is a cover of X by coordinate neighborhoods with local charts $z_\alpha \colon U_\alpha \to V_\alpha \subset \mathbb{C}$, then let $\omega|_{U_\alpha} = f_\alpha\, dz_\alpha$. We claim that the collection of functions $1/f_\alpha$ determines a nonvanishing vector field on $X \setminus S$.

Indeed, on each $V_\alpha \setminus S$ (more exactly, $V_\alpha \setminus z_\alpha(S \cap U_\alpha)$) consider the vector field defined by assigning the vector $1/f_\alpha(t)$ to each point t. Applying the diffeomorphism z_α^{-1}, we obtain a vector field on $U_\alpha \setminus S$, denote it by v_α. However, on $U_\alpha \cap U_\beta$ the vector fields v_α and v_β coincide: this follows from the equality $1/f_\beta = (1/f_\alpha)(dz_\beta/dz_\alpha)$, which, in turn, follows from the equality $f_\beta = f_\alpha \cdot (dz_\alpha/dz_\beta)$.

Denote the obtained vector field by v. By Lemma 13.42, the index of every point $p \in S$ with respect to v is equal to $-\mathrm{ord}_p\, \omega$. By Proposition 13.43, we now have

$$2 - 2g = \sum_{p \in S}(-\mathrm{ord}_p\, \omega),$$

whence $\sum\limits_{p \in S} \mathrm{ord}_p\, \omega = 2g - 2$. $\qquad\qquad\Box$

Holomorphic forms can be integrated over piecewise smooth paths. For the reader's convenience, I will give the definition of such integrals as applied to the case we are interested in. This is done in two steps.

Definition 13.44 Let $\gamma \colon [p;q] \to X$ be a piecewise smooth path on a Riemann surface X and ω be a holomorphic form on X. If $\gamma([p;q])$ lies in a coordinate neighborhood U_α with local chart $z_\alpha \colon U_\alpha \to \mathbb{C}$ and $\omega = f_\alpha\, dz_\alpha$ in this local coordinate, then the integral of ω over γ is the number

$$\int_\gamma \omega = \int_{z_\alpha \circ \gamma} (f_\alpha \circ z_\alpha^{-1})\, dz_\alpha.$$

If $\gamma([p;q])$ lies also in a coordinate neighborhood U_β with local chart z_β and $\omega = f_\beta\, dz_\beta$ in this local coordinate, then

$$\int_{z_\alpha \circ \gamma} (f_\alpha \circ z_\alpha^{-1})\, dz_\alpha = \int_{z_\beta \circ \gamma} (f_\beta \circ z_\beta^{-1})\, dz_\beta$$

in view of the relation $f_\beta = f_\alpha \cdot (dz_\alpha/dz_\beta)$ and the change of variable formula, so the integral is well defined.

In the general case, the integral is defined as follows.

Definition 13.45 Let $\gamma : [p; q] \to X$ be a piecewise smooth path on a Riemann surface X and ω be a holomorphic form on X. If the interval $[p; q]$ is divided by points $p = p_0 < p_1 < \ldots < p_n = q$ into subintervals such that for every j the set $\gamma([p_j; p_{j+1}])$ is entirely contained in a coordinate neighborhood, then we set

$$\int_\gamma \omega = \sum_{j=0}^{n-1} \int_{\gamma|_{[p_j : p_{j+1}]}} \omega.$$

It is easy to see that the integral thus defined does not depend on the choice of a partition.

If f is a holomorphic function on an open subset of the complex plane, then the integral $\int_\gamma f(z)\, dz$ coincides with the integral of the form $f(z)\, dz$ over the path γ we have just defined.

Remark 13.46 For the reader familiar with the corresponding definition, I mention that holomorphic forms on a Riemann surface X can be regarded as smooth differential forms of degree 1 on the same X regarded as a smooth two-dimensional manifold (we obtain complex-valued differential forms rather than real-valued ones, but this causes no problems). Then the integral we have defined is a special case of the integral of a 1-form over a path.

Now we extend the notion of residue to the case of Riemann surfaces. For a function on a Riemann surface, residue can be defined neither as a Laurent coefficient, nor as an integral: the coefficients of a Laurent series depend on the choice of local coordinates, and the integral of a function on a Riemann surface over a curve cannot be consistently defined at all. Thus, neither of our definitions of residue can be generalized to Riemann surfaces. However, it turns out that residues can be defined for differential forms.

To begin with, note that if $f : V' \to V''$ is a conformal isomorphism between open subsets of the complex plane, $p \in V'$, and γ is a closed piecewise smooth path in V' that does not pass through p, then $\mathrm{Ind}_p\, \gamma = \mathrm{Ind}_{f(p)}(f \circ \gamma)$ (the proof may be safely left to the reader). Thus, if X is an arbitrary Riemann surface, $U \subset X$ is an open subset of the complex plane isomorphic to the disk, $p \in U$ is a point, and γ is a closed piecewise smooth path that does not pass through p, then the index $\mathrm{Ind}_p\, \gamma$ is well defined: if $\varphi : U \to V \subset \mathbb{C}$ is an isomorphism, then it can be defined as $\mathrm{Ind}_{\varphi(p)}(\varphi \circ \gamma)$, and this number does not depend on the choice of the pair (φ, V) by the above remark.

Now we can define the residue of a differential form on a Riemann surface at an isolated singularity.

Let X be a Riemann surface, $p \in X$, $W \ni p$ be a neighborhood of p, and ω be a holomorphic form on $W \setminus \{p\}$.

Also, let $U \ni p$, $U \subset W$ be a neighborhood isomorphic to a disk in $\mathbb{C}$, and let γ be a closed curve in U such that $\mathrm{Ind}_p \, \gamma = 1$.

Definition 13.47 The *residue* of the form ω at the point p is the number

$$\mathrm{Res}_p \, \omega = \frac{1}{2\pi i} \int_\gamma \omega.$$

When finding residues on Riemann surfaces, it is also convenient to use Laurent series. The following proposition obviously follows from the definition of residue and Proposition 8.2.

Proposition 13.48 *Let ω be a holomorphic form in a punctured neighborhood of a point p on a Riemann surface. Assume that z is a local chart in a neighborhood of p, that $z(p) = a$, and that $\omega = f(z)\,dz$ in this neighborhood. Then the residue $\mathrm{Res}_p \, \omega$ is equal to the coefficient of $(z - a)^{-1}$ in the Laurent expansion of f at the point a.*

Example 13.49 The meromorphic form dz/z on $\mathbb{C}$ can be redarded as a form with isolated singularities on $\overline{\mathbb{C}}$; let us find its residue at ∞. Taking the function $z = 1/w$ for a local chart in a neighborhood of ∞, we obtain $dz = d(1/w) = -dw/w^2$, whence

$$\frac{dz}{z} = \frac{-dw}{w^2} \bigg/ \frac{1}{w} = -\frac{dw}{w}.$$

Thus, in view of Proposition 13.48, we have $\mathrm{Res}_\infty(dz/z) = -1$.

In the previous example, the sum of the residues of the form dz/z on $\overline{\mathbb{C}}$ vanishes; this is a manifestation of a general fact similar to Proposition 8.3.

Proposition 13.50 *If X is a compact Riemann surface and ω is a form on X that is holomorphic everywhere except finitely many points, then the sum of the residues of this form at all its singular points vanishes.*

By singular points of ω we mean points on X at which ω is not holomorphic.

Proof Let ω be a form with finitely many singularities on a compact Riemann surface X. Consider a triangulation of X such that the set of its vertices contains all points at which ω is not holomorphic and every closed triangle is entirely contained in some coordinate neighborhood.

Now, for each singular point consider a disk around it (containing no other singular points) and remove the interiors of all these disks from X; thus, the resulting surface with boundary X' is divided into curvilinear polygons. The resulting figure looks like Fig. 5.3 (p. 55), except that there is no outer curve γ (a compact Riemann surface has no boundary).

Endow the boundary of each of these polygons with the orientation induced by the orientation of X. The integral of ω over each of these boundaries vanishes by Definition 13.44 and Cauchy's theorem. If we sum up all these vanishing integrals, then the integrals over the inner parts of boundaries cancel, and the sum of the integrals over the parts of the disk boundaries equals the sum of the integrals over the boundaries of these disks; thus, this sum vanishes, and it differs only by the factor $2\pi i$ from the sum of the residues. $\qquad\square$

As you see, Propositions 13.50 and 8.3 are similar not only in statements, but also in proofs.

As the first application of Proposition 13.50, we give a proof of Proposition 13.33 that does not rely on the theory of ramified coverings. So, let f be a meromorphic function on a compact and connected Riemann surface X that is not identically zero; we want to prove that $\sum\limits_{p \in X} \mathrm{ord}_p(f) = 0$. To this end, consider the meromorphic form df/f (the "logarithmic differential" of f) on X. Obviously, it is holomorphic at points different from zeros and poles of f, and if p is a zero or pole of f, then Lemma 8.4 shows that $\mathrm{Res}_p(df/f) = \mathrm{ord}_p(f)$. By what we have just proved, the sum of all residues of the form df/f vanishes, hence the sum of the orders of f at all its zeros and poles vanishes too.

13.5 On Riemann's Existence Theorem

In the theory of compact Riemann surfaces there are several quite important results that share a common property: they are easy to state, but in order to prove them, one needs to develop a theory that goes far beyond the scope of this book. In the remainder of this chapter, I introduce the reader to these fundamental facts. All theorems (with one exception) will be stated in full, but the proofs will be given in the simplest nontrivial case, that of elliptic curves, for which there is no need to develop general theories, because you can "feel everything with your hands." (The Riemann sphere is an even simpler example of a compact Riemann surface, but this case can hardly be called nontrivial.)

We begin with Riemann's existence theorem, which has been already mentioned several times. Recall the statement of this theorem.

Theorem 13.51 *Every compact and connected Riemann surface is isomorphic to a Riemann surface corresponding to an irreducible polynomial in two variables via the construction from Sec. 13.2.*

We will prove this theorem in the special case where the Riemann surface in question is an elliptic curve.

So, let $X = \mathbb{C}/\Gamma$ where $\Gamma \subset \mathbb{C}$ is a lattice generated by numbers ω_1 and ω_2. Let $\wp$ be the Weierstrass function corresponding to Γ (see Sec. 11.2); again, set $e_1 = \wp(\omega_1/2)$, $e_2 = \wp((\omega_1 + \omega_2)/2)$, $e_3 = \wp(\omega_2/2)$.

Proposition 13.52 *The elliptic curve $\mathbb{C}/\Gamma$ is isomorphic to the Riemann surface corresponding to the equation*

$$w^2 = 4(z - e_1)(z - e_2)(z - e_3). \tag{13.4}$$

Proof To begin with, recall that the numbers e_1, e_2, and e_3 are distinct, see Exercise 11.5. To avoid referring to an exercise, we prove this fact directly. Namely, regarding $\wp$ as a meromorphic function on X, i.e., as a holomorphic map from X to $\overline{\mathbb{C}}$, we have, since $\wp$ has poles only at the points of the lattice, $f_\wp^{-1}(\infty) = \{0\}$ (hereafter, we identify complex numbers with their images in $X = \mathbb{C}/\Gamma$; this will not cause any confusion); and since these poles are of order 2, the ramification index of the map $\wp$ at 0 is equal to 2. Therefore, $\deg \wp = 2$ (by Proposition 13.31). Now observe that $\wp'$ has a simple zero at each of the points $\omega_1/2$, $(\omega_1 + \omega_2)/2$, and $\omega_2/2$ (see the proof of Proposition 11.17), so the map $\wp$ has ramification index 2 at each of these points. But then the images in $\overline{\mathbb{C}}$ of any two of these points are distinct from each other: if over one point in $\overline{\mathbb{C}}$ there are two points with ramification index 2, then, by the same Proposition 13.31, the degree of $\wp$ is at least $2 + 2 = 4$, a contradiction.

Now denote the compact Riemann surface corresponding to equation (13.4) by Y. Applying the construction from Sec. 13.2, we see that Y is obtained by adding to the Riemann surface

$$Y_0 = \{(z, w) \in \mathbb{C}^2 : w^2 = 4(z - e_1)(z - e_2)(z - e_3), \, z \neq e_1, e_2, e_3\}$$

points lying over e_1, e_2, e_3, and ∞.

As in Example 13.21, we deduce that exactly one point is added over each point e_j. Denote these new points by $\tilde{e}_j \in Y$. Over ∞, the picture is different from that in the above-mentioned example: if in a neighborhood of the point $\infty \in \overline{\mathbb{C}}$ we choose the local coordinate $s = 1/z$, then equation (13.4) can be rewritten in the form

$$w^2 = \frac{4(1 - e_1 s)(1 - e_2 s)(1 - e_3 s)}{s^3};$$

set

$$\psi(s) = \frac{2\sqrt{(1 - e_1 s)(1 - e_2 s)(1 - e_3 s)}}{s},$$

where $\sqrt{(1 - e_1 s)(1 - e_2 s)(1 - e_3 s)}$ is a single-valued branch of the root, which exists for all sufficiently small s. Denote by $p : Y_0 \to \overline{\mathbb{C}} \setminus \{e_1, e_2, e_3, \infty\}$ the projection $(z, w) \mapsto z$. Then the map $(s, w) \mapsto (s, w/\psi(s))$ defines an isomorphism between the covering $p^{-1}(U) \to U$, where

$$U = \left\{ z : |z| > \frac{1}{\varepsilon} \right\} = \{s : 0 < |s| < \varepsilon\},$$

and the covering $p': V \to U$, where $V = \{(a, s): a^2 = 1/s, 0 < |s| < \varepsilon\}$, $p': (a, s) \mapsto s$. Thus, $p^{-1}(U)$ is connected, and over ∞ there is also only one new point, denote it by $\tilde{\infty}$. Schematically, all this is shown in Fig. 13.3.

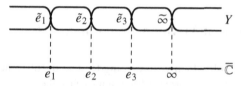

Fig. 13.3 $w^2 = 4(z - e_1)(z - e_2)(z - e_3)$

Now set $X_0 = X \setminus \{0, \omega_1/2, (\omega_1 + \omega_2)/2, \omega_2/2\}$ and define a map $f_0: X_0 \to Y_0$ by

$$f_0: t \mapsto (\wp(t), \wp'(t)).$$

We have $f_0(X_0) \subset Y_0$ by (11.11); the map f_0 is holomorphic, since for a local coordinate on the whole of Y_0 we can take z, but $z = \wp(t)$. Extend f_0 to a map $f: X \to Y$ by setting

$$f(0) = \tilde{\infty}, \quad f(\omega_1/2) = \tilde{e}_1, \quad f((\omega_1 + \omega_2)/2) = \tilde{e}_2, \quad f(\omega_2/2) = \tilde{e}_2.$$

Note that f is continuous; this follows from the fact that for a fundamental system of neighborhoods of each $\tilde{e}_j$ we can take the sets of the form $p^{-1}(\{z: |z - e_j| < \varepsilon\})$ where $p: Y \to \overline{\mathbb{C}}$ is the map induced by the projection $(z, w) \mapsto z$, and for a fundamental system of neighborhoods of $\tilde{\infty}$ we can take the preimages of the neighborhoods of ∞ in $\overline{\mathbb{C}}$. Therefore, the map f is holomorphic at the four new points of Y too; it suffices to verify this in local coordinates, and then this follows from the Riemann removable singularity theorem (if a function can be extended by continuity to a singular point, then it is certainly bounded in a neighborhood of this point).

So, $f: X \to Y$ is a holomorphic (and, obviously, nonconstant) map. We will show that it has degree 1. For this, it suffices to check that every point from Y_0 has exactly one preimage.

Indeed, if $f(t) = (z, w)$ where $z \notin \{e_1, e_2, e_3, \infty\}$, then $\wp(t) = z$. Since the function $\wp$ has degree 2 as a map from X to $\overline{\mathbb{C}}$, and the function $\wp$ is even, the curve $X = \mathbb{C}/\Gamma$ contains exactly one additional point mapped by $\wp$ to z, and this is the point $-t$ (note that $t \neq -t$ on X: otherwise, t is congruent modulo Γ to one of the "semiperiods" $\omega_1/2$, $(\omega_1 + \omega_2)/2$, or $\omega_2/2$, but this cannot happen, because $z \neq e_j$). Thus, the preimage of the point $(z, w) \in Y_0 \subset Y$ can contain only t or $-t$. Both these points cannot lie in the preimage simultaneously: if $f(t) = (z, w)$, then $f(-t) = (z, -w)$, because $\wp$ is even and $\wp'$ is odd, but $w \neq -w$, because $w \neq 0$ by (13.4). So, $\deg f = 1$. Thus, the holomorphic map f is bijective. By the compactness of X and Y, it is a homeomorphism; but then the inverse map is also holomorphic: again, it suffices to check this in local coordinates, and then this follows from Proposition 9.9.

So, the map $f: X \to Y$ is an isomorphism, and the proposition is proved. $\qquad\square$

In conclusion, note that the converse is also true: every compact Riemann surface of genus 1 is isomorphic to an elliptic curve, i.e., a Riemann surface of the form $\mathbb{C}/\Gamma$ where Γ is a lattice. However, this statement will not be proved here.

13.6 On the Field of Meromorphic Functions

The next classical result which we are going to state in full generality but prove in a special case says that on a compact Riemann surface there are "few" meromorphic functions (in a strictly defined sense).

So, let X be a connected Riemann surface. As we have already observed, the sum, difference, product, and quotient (if the divisor is not identically zero) of meromorphic functions on X is also a meromorphic function on X, so the set of all meromorphic functions on X is a field.

Theorem 13.53 *The field of meromorphic functions on a compact and connected Riemann surface X is generated over $\mathbb{C}$ by two elements satisfying an algebraic relation.*

The conclusion of the theorem means that on X there are meromorphic functions f and g such that every meromorphic function on X can be written in the form $R(f, g)$ where $R \in \mathbb{C}(T_1, T_2)$ (i.e., R is the quotient of polynomials with complex coefficients in two variables), and there exists a nonzero polynomial P in two variables with complex coefficients such that $P(f, g)$ is identically zero.

If X is not compact, then the analogous result does not hold. Indeed, Theorem 13.53 implies that the field of meromorphic functions on a compact Riemann surface has transcendence degree 1 over $\mathbb{C}$, so any two meromorphic functions satisfy an algebraic relation, while already on $\mathbb{C}$ we can give an example of an infinite family of algebraically independent functions, say, e^z, e^{e^z}, $e^{e^{e^z}}$, ..., see Exercise 13.22. Apparently, the point here is that the exponential and its iterations are "transcendental" functions, and it is exactly the existence of such functions that makes it possible to construct many algebraically independent functions. But on a compact Riemann surface all meromorphic functions are algebraic (this claim can also be given a precise meaning), and that is why there are relatively few of them.

We cannot prove Theorem 13.53 in full generality (it can be deduced from Riemann's existence theorem), but we will at least verify that it is true in two cases: the trivial one and the simplest nontrivial one.

The trivial case is, of course, $\overline{\mathbb{C}}$. As we know from Proposition 7.29, every meromorphic function on $\overline{\mathbb{C}}$ is a rational function, so the field of meromorphic functions on $\overline{\mathbb{C}}$ is generated by a single function z. (For pedants: for the second generating function, one can take the same function z. The algebraic relation between the functions $f(z) = z$ and $g(z) = z$ has the form $f = g$.)

The simplest nontrivial case is still the case of elliptic curves.

Proposition 13.54 *Let $\Gamma \subset \mathbb{C}$ be a lattice and $\wp$ be the corresponding Weierstrass function. Then every meromorphic function on $X = \mathbb{C}/\Gamma$ can be written as a ra-*

tional function of $\wp$ and $\wp'$. Moreover, the functions $\wp$ and $\wp'$ satisfy the algebraic relation (11.11).

Proof The last assertion is already proved in Chap. 11. It remains to prove the assertion concerning the field. In further reasoning, we do not distinguish between meromorphic functions on X and functions elliptic with respect to the lattice Γ (which we call simply elliptic, since we will not encounter any other lattices).

Lemma 13.55 *Every even elliptic function can be written as a rational function of $\wp$.*   □

Proof Let f be an even elliptic function. Construct a map $R\colon \overline{\mathbb{C}} \to \overline{\mathbb{C}}$ as follows. Let $z \in \mathbb{C}$. Then, as we have seen in the proof of Proposition 13.52, $\wp^{-1}(z) = \{t, -t\}$ for some point $t \in X$ (again, we identify elements of $\mathbb{C}$ with their images in X). Now set $R(z) = f(t) = f(-t)$ (recall that f is an even function); if f has a pole at t (and hence at $-t$), we set $R(z) = \infty$.

We will show that R is a holomorphic map from $\overline{\mathbb{C}}$ to $\overline{\mathbb{C}}$. Indeed, if $z \in \overline{\mathbb{C}}$ does not coincide with a branch point of the map $\wp\colon X \to \overline{\mathbb{C}}$ and $\wp(t) = z$, then, by Proposition 13.29, there exist neighborhoods $U \ni t$ and $V \ni z$ such that the restriction $\wp|_U$ is an isomorphism from U onto V. Since for every $z \in V$ we can set $R(z) = f((\wp|_U)^{-1}(z))$, on the neighborhood $V \ni z$ the map R can be represented as the composition of the holomorphic maps $(\wp|_U)^{-1}$ and f, so it is holomorphic at the point z. One can easily see from construction that the map R is continuous also at the branch points of the map $\wp\colon X \to \overline{\mathbb{C}}$; from this, using the Riemann removable singularity theorem (as in the proof of Proposition 13.52), one can deduce that R is holomorphic everywhere.

So, R is a holomorphic map from $\overline{\mathbb{C}}$ to $\overline{\mathbb{C}}$, that is, a meromorphic function on $\overline{\mathbb{C}}$, i.e., a rational function. By construction, $f(t) = R(\wp(t))$ for every $t \in X$, whence $f = R(\wp)$.   □

So, we have shown that every even elliptic function is a rational function of $\wp$. Now we will show that every odd elliptic function is a rational function of $\wp$ and $\wp'$. Indeed, the function $\wp'$ is odd; if g is an arbitrary odd elliptic function, then the function $g/\wp'$ is elliptic and odd, so, by the above lemma, $g/\wp' = R(\wp)$ where R is a rational function. Therefore, $g = \wp' \cdot R(\wp)$, and the right-hand side of this equality is a rational function of $\wp$ and $\wp'$.

Finally, let h be an arbitrary elliptic function. Then we have the identity

$$h(z) = \frac{h(z) + h(-z)}{2} + \frac{h(z) - h(-z)}{2}.$$

Both terms in the right-hand side are, obviously, elliptic functions, the first of them is even, and the second one is odd. By the above, each of these terms can be written as a rational function of $\wp$ and $\wp'$, so the same holds for their sum.   □

13.7 On the Riemann–Roch Theorem

In this section, we discuss to what extent the Mittag-Leffler theorem on the existence of a function with given principal parts (Theorem 11.31) can be extended from open subsets in $\mathbb{C}$ to Riemann surfaces, and especially to compact Riemann surfaces.

To begin with, we define principal parts in the context of Riemann surfaces. This generalization is quite banal:

Definition 13.56 Let X be a Riemann surface and $p \in X$. A *principal part* at p is a function φ that is holomorphic in a punctured neighborhood of p and satisfies the following property: for some local chart z defined in a neighborhood of p,

$$\varphi(z) = \sum_{n=1}^{\infty} c_n (z - a)^{-n},$$

where $a = z(p)$ and the series converges in a punctured neighborhood of a.

If f is a function holomorphic in a punctured neighborhood of a and φ is a principal part at a, one says that f has principal part φ at a if $f - \varphi$ can be extended to a function holomorphic at a.

Now consider the following situation: X is a Riemann surface, $S \subset X$ is a subset without accumulation points, and at each point $a \in S$ a principal part φ_a is given. The question is whether there exists a holomorphic function $f: X \setminus S \to \mathbb{C}$ that has principal part φ_a at each point $a \in S$.

This question is called the *Mittag-Leffler problem*.

Let X be compact (without any loss, we may and will assume it to be connected). In this case, a subset $S \subset X$ without accumulation points is necessarily finite.

In the trivial case $X = \overline{\mathbb{C}}$, every Mittag-Leffler problem has a solution. Indeed, in this case every principal part at a point $a \in \overline{\mathbb{C}}$ determines a function on $\overline{\mathbb{C}}$ with one singular point (an entire function if $a = \infty$, and a function of the form $f(1/(z - a))$ with f entire if a is a finite point); taking the (finite) sum of these functions, we obtain a desired function with given principal parts.

If, however, the genus of X is not zero, then, necessarily, not every Mittag-Leffler problem has a solution. Assume, for example, that we are given only one principal part which in local coordinates has the form $1/(z - a)$ (a simple pole). A solution f to this Mittag-Leffler problem would be a meromorphic function on X with exactly one, and simple, pole. If we regard it as a holomorphic map $f: X \to \overline{\mathbb{C}}$, then $f^{-1}(\infty)$ consists of a single point at which f is unramified, hence $\deg f = 1$ and f defines an isomorphism between X and $\overline{\mathbb{C}}$, which is impossible unless the genus of X is zero.

So, in the general case the Mittag-Leffler problem on a compact Riemann surface is nontrivial. To state an exact criterion for its solvability, we take recourse to residues.

Note that if ω is a holomorphic form on a Riemann surface X and φ is a principal part at a point $a \in X$, then the product $\varphi \cdot \omega$ is a holomorphic form in a punctured neighborhood of a, so the residue $\mathrm{Res}_a(\varphi \cdot \omega)$ is well defined.

Theorem 13.57 (Riemann–Roch theorem) *Let X be a compact Riemann surface, $S \subset X$ be a finite subset, and assume that for each point $a \in S$ a principal part φ_a at a is given. Then the following two conditions are equivalent:*

(1) there exists a function f holomorphic on $X \setminus S$ that has principal part a at each point $a \in S$;

(2) for every holomorphic form ω on X,

$$\sum_{a \in S} \mathrm{Res}_a(\varphi_a \cdot \omega) = 0. \tag{13.5}$$

It is quite easy to verify that condition (2) is necessary for the Mittag-Leffler problem to be solvable. Indeed, if f is holomorphic on $X \setminus S$ and has principal part φ_a at each point $a \in S$, then the function $f - \varphi_a = h_a$ is holomorphic in a neighborhood of a, hence for every holomorphic form ω on X we have

$$\mathrm{Res}_a(\varphi_a \cdot \omega) = \mathrm{Res}_a(f\omega) - \mathrm{Res}_a(h_a\omega) = \mathrm{Res}_a(f\omega),$$

since the form $h_a\omega$ is holomorphic in a neighborhood of a. Summing these equalities over all $a \in S$, we obtain

$$\sum_{a \in S} \mathrm{Res}_a(\varphi_a \cdot \omega) = \sum_{a \in S} \mathrm{Res}_a(f\omega) = 0$$

by Proposition 13.50.

So, the implication $(1) \Rightarrow (2)$ in Theorem 13.57 is established. Note that since the space of holomorphic forms on a compact Riemann surface is finite-dimensional, condition (2) is quite suitable for practical use: it suffices to check relation (13.5) for all elements of a basis of the space of holomorphic forms.

In contrast, the converse implication $(2) \Rightarrow (1)$ is a very profound result. Its proof can be found in other manuals. In many textbooks, the Riemann–Roch theorem for compact Riemann surfaces is stated differently. To establish the equivalence of these statements is a useful exercise. I recommend to verify this equivalence first for the case where all principal parts correspond to poles but not essential singularities (this case is much simpler, and it is of special interest for algebraic geometry).

But we will prove the Riemann–Roch theorem for the special case where X is an elliptic curve and all principal parts correspond to poles.

As we have seen in Example 13.38, every holomorphic form on an elliptic curve $X = \mathbb{C}/\Gamma$ has the form $c\,dz$ where c is a constant. Consequently, to check the Riemann–Roch theorem for elliptic curves, it suffices to show that the Mittag-Leffler problem is solvable if relation (13.5) holds for $\omega = dz$. Note also that if f is a meromorphic function on $\mathbb{C}/\Gamma$ regarded as an elliptic function with respect to Γ, then the residue $\mathrm{Res}_a(f\,dz)$ is equal to the coefficient of $(z - a)^{-1}$ in the Laurent expansion of f at the point a (we again identify points from $\mathbb{C}$ with their images in $\mathbb{C}/\Gamma$).

Proposition 13.58 *Let $X = \mathbb{C}/\Gamma$ be an elliptic curve, and assume that we are given principal parts $\varphi_1, \ldots, \varphi_n$ at points $a_1, \ldots, a_n$. Assume that all principal parts φ_j correspond to poles but not essential singularities, and let*

$$\sum_{j=1}^{n} \text{Res}_{a_j}(\varphi_j \, dz) = 0.$$

Then on X there exists a meromorphic function f that has principal part φ_j at each point a_j and has no singularities outside the points $a_1, \ldots, a_n$.

Proof We will construct a solution to this problem from $\wp$-functions and their derivatives. Obviously, the proposition will be proved as soon as we establish the following two facts.

(1) For every point $a \in X$ and every integer $k \geq 2$ there exists a meromorphic function on X that has principal part $1/(z-a)^k$ at a and no poles other than a.

(2) For any two different points $a, b \in X$ there exists a meromorphic function on X that has principal part $1/(z-a)$ at a, principal part $-1/(z-b)$ at b, and no poles other than a and b.

Indeed, if the data of the Mittag-Leffler problem satisfy condition (13.5) for $\omega = dz$, then its solution can be obtained by taking an appropriate linear combination of functions of the form (1) and (2).

To verify claim (1), note that the function $\wp$ has principal part $1/z^2$ at the origin (since, by definition, it is the sum of $1/z^2$ and a function holomorphic in a neighborhood of the origin). Hence, the $(k-2)$th derivative of $\wp$ at the origin has principal part of the form const/z^k (with a nonzero constant) and no poles outside the origin; multiplying by an appropriate constant and replacing $\wp(z)$ with $\wp(z-a)$, we have everything we need.

To verify claim (2), consider a number $c \in \mathbb{C}$ such that $2c \notin \Gamma$ and set $\lambda = \wp(c) = \wp(-c)$. By the choice of c, the points c and $-c$ do not coincide not only on $\mathbb{C}$, but also on $\mathbb{C}/\Gamma$. Since $\wp$ regarded as a holomorphic map from $\mathbb{C}/\Gamma$ to $\overline{\mathbb{C}}$ has degree 2 (see the beginning of the proof of Proposition 13.52), $\wp(z) = \lambda$ if and only if $z = c$ or $z = -c$ (here we regard z, c, and $-c$ as elements of $\mathbb{C}/\Gamma$). Thus, the function $f(z) = 1/(\wp(z) - \lambda)$ has poles of order 1 at the points c and $-c$ and no other poles; by Proposition 13.50, the residues of the form $f \, dz$ at the points c and $-c$ are opposite to each other, so, multiplying f by an appropriate constant, we may assume that the principal parts of f at the points c and $-c$ have the form $1/(z-c)$ and $-1/(z+c)$, respectively. Thus, claim (2) is proved for the case where $a = c, b = -c$. In the general case, given points a and b, find elements $c, t \in \mathbb{C}/\Gamma$ such that $a = c - t, b = -c - t$: it suffices to take any solution of the equation $2c = a - b$ for c, and any solution of the equation $2t = -(a+b)$ for t; note that $2c \notin \Gamma$, since $a \not\equiv b \pmod{\Gamma}$, so $2c \notin \Gamma$ and the previous construction applies to c; now the function $z \mapsto f(z-t)$, obviously, satisfies all the required properties. $\square$

On a connected and *non*compact Riemann surface, every Mittag-Leffler problem has a solution: perhaps, it is this statement that should be regarded as the most direct generalization of Theorem 11.31. The proof of this result is also beyond the scope of this book.

13.8 On Abel's Theorem

In this concluding section, we discuss the extension to Riemann surfaces of Weierstrass' theorem on the existence of holomorphic functions with given zeros (Theorem 11.34). The literal generalization of Theorem 11.34 to compact Riemann surfaces necessarily fails, since on such a surface there is no nonconstant holomorphic function. To obtain a meaningful problem, let us try to generalize Corollary 11.38 (completely equivalent to Theorem 11.34) about the existence of a meromorphic function with given zeros and poles.

First of all, we introduce the notion of divisor, which plays an important role in the further development of the theory; we do not make use of the full force of this notion, it serves solely to simplify statements.

Definition 13.59 A *divisor* on a compact and connected Riemann surface X is a formal expression

$$D = m_1 a_1 + \ldots + m_n a_n$$

where $a_j \in X$ and m_j are integers (n is arbitrary).

Divisors can be added and subtracted by collecting like terms.

Definition 13.60 The *degree* of a divisor $D = \sum_{i=1}^{n} m_i a_i$ is the number $\deg D = \sum_{i=1}^{n} m_i$.

Definition 13.61 If f is a meromorphic function on a compact Riemann surface X, then (f) stands for the divisor of the form $\sum_{a \in X} \operatorname{ord}_a(f) \cdot a$ (the sum is finite, because $\operatorname{ord}_a(f) = 0$ if a is neither zero nor pole of f).

Every divisor of the form (f) where f is a meromorphic function is called a *principal* divisor.

Using the notion of divisor, the problem of generalizing Weierstrass' theorem to compact Riemann surfaces can be stated as follows. Let D be a divisor on a compact Riemann surface. When is it principal, i.e., when does there exist a meromorphic function f such that $(f) = D$?

One necessary condition is seen immediately. Indeed, Proposition 13.33 can be restated in the language of divisors as follows: if f is a meromorphic function on a compact Riemann surface, then $\deg(f) = 0$. Therefore, if $\deg D \neq 0$, then a meromorphic function f with zeros and poles prescribed by the divisor D cannot exist. A genuinely interesting question arises if we assume additionally that $\deg D = 0$.

On the Riemann sphere, everything is still simple: every divisor of degree 0 on $\overline{\mathbb{C}}$ is principal. Indeed, if

$$D = (a_1 + \ldots + a_n) - (b_1 + \ldots + b_n)$$

where all points a_i and b_j are different from ∞ (there may be repeated points among $a_1, \ldots, a_n$ and $b_1, \ldots, b_n$), then $D = (f)$ where

$$f(z) = \frac{(z - a_1) \dots (z - a_n)}{(z - b_1) \dots (z - b_n)}$$

(observe that $f(\infty) = 1$, so at infinity f has neither zero nor pole). The case where the divisor D contains the point ∞ is left to the reader.

On Riemann surfaces of nonzero genus, not every divisor of degree 0 is principal. Here is an example.

Remark 13.62 Let X be a compact Riemann surface of nonzero genus, and let $a, b \in X$ be two different points. Then the divisor $a - b$ is not principal. Indeed, if $a - b = (f)$, then f is a meromorphic function on X with one and simple zero at the point a and one and simple pole at the point b. If we regard f as a holomorphic map from X to the Riemann sphere, then $\deg f = 1$ and f is an isomorphism, which is absurd.

The classical Abel's theorem gives a necessary and sufficient condition for a given divisor of degree 0 on a compact Riemann surface to be principal. In contrast to the three other classical theorems, I will not even state Abel's theorem in full generality: this requires introducing a series of new definitions, which would bring us too far. Instead, I will state and prove the special case of this theorem for elliptic curves: this does not require introducing new notions, and the theorem in this case can be proved "with bare hands."

So, consider an elliptic curve $X = \mathbb{C}/\Gamma$ where $\Gamma \subset \mathbb{C}$ is a lattice. Since both $\mathbb{C}$ and Γ are additive groups, X is also equipped with an addition as the quotient of $\mathbb{C}$ with respect to Γ.

Proposition 13.63 (Abel–Jacobi theorem for elliptic curves) *Let*

$$D = p_1 a_1 + \dots + p_n a_n$$

be a divisor on the elliptic curve $X = \mathbb{C}/\Gamma$. There exists a meromorphic function f such that $(f) = D$ if and only if the following two conditions are satisfied:

(1) $p_1 + \dots + p_n = 0$ *(in other words, $\deg D = 0$);*

(2) $p_1 a_1 + \dots + p_n a_n = 0$, *where addition is understood in the sense of the quotient group $\mathbb{C}/\Gamma$.*

We will give two proofs of this theorem. In the first one, as in the proof of the Riemann–Roch theorem in the previous section, the key role is played by the Weierstrass $\wp$-function.

First proof of Proposition 13.63 Throughout the proof, we denote the addition and subtraction of points on $\mathbb{C}/\Gamma$ by $\oplus$ and $\ominus$ (in order not to confuse them with the formal addition and subtraction of points in the notation for divisors).

We begin with the "if" part.

Let $D = a_1 + \dots + a_n - b_1 - \dots - b_n$ where $a_1, \dots, a_n, b_1, \dots, b_n \in \mathbb{C}/\Gamma$ and $a_1 \oplus \dots \oplus a_n = b_1 \oplus \dots \oplus b_n$. We must prove that D is a principal divisor. Let us use induction on n.

For $n = 1$, the claim is trivial, since it follows from the assumption that $D = 0 = (f)$ where f is a nonzero constant.

For $n = 2$, we have $D = a + b - c - d$ where $a \oplus b = c \oplus d$. To show that the divisor $a + b - c - d$ is principal, we proceed as follows. Let us regard the Weierstrass $\wp$-function corresponding to the lattice Γ as a meromorphic function on $\mathbb{C}/\Gamma$. As we have already mentioned several times, $\wp(u) = \wp(v)$ if and only if $u = v$ or $u = \ominus v$. Since $\wp$ has a pole of order 2 at the origin, we see that for every point $u \in \mathbb{C}/\Gamma \setminus \{0\}$ the divisor $u + (\ominus u) - 2 \cdot 0$ is principal: it has the form (f) where $f(z) = \wp(z) - \wp(u)$. Therefore, for any points $u, x \in E$ the divisor $(x \oplus u) + (x \ominus u) - 2 \cdot x$ is also principal, since it has the form (f) where $f(z) = \wp(z - x) - \wp(u)$. In the group $E = \mathbb{C}/\Gamma$, for any two points $a, b \in E$ we can choose $u, x \in E$ such that $a = x \oplus u$, $b = x \ominus u$ (take "half" $a \oplus b$ for x and "half" $a \ominus b$ for y); since $a \oplus b = x \oplus x$, we see that the divisor $a + b - 2x$ is principal; since $c \oplus d = a \oplus b$, the divisor $c + d - 2x$ is also principal; further, the difference of two principal divisors is a principal divisor too: if $D_1 = (f_1)$, $D_2 = (f_2)$, then, obviously, $D_1 - D_2 = (f_1/f_2)$. Therefore, the divisor $(a + b) - (c + d) = (a + b - 2x) - (c + d - 2x)$ is principal.

Finally, we prove the induction step from $n - 1 \geq 2$ to n. Let $a_1 \oplus \ldots \oplus a_n = b_1 \oplus \ldots \oplus b_n$; setting $c = a_{n-1} \oplus a_n \ominus b_n$, we have

$$a_{n-1} \oplus a_n = b_n \oplus c,$$
$$a_1 \oplus \ldots \oplus a_{n-2} \oplus c = b_1 \oplus \ldots \oplus b_{n-1}.$$

Hence, by the induction hypothesis, the divisors $D' = a_{n-1} + a_n - b_n - c$ and $D'' = a_1 + \ldots + a_{n-2} + c - b_1 - \ldots - b_{n-1}$ are principal. However, the sum of two principal divisors is principal too (if $D_1 = (f_1)$, $D_2 = (f_2)$, then $D_1 + D_2 = (f_1 f_2)$), so the divisor

$$a_1 + \ldots + a_n - b_1 - \ldots - b_n = D' + D''$$

is principal, and the "if" part is proved by induction.

Now we prove the "only if" part. Let $D = a_1 + \ldots + a_n - b_1 - \ldots - b_n$ be a principal divisor; by Proposition 13.33, we may assume that $D = a_1 + \ldots + a_n - b_1 - \ldots - b_n$ where $a_i, b_j \in X$; we must prove that $a_1 \oplus \ldots \oplus a_n = b_1 \oplus \ldots \oplus b_n$. Set

$$c = a_1 \oplus \ldots \oplus a_n \ominus b_1 \ldots \ominus b_{n-1}.$$

Then

$$a_1 \oplus \ldots \oplus a_n = b_1 \oplus \ldots \oplus b_{n-1} \oplus c, \tag{13.6}$$

so the divisor $D' = a_1 + \ldots + a_n - b_1 - \ldots - b_{n-1} - c$ is principal by what we have already proved. Hence, the divisor $D' - D = b_n - c$ is principal too, but, in view of Remark 13.62, this can happen only if $b_n = c$. Now it follows from (13.6) that $a_1 \oplus \ldots \oplus a_n = b_1 \oplus \ldots \oplus b_n$, as required. $\square$

The second proof of the Abel–Jacobi theorem is remarkable in that it can be directly transferred to the case of arbitrary genus.

Second proof of Proposition 13.63 Again, let $X = \mathbb{C}/\Gamma$ where $\Gamma \subset \mathbb{C}$ is a lattice with basis $\langle \omega_1, \omega_2 \rangle$, and let Π be a fundamental parallelogram of Γ with vertices $b, b + \omega_1, b + \omega_1 + \omega_2, b + \omega_2$.

Lemma 13.64 *If g is an elliptic function with respect to Γ that has no poles on the boundary $\partial\Pi$, then*

$$\int_{\partial\Pi} zg(z)\,dz = \omega_1 \int_b^{b+\omega_2} g(z)\,dz - \omega_2 \int_b^{b+\omega_1} g(z)\,dz \qquad (13.7)$$

where $\partial\Pi$ is the positively oriented boundary of the parallelogram Π and $\int_A^B$ stands for the integral over the segment between A and B directed from A to B. □

Proof Denoting the left-hand side of (13.7) by I, we have

$$I = \underbrace{\left(\int_b^{b+\omega_1} zg(z)\,dz + \int_{b+\omega_1+\omega_2}^{b+\omega_2} zg(z)\,dz \right)}_{J_1}$$

$$+ \underbrace{\left(\int_{b+\omega_1}^{b+\omega_1+\omega_2} zg(z)\,dz + \int_{b+\omega_2}^{b} zg(z)\,dz \right)}_{J_2}. \qquad (13.8)$$

Since the function g has period ω_2, we have

$$J_1 = \int_b^{b+\omega_1} zg(z)\,dz + \int_{b+\omega_1+\omega_2}^{b+\omega_2} zg(z)\,dz = \int_b^{b+\omega_1} zg(z)\,dz - \int_{b+\omega_2}^{b+\omega_1+\omega_2} zg(z)\,dz$$

$$= \int_b^{b+\omega_1} zg(z)\,dz - \int_b^{b+\omega_1} (z+\omega_2)g(z)\,dz = -\omega_2 \int_b^{b+\omega_1} g(z)\,dz.$$

A similar calculation shows that $J_2 = \omega_1 \int_b^{b+\omega_2} g(z)\,dz$; substituting this into (13.8) yields the required result. □

Now we prove the "only if" part of the Abel–Jacobi theorem. Let $D = \sum m_i a_i = (f)$ be a principal divisor on X. Condition (1) is still satisfied by Proposition 13.33. To establish condition (2), choose a point $b \in \mathbb{C}$ such that the preimages of the points a_j do not lie on a side of the fundamental parallelogram Π. Also, denote by $\tilde{a}_1, \ldots, \tilde{a}_n$

the preimages of the points $a_1, \ldots, a_n \in \mathbb{C}/\Gamma$ lying in Π, and regard the meromorphic function f for which $(f) = D$ as an elliptic function with respect to Γ.

Set $g(z) = f'(z)/f(z)$ in (13.7). The logarithmic residue theorem (Lemma 8.4) implies that the principal part of g at the point $\tilde{a}_j$ is equal to $p_j/(z - \tilde{a}_j)$, whence

$$\operatorname{Res}_{\tilde{a}_j} zg(z) = m_j \tilde{a}_j;$$

thus, the left-hand side of (13.7) (divided by $2\pi i$) is equal to

$$\frac{1}{2\pi i} \int_{\partial \Pi} zg(z)\, dz = \sum m_j \tilde{a}_j. \tag{13.9}$$

On the other hand, the function f has neither zeros nor poles on the segment $[b; b + \omega_2]$, hence $\log f$ is well defined in some neighborhood of this segment; we have $(\log f)' = f'/f$, so

$$\int_b^{b+\omega_1} g(z)\, dz = \int_b^{b+\omega_1} \frac{f'(z)\, dz}{f(z)} = \log f(b + \omega_1) - \log f(b).$$

Since $f(b + \omega_1) = f(b)$, the difference of the logarithms of f at these points is a multiple of $2\pi i$, so $\int_b^{b+\omega_1} g(z)\, dz = 2\pi i k_1$ for some $k_1 \in \mathbb{Z}$. In a similar way, $\int_b^{b+\omega_1} g(z)\, dz = 2\pi i k_2$ for some $k_2 \in \mathbb{Z}$. Comparing these relations with (13.7) and (13.9), we obtain

$$\sum m_j \tilde{a}_j = k_2 \omega_1 - k_1 \omega_2 \in \Gamma,$$

as required.

Now we prove the "if" part. Let $D = \sum_{i=1}^{k} n_i a_i$ be a divisor on $X = \mathbb{C}/\Gamma$ such that $\sum n_i = 0$ and $\sum n_i a_i = 0$ in the group $\mathbb{C}/\Gamma$. We must prove that there exists a function f such that $(f) = D$. To this end, we first construct the logarithmic derivative of f (more exactly, of the corresponding elliptic function).

Namely, since $\sum n_i = 0$, the Riemann–Roch theorem for elliptic curves (Proposition 13.58) on $\mathbb{C}/\Gamma$ implies that there exists a meromorphic function g on $\mathbb{C}/\Gamma$ that has a simple pole with $\operatorname{Res}_{a_i} g\, dz = n_i$ at each point a_i, where $1 \leq i \leq k$, and no other poles. As above, choose a fundamental parallelogram Π in such a way that the preimages of the points a_i in $\mathbb{C}$ do not lie on its boundary; we denote by $\tilde{a}_1, \ldots, \tilde{a}_n$ the preimages of the points $a_1, \ldots, a_n \in \mathbb{C}/\Gamma$ lying in Π and do not distinguish between meromorphic functions on $\mathbb{C}/\Gamma$ and elliptic functions on $\mathbb{C}$ with respect to Γ.

Since $\operatorname{Res}_{\tilde{a}_i} zg(z) = n_i \tilde{a}_i$, identity (13.7) implies that

$$\sum_{i=1}^{k} n_i a_i = B\omega_1 - A\omega_2, \tag{13.10}$$

where

$$A = \frac{1}{2\pi i} \int_b^{b+\omega_1} g(z)\, dz, \quad B = \frac{1}{2\pi i} \int_b^{b+\omega_2} g(z)\, dz.$$

Since $\sum n_i \tilde{a}_i \equiv 0 \pmod{\Gamma}$, it follows from (13.10) that

$$B\omega_1 - A\omega_2 = p\omega_1 + q\omega_2, \quad p, q \in \mathbb{Z}. \tag{13.11}$$

Adding a constant to g does not affect its poles and residues; let us show that this constant can be chosen in such a way that the integrals of the new function over the segments $[b; b + \omega_1]$ and $[b; b + \omega_2]$ become integer multiples of $2\pi i$. Indeed, setting

$$g_1 = g - 2\pi i \frac{A+q}{\omega_1},$$

we have

$$\frac{1}{2\pi i} \int_b^{b+\omega_1} g_1(z)\, dz = A - \omega_1 \frac{A+q}{\omega_1} = -q,$$

$$\frac{1}{2\pi i} \int_b^{b+\omega_2} g_1(z)\, dz = B - \omega_2 \frac{A+q}{\omega_1} = p, \tag{13.12}$$

where the last equality can be obtained if we express $\frac{(A+q)\omega_2}{\omega_1}$ from (13.11).

Now, proceeding from g_1, we construct a function f. Since g_1 is a meromorphic function whose residues at all poles are integers, the function

$$f(z) = e^{\int_b^z g_1(\zeta)\, d\zeta},$$

where the integral is over an arbitrary path that does not pass through the points $\tilde{a}_j + \gamma$, $\gamma \in \Gamma$, is a well-defined holomorphic function on $\mathbb{C} \setminus \{\tilde{a}_j + \gamma\}_{\gamma \in \Gamma}$: the integrals over two paths connecting b and z differ by a finite sum of integrals along small circles around $\tilde{a}_j + \gamma$, and upon exponentiating this difference cancels out since the residues are integers. Further, since $\mathrm{Res}_{\tilde{a}_j + \gamma}\, g_1 = n_j \in \mathbb{Z}$ for every $\gamma \in \Gamma$, the function f extends to a meromorphic function on $\mathbb{C}$ with $\mathrm{ord}_{\tilde{a}_j + \gamma}\, f = n_j$ that has no poles and zeros outside $\tilde{a}_j + \gamma$. Finally, it follows from (13.12) that f is periodic with respect to Γ. Thus, f induces a meromorphic function on $\mathbb{C}/\Gamma$ such that $(f) = \sum n_j a_j$, as required. $\square$

Let us conclude by saying that on a *noncompact* and connected Riemann surface X, every "divisor" is principal: if $S \subset X$ is a (not necessarily finite) subset without accumulation points, and if each point $a \in S$ is assigned an integer n_a, then on X there exists a meromorphic function f such that $\mathrm{ord}_a(f) = n_a$ for every $a \in S$

and f has neither zeros nor poles outside S. This is a direct generalization of the Weierstrass theorem and Corollary 11.38.

Exercises

13.1. Show that if $f: X \to Y$ is a holomorphic and bijective map between Riemann surfaces, then the inverse map is automatically holomorphic.

13.2. Let $X = \{(z, w) \in \mathbb{C}^2 : z^2 = w^3\}$. Show that X does not coincide with the set of zeros of any polynomial satisfying the conditions from Example 13.7. (*Hint.* Look at a neighborhood of the origin.)

13.3. Let $X = \{(z, w): z^3 + w^3 = 2\} \subset \mathbb{C}^2$; define a function $f: X \to \mathbb{C}$ by the formula $f: (z, w) \mapsto zw - 1$. Find $\mathrm{ord}_p(f)$ where $p = (1, 1)$. (*Hint.* Consider the power series expansion.)

13.4. Show that on the Riemann surface $V(F)$ from Example 13.7, every bounded holomorphic function is a constant. (*Hint.* See Sec. 13.2.)

13.5. Let $\Gamma \subset \mathbb{C}$ be a lattice and $\wp$ be the corresponding Weierstrass function. Regarding $\wp$ as a holomorphic map from $\mathbb{C}/\Gamma$ to $\bar{\mathbb{C}}$, find its ramification points.

For each of the algebraic equations listed below, denote by X the compact Riemann surface corresponding to it in the sense of Sec. 13.2. For each of these Riemann surfaces, find the following.

(a) The ramification points (and ramification indices) of the map $X \to \bar{\mathbb{C}}$ induced by the map $(z, w) \mapsto z$.

(b) The genus.

(c) The zeros and poles of the meromorphic function w.

If the genus g of the surface X is positive, construct g linearly independent holomorphic differential forms on X.

13.6. $w^2 = (z - a_1)(z - a_2)(z - a_3)$, all a_j are distinct. (*Hint.* Much work has already been done in the text of the chapter.)

13.7. $w^2 = (z - a_1) \cdot \ldots \cdot (z - a_n)$, all a_j are distinct. (*Hint.* The course of action depends on the parity of n; to construct holomorphic forms, begin with dz/w.)

13.8. $w^2 = (z - a_1)^2(z - a_2)(z - a_3)(z - a_4)$, all a_j are distinct.

13.9. $z^4 + w^4 = 1$.

13.10. $zw^4 = z^3 + 2z + 1$.

13.11. $w^2 = z^3$.

13.12. Let X and Y be compact Riemann surfaces with the genus of X less than that of Y. Show that every holomorphic map from X to Y is constant.

13.13. Show that there does not exist a nonconstant holomorphic map from a compact Riemann surface of genus 8 to a compact Riemann surface of genus 5.

13.14. Show that there does not exist a nonconstant holomorphic map $f: X \to \bar{\mathbb{C}}$, where X is a compact Riemann surface, that is ramified over exactly one point.

13.15. Assume that there exists a nonconstant holomorphic map from a compact Riemann surface X to the Riemann sphere $\overline{\mathbb{C}}$ that is ramified over exactly two points; what can be said about the surface X?

13.16. Let $f: X \rightarrow Y$ be a holomorphic map between compact Riemann surfaces, and let $\deg f = 2$. Can f be ramified over exactly 2015 points?

13.17. Show that on the Riemann sphere there are no nonzero holomorphic forms.

13.18. Show that the space of holomorphic forms on a compact Riemann surface is finite-dimensional. (*Hint.* If nonzero holomorphic forms do not exist at all, then there is nothing to prove; if there is at least one such form, denote it by ω, then the space in question is isomorphic to the space of meromorphic functions f such that the form $f\omega$ is also holomorphic.)

13.19. Let f be a function holomorphic on all of $\mathbb{C}$ except finitely many isolated singularities. Regarding $f(z)dz$ as a form with isolated singularities on $\overline{\mathbb{C}}$, show that $\mathrm{Res}_\infty (f(z)\,dz) = -c_{-1}$, where c_{-1} is the coefficient of z^{-1} in the Laurent series of f in a punctured neighborhood of infinity. (Thus, the definition of residue at infinity given in Exercise 8.23 is justified.)

13.20. Let X be the compact Riemann surface corresponding to the equation

$$w^3 = z^3 + z - 2.$$

Consider the meromorphic differential form $\omega = dz/w^2$ on X. Find all zeros and poles of ω, their multiplicities, and the residues of ω at all its poles.

13.21. Show that if a meromorphic form on a compact Riemann surface has exactly one pole, then this pole cannot be simple.

13.22. (a) Show that the functions z and e^z on $\mathbb{C}$ are algebraically independent, i.e., there is no nonzero polynomial P with complex coefficients in two variables such that $P(z, e^z) = 0$ identically. (*Hint.* See Lemma 13.20.)

(b) Show that the functions e^z, e^{e^z}, $e^{e^{e^z}}$, ... are algebraically independent over $\mathbb{C}$.

13.23. Let $\Gamma \subset \mathbb{C}$ be a lattice with basis $\langle 1, i \rangle$. Show that the elliptic curve $\mathbb{C}/\Gamma$ is isomorphic to the compact Riemann surface corresponding to the equation $w^2 = z^3 - z$. (*Hint.* It can be seen from the symmetry of this lattice that the Eisenstein series G_3 is zero.)

13.24. Give an example of a lattice $\Gamma \subset \mathbb{C}$ for which $\mathbb{C}/\Gamma$ is isomorphic to the compact Riemann surface corresponding to the equation $w^2 = z^3 - 1$.

References

1. Gamelin, T. W.: Complex Analysis. Springer, New York (2001).
2. Lang, S.: Complex Analysis. Springer, New York (1999).
3. Privalov, I. I.: Introduction to the Theory of Functions of a Complex Variable [in Russian]. Leningrad (1948).
4. Shabat, B. V.: Introduction to Complex Analysis [in Russian], Nauka, Moscow (1985).
5. Zorich, V. A.: Mathematical Analysis I. Springer, Berlin (2015).
6. Zorich, V. A.: Mathematical Analysis II. Springer, Heldelberg (2016).

© Springer Nature Switzerland AG 2020
S. Lvovski, *Principles of Complex Analysis*, Moscow Lectures 6,
https://doi.org/10.1007/978-3-030-59365-0

Index

© Springer Nature Switzerland AG 2020
S. Lvovski, *Principles of Complex Analysis*, Moscow Lectures 6,
https://doi.org/10.1007/978-3-030-59365-0

Printed in the United States
by Baker & Taylor Publisher Services